Studies in Rhythm Engineering

Series Editors

Anirban Bandyopadhyay, Senior Researcher, ANCC, National Institute for Materials Science, Tsukuba, Ibaraki, Japan

Kanad Ray, Amity School of Applied Sciences, Amity University Rajasthan, Jaipur, Rajasthan, India

Chi-Sang Poon, Department of Health Sciences and Technology, Massachusetts Institute of Technology, Cambridge, MA, USA

This is a multi-disciplinary book series, ranging from astrophysics to biology, chemistry, mathematics, geophysics and materials science. Its primary scope is the fundamental science and associated engineering wherever cyclic and rhythmic oscillations are observed.

Time neither being an entity nor a process is unmeasurable and undefined, although a clock only measures the passage of time. The clock drove recurring processes are observed in the biological rhythms, astrophysical and geophysical environments. Always, the clocks are nested, arranged in a geometric shape to govern a phenomenon in nature. These clocks are made of atoms, molecules, their circuits, and complex networks. From biology to astrophysics, the clocks have enriched the science and engineering and the series would act as a catalyst and capture the forthcoming revolution of time cycles expected to unfold in the 21st Century. From the cyclic universe to the time crystal, the book series makes a journey through time to explore the path that time follows.

The series publishes monographs and edited volumes.

Tanusree Dutta · Anirban Bandyopadhyay

Emotion, Cognition and Silent Communication: Unsolved Mysteries

Springer

Tanusree Dutta
Indian Institute of Management Ranchi
Ranchi, Jharkhand, India

Anirban Bandyopadhyay
International Center for Materials
and Nanoarchitectronics (MANA)
National Institute for Materials Science
(NIMS)
Tsukuba, Japan

ISSN 2524-5546 ISSN 2524-5554 (electronic)
Studies in Rhythm Engineering
ISBN 978-981-99-9333-8 ISBN 978-981-99-9334-5 (eBook)
https://doi.org/10.1007/978-981-99-9334-5

This Springer imprint is published by the registered company Springer Nature Singapore Pte Ltd.
The registered company address is: 152 Beach Road, #21-01/04 Gateway East, Singapore 189721, Singapore

Preface

Imagine if we could peek inside our brains to uncover the secrets of how we learn and make decisions in organizations, understand the mind, its hidden treasures. That's exactly what this book aims to do! Our brains are like super interconnected networks, and when they sync up, amazing things happen, from effective marketing to the treatment of neurological disorders, it is therefore utmost essential to read the mind right. Do we have the right tool? Think of it like an avalanche in the mountains, but instead of snow, its brain cells working together. When this happens under the right conditions, it helps us learn complicated stuff by teaming up. Now, here's the cool part: When our brain goes through this syncing-up process, it breaks into smaller groups, kind of like puzzle pieces. These pieces have a special pattern that looks like those intricate fractal designs you see in art. We wanted to measure this special "fractal time" during different brain activities, so we built a state-of-the-art lab at IIM Ranchi for consciousness research, thanks to Government of India SPARC project. Brain science is the greatest challenge of 21st century, and we expect this book would remain as a classic masterpiece in decades to come.

Here's the big mystery that the book has investigated: We don't know exactly how our brain's learning skills connect to make big decisions. So, in this book, we're going to deep dive into some unprecedented brain experiments. We'll see if this fractal time stuff relates to the way we make personal decisions and decisions in big groups. By comparing how our brains handle complex decisions, we hope to find some patterns. These patterns spread over a wide time range might show us how we learn and make efficient choices. It's like finding the best way to organize our thoughts and share information in our brain network. And why is this exciting? Well, if we figure it out, it could help us learn and take better decisions! Think about how handy that would be for managing our time and making important choices for those who are suffering from neurological disorders.

Additionally, this knowledge isn't just useful for brain science. It can help in all sorts of fields, like scientific training and teaching, understanding human behavior, and how companies are run via neuromanagement. So, stay tuned for some brainy discoveries that could change the way we learn and make decisions in the real world!

Most research on human consciousness and psychology has primarily remained theoretical, with limited experimental exploration involving multiple layers of instruments. Surprisingly, there has been minimal effort to simultaneously examine human consciousness using a variety of instruments. However, we are now excited to present a groundbreaking development resulting from the MHRD project: the creation of the Dodecanogram, DDG, an advanced iteration of electroencephalography (EEG). This represents a significant milestone, considering that EEG technology has remained largely unchanged since its inception in 1875, spanning over 150 years. Traditionally, EEG was confined to measuring brain activity within the frequency range of 1–40 Hz. In contrast, our book introduces a new EEG capable of measuring brain activity up to terahertz frequencies, we have given the name as DDG. These remarkable nine orders' advancement in frequency bandwidth opens a wide spectrum of unexplored dimensions within the human mind, which has hitherto remained hidden from scientific scrutiny. This book is a detailed review of 150 years of advancement that has led to this invention.

We acknowledge the project team comprising of Dr. Stuart R. Hameroff, Prof. Manas Kumar Mandal, Prof. Soumya Sarkar, Mr. Anindya Pattanayak, Dr. Piyush Pranjal, Dr. Pushpendra Singh, Mr. Radha Govind, and Dr. Komal Saxena, along with a large number of students who worked day and night helping us. We extend our thanks to Prof. Soami Daya Krisnananda, Martin Timms for their help in our experiments and building user interface for automation from 2019 to 2022. Josh Nino from Monarch Research Organization, USA, helped us in DDG experiments.

We have worked primarily in the Indian Institute of Management, Ranchi, where state-of-the-art lab was made to carry out an unprecedented experiment. We took ethical committee clearance of IIM Ranchi and the written consent from all 289 participants. The student volunteers who also helped us and put their best efforts for the completion of the project include Dr. Jhimli Sarkar, Dr. Sudipa Nag, Mr. Saunak Sarkar, Ms. Shanthi Banishetty, Ms. Chahat Maru, Ms. Abhinanda Mallick, Ms. Priya Katiyar, Mr. D Tarun Karthik, Mr. Kislay Kishore Singh, Ms. Pritika Paul, Dr. Ankur Shukla, Mr. Rajiv Ranjan, Mr. Raunak Srivastav, Mr. Bishnu Prasad Dash, Mr. Manas Jyoti Borah, and some students of the MBA, IPM, and Ph.D. scholars. From Japan, Dr. Pathik Sahoo and Dr. Pushpendra Singh worked primarily in 2019. We also acknowledge the help extended by Dr. Nitesh Bhatia, Central University of Jharkhand, for making arrangements for the participants of the experiment. We apologize to a large number of other volunteers who worked relentlessly for more than 16 h a day, and without their coherent effort, this book would never have seen the light.

Authors acknowledge the help, support, and encouragement received from Prof. Deepak Srivastava, Director, IIM Ranchi; Dr. Satinder Singh Rekhi, Founder, Rekhi Foundation, USA, for his insights on the experience of awe. The authors also acknowledge the support of Prof. Shailendra Singh and Prof. P. K. Bala.

Authors acknowledge the Asian Office of Aerospace R&D (AOARD), a part of the United States Air Force (USAF), for the Grant no. FA2386-16-1-0003 (2016–2019) on the electromagnetic resonance-based communication and intelligence of biomaterials. Authors also acknowledge the financial assistance of Scheme for Promotion of Academic and Research Collaboration (SPARC) an MHRD, Government of India initiative, for the project titled 'Management of Fractal Time in High-level Decision Making' (Government of India, MHRD; project number P 524; Start date: 15.03. 2019). We sincerely acknowledge the funds and IIM Ranchi management for all official cooperation between 2019 and 2023.

Ranchi, India
Tsukuba, Japan

Tanusree Dutta
Anirban Bandyopadhyay

Contents

Summary of the Chapters

Chapter 1: Our book embarks on a comprehensive exploration of consciousness and the fractal brain theory, initially conceived by multiple esteemed scientists, and subsequently advances were made in this theory through experimental investigations involving human subjects. Within the intricate realm of brain function, we unveil a noteworthy revelation: the dual facets of noise as a signal and the disruption of consciousness caused by ordered signals. We provide a concise overview of these findings while delving into various metaanalysis methodologies employed by psychologists and neuroscientists to unravel the origins of human cognitive experiences. Furthermore, we scrutinize the constraints inherent in EEG and group study techniques, foreshadowing their in-depth examination in forthcoming chapters. In essence, this chapter serves as the foundational cornerstone for the comprehensive journey delineated across the subsequent nine chapters of our monograph.

Chapter 2: The mirror drawing task, originating in the 1930s, initially sought to explore interhemispheric signal transfer between the left and right brain. Over time, this historical tool evolved into a powerful instrument for contemporary research. By incorporating EEG technology, we now witness real-time brain changes in human subjects, yielding invaluable insights. Through numerous studies, we've discovered a fascinating aspect of skill acquisition: The time it takes for the human brain to master a skill can range from a mere 30 s to a substantial 40 min, highlighting the uniqueness of each individual's brain. Managing time within the brain is intricately tied to emotional and cognitive traits, influencing social interactions, competitive performance, and creative problem-solving. This chapter advocates for a paradigm shift in psychological research, emphasizing the diversity among human subjects. Moreover, it underscores the need to reform our education system and society to accommodate the diverse hardware of the human brain, fostering a more inclusive and equitable environment that taps into the full spectrum of human potential.

Chapter 3: In Chap. 2, we tackled the issue of assuming equal time management abilities among all humans. In this chapter, we delve even deeper into the complexities of human expression, specifically how expression of emotions varies widely from person to person. Through extensive studies involving human subjects, we've revealed that emotions like love, pain, anger, or happiness are managed differently in each individual. For instance, drawing on half a century of research on blood flow through facial vessels and utilizing thermal cameras, we've uncovered a fascinating phenomenon. These studies consistently demonstrate that blood vessels beneath the facial skin transport blood in varying volumes and pressures, creating a unique heat distribution map that can be captured by thermal cameras. Similar to the mirror drawing task described in Chap. 2, we've categorized human subjects based on their distinctive thermal expressions of emotions and cognitive responses. This chapter introduces an automated emotion recognition system using thermal cameras, which proves unequivocally superior to optical image-based artificial intelligence tools.

Chapter 4: When diving into the classic brain science articles, which connect fMRI, EEG, and PET scans to human cognitive tasks, it may give the impression of a direct one-to-one relationship. This suggests that these texts portray a clear understanding of how specific brain regions trigger particular cognitive and emotional responses. However, this impression is deceptive. Despite the abundance of such correlations in research papers and textbooks, experimental psychologists have consistently indicated otherwise. We conducted experiments involving straightforward cognitive tasks like mathematical calculations and pattern recognition, anticipating clear evidence of interhemispheric signal transfer. Yet, the brain's signal transmission pathways proved to be intricate, defying simple routes from the left brain to the right and vice versa. This complexity highlights that there are no universal rules governing signal transmission in the brains of all individuals. Nevertheless, we made a significant discovery applicable to all subjects we studied: ASMR and random noise enhance the efficiency of various cognitive tasks, while periodic or repetitive sounds diminish task efficiency universally. This intriguing link between human consciousness and noise underscores the fundamental role of noise in shaping cognitive, perceptual, and emotional responses. Thus, Chap. 4 explores the downside of not finding a predefined signal transfer pattern in the brain while illuminating the positive aspect of noise as the genuine signal of the brain's conscious processes.

Chapter 5: The EEG machine, with its origins dating back to 1875, has played a crucial role in understanding brain signals, treating neurodegenerative disorders, and granting us a window into the inner workings of the brain for over a century and a half. However, it has certain limitations: It filters the original signal, disregards sudden changes, operates within a millisecond time domain (ranging from 1–300 Hz), takes averages of multiple signals, and interprets them as accurate representations of neuron firing in the cortex region. In this chapter, we introduce a groundbreaking advancement in EEG technology, known as Dodecahedron-driven EEG (DDG), which functions in two distinct modes. Firstly, it emits streams of picoseconds or nanoseconds pulses and employs a logic analyzer to precisely measure the duration and magnitude of potential bursts spanning from picoseconds to milliseconds. Thus, we delve into how human cognition, perception, emotions, and all conscious experiences occur within the astonishingly brief time frame of 200–700 nanoseconds, a realm that falls well below the traditional EEG's time domain. In its second mode, DDG delivers electromagnetic, magnetic, or mechanical noise bursts to the brain and gauges the resulting electromagnetic radiation band or frequency distribution. Consequently,

DDG challenges the notion that we've been glimpsing only snapshots of brain operation, whereas there exists an entire movie of information processing unfolding across the entire spectrum from picoseconds to seconds. Chapter 5 calls for a revolution in the fields of neuroscience, brain science, as well as both artificial and natural intelligence, in light of these transformative insights.

Chapter 6: In Chap. 5, we introduced a groundbreaking invention, DDG, designed to map the realm of the subconscious mind. The pivotal question that arises is how this innovation could potentially reshape the entire landscape of neuroscience and our comprehension of consciousness. Within this chapter, we delved into the intricacies of group studies, expanding from a single subject to eight, revealing a profound revelation: The brain doesn't function like a Turing machine but operates under a unique framework known as Fractal Information Theory (FIT). Moreover, we explored the concept of collective consciousness, where one's consciousness is intricately linked to others. In a dimly lit room, during moments of collective silence among human subjects, we observed remarkable responses that allowed us to reevaluate various theories of consciousness. Ultimately, we found substantial support for the Self-Operating Mathematical Universe (SOMU) theory, which posits a mathematical foundation for consciousness. As a result, Chap. 6 embarks on an exploration of higher-level consciousness emergence through group studies, presenting a compelling case for reevaluating all existing theories of consciousness through the lens of DDG.

Chapter 7: In the preceding two chapters, Chaps. 5 and 6, we've explored a groundbreaking concept: That time within the brain isn't linear or nonlinear but rather an intricate architecture of interconnected clocks. To validate this unique notion of time, we conducted experiments involving human subjects tasked with deciphering simple patterns and drawings guided by a 3D assembly of clocks. This chapter also delves into a common psychological test where individuals draw images based on hints, which themselves are questions. Through the expansion of these patterns, we've constructed a framework for understanding the architecture of time that represents the subconscious mind. With the insights gained from human subjects undergoing DDG tests, mirror drawing experiments, and thermal camera assessments of emotional transitions, we've been able to unveil the underlying structure of the subconscious mind, further classifying these insights into multiple categories, building upon the ideas explored in the preceding chapters, namely Chaps. 2, 3, and 4.

Chapter 8: DDG leads us to a fresh perspective on consciousness, challenging the conventional pursuit of happiness through rewards and advocating the exploration of pain as a pathway to true happiness. While the prevalent belief holds that happiness is a broad, overarching state and pain a localized phenomenon within the brain, this chapter sheds light on a different understanding. Throughout history, mental engineers have extolled the virtues of happiness as a means to alleviate suffering and pain. However, we take a novel approach by profiling brain waves associated with two distinct emotions in various contexts, discerning their geometric invariants within the brainwave patterns. Surprisingly, we find that pain transcends localized boundaries and manifests as a global phenomenon in the brain, while happiness emerges as a subset of the governing invariants of pain. By comparing the activation pathways of spontaneously generated emotions with induced emotional states, we establish the significance of latency, non-responsiveness to pain, and a universal connection to elements like children and flowers. These factors underscore the intricate hardware and software relationships between happiness and pain. This chapter provides a comprehensive review of the exploration of the path to finding happiness within the realm of pain.

Chapter 9: The use of awareness campaigns stands as a powerful tool in the arsenal of moral advocates aiming to foster a society rooted in peace and security. Its effectiveness remains unchallenged, and extensive research has been dedicated to discovering the most impactful ways to instill awareness in the human mind. In this context, our exploration involves mapping the emotional responses generated by an awareness campaign focused on combatting sexual harassment in the workplace. What we've uncovered is quite remarkable: A profound sense of pure positive awe has the capacity to entirely counteract emotional arousal triggered by awareness videos. To the extent that not even a negative sense of awe can sway this response. The discussions in this chapter redefine our fundamental understanding of crafting effective awareness campaign materials. It becomes evident that introducing an entirely unrelated and deeply positive awe experience during a campaign has the potential to redirect a mind that may have become misguided due to the sensationalization of pressing issues. Pure positive awe emerges as a universally compelling force, capable of resonating with minds that might otherwise be resistant to the messages of awareness campaigns.

Chapter 10: In the concluding chapter, we undertake a comprehensive review of all eight preceding chapters (Chaps. 2–9) through the lens of our groundbreaking consciousness theory, aptly named the "Self-Operating Mathematical Universe" or SOMU for brevity. Within this chapter, we revisit numerous consciousness theories and concepts related to human perception and cognition by means of a series of carefully designed experiments. Our focus centers on three pivotal points: firstly, the necessity of classifying human subjects based on their time management and emotional expression; secondly, the revelation that human consciousness is a complex map of signal processing across multiple time domains concurrently, challenging the conventional belief that millisecond neuron firing alone defines consciousness; and lastly, the assertion that an architecture of time, represented by interconnected clocks and governed by geometric invariants, forms the fundamental basis for human cognition, emotion, and perception. These key insights serve as a concise summary of our extensive exploration into the measurement of consciousness employing a diverse array of instruments.

About the Authors

Dr. Tanusree Dutta is presently working as Associate Professor in the Indian Institute of Management, Ranchi. Dr. Dutta obtained her Postgraduate and Doctorate degrees from Banaras Hindu University and Indian Institute of Technology, Kharagpur, in 2000 and 2004, respectively. She received the M. M. Sinha Endowment Scholarship in 1998-99, B.H.U. Merit and Prize Award in 1999, and a gold medal in Masters in Psychology in 2000. Dr. Dutta was Assistant Professor at Banaras Hindu University in the Psychology Section. She has also served on lien as Assistant Professor in the Management Department of B.I.T., Mesra Indian Institute of Technology Rajasthan. She has research publications in peer-reviewed national and international journals which have been cited in manuscripts and books of national and international repute. She has co-authored a book titled *Neuromarketing in India: Understanding the Indian Consumer* and has a co-edited book titled *Bias in Human Behaviour*.

Anirban Bandyopadhyay is principal research scientist at the National Institute for Materials Science (NIMS), Tsukuba, Japan. He has done Ph.D. in Supramolecular Electronics at the Indian Association for the Cultivation of Science (IACS), Kolkata, 2005. From 2005 to 2008, he was ICYS Research Fellow at the ICYS NIMS Japan, and worked on the brain-like bio-processor. In 2008, he joined as Permanent Scientist at NIMS working on the time crystal model of human brain and design-synthesis of brain-like organic jelly, and he wrote a book *Nanobrain: The Making of an Artificial Brain from a Time Crystal*, 2020. From 2013 to 2014, he was Visiting Scientist at the Massachusetts Institute of Technology (MIT), USA. He received Hitachi Science and Technology award 2010, Inamori Foundation award 2011–2012, Kurata Foundation Award, Inamori Foundation Fellow (2011–), and Sewa Society international member, Japan.

Chapter 1
Unsolved Mysteries of the Mind and the Brain: Fractal Brain Hypothesis

1.1 Introduction

1.1.1 History of Studies of Consciousness

1.1.1.1 Vedic Consciousness and Universal Consciousness

In the realm of consciousness theories, various perspectives abound. Yet, in light of its statistical-mathematical character and its extensive incorporation of fractal attributes, we have elected to commence our discourse from the vantage point of the Vedic viewpoint [38, 69]. The Vedic theory of consciousness is a holistic approach to understanding the nature of consciousness (Kalpa, Ulbastra; [17]). It views consciousness as an essential aspect of statistical mathematics (Sankhya Darshan), rather than as an epiphenomenon of the physical world. The Western debate that differentiates physicality with the non-physical nature of the universe does not exist in the Vedic tenet of scholarly debate. The Vedas, the ancient scriptures of Hindus, describe consciousness as a field of pure awareness that underlies and permeates all of existence, where a right matrix (aadhaar) is created. Thousands of years of debate exists on how to make a proper matrix. A right matrix is a mathematical condition, and we would explore in this book how experimentally one could determine these unquantified parameters [17]. Given the nature of readers that we address, higher-dimensional tensor mathematics has been removed from this book, instead their physical significance has been incorporated in lucid language.

Through these ideas, researchers proposed a series of concepts [74]:

1. The sensor detected world or Prakriti doesn't physically exist. The world's apparent existence is a product of the mind.
2. The mind itself doesn't physically exist. The mind's apparent existence is rooted in the 'I-thought.'

T. Dutta and A. Bandyopadhyay, *Emotion, Cognition and Silent Communication: Unsolved Mysteries*, Studies in Rhythm Engineering,
https://doi.org/10.1007/978-981-99-9334-5_1

3. This 'I' is also not a physical existence. Only pure consciousness or Purusha is real.
4. Consequently, the reality of both the 'I' and the world exists solely within consciousness itself.

To aid us in this endeavor, scholars employ the Hindu concept of Triputi, which revolves around the relationship between the seer, seeing, and the seen, as well as the abstract exercise of the topological metaphor. Through the latter, we create a symbolic cosmogony that strives to account for such a worldview, with the hope of encouraging individuals to surpass its fragile construction.

The Vedic theory of consciousness has been influenced by the philosophy of monism, which holds that the universe is ultimately one and undivided (Virata Purusha). There is a fractal network of life [8], it means the existence of an endless chain of life or consciousness. According to monism, there is no separation between the individual self and the universal Self, one is a subset of the other. This means that all beings are fundamentally interconnected and that consciousness is shared by all [66]. What is least understood about this eastern school of thoughts is that there is no endless linear chain of elements, every single cognitive, perceptive and emotional experience is fractal in nature [8], one has to zoom inside an experience of mind to find its constituent mental states. And the journey would never stop, one goes above or within. As Carl Sagan argued, "if time is circular, the space is also circular" [27]. Time is cyclic and Hindu scale of time plays crucial role in the advent of consciousness [27, 52]. There are several traditions Hinduism that worked on structure of time [47].

The most misunderstood part in the west regarding this concept is that Vedic consciousness and Panpsychism are similar, it is very untrue. There is a consciousness that pervades everything is Panpsychism, which was mis-inspired by Hinduism in the fifteenth century. In contrast, in Vedic wisdom, there is one consciousness that is made of several conscious constituents (Mandukya Upanishad [50]), and we can enter inside one of these constituents and continue the journey [4]. The problem to understand this concept is the nature of integration, to a conscious entity, it is difficult to sense that it is an integral part of the whole, and at the same time, it is whole of its constituent parts. At a given time, it only understands its own world, the world above and the worlds within are not sensed.

Thus, came the concept of infinity, and changing the body sensors such that it can sense the infinite network of worlds. The Vedic theory of consciousness therefore emphasizes the importance of physical transformation of a human body with 108 rhythms of music, 108 forms of dances [13], 108 emotional and cognitive states (Raga), 108 tastes and 108 smells. All bodily transformations are directed to the unification with infinity (Yogic) practices for cultivating and expanding consciousness. These practices are seen as a way to connect with the underlying field of awareness and to experience a deeper level of reality.

The Vedic theory of consciousness is mathematically complex (indefinability of Naya school [15]), since it has to process information as a network of imaginary worlds interconnected fractally and rhythmic engineering view of consciousness has become profound to hold the continuity of the chain that connects all imaginary worlds (loka). The scholarly debates consider that universe is made of three elementary properties, bonding (T), transformation (R) and acquiring feedback from infinity (S), not matter or field or any entity, its SRT triplet that makes Prakriti [63]. This is amazing that when we move orthogonal to the Western view of a linear or nonlinear universe, instead of self-assembling matter, we self-assemble rhythms, instead of engineering the circuits, we engineer the architecture of properties, instead of facts and quantity of information, and we need to worry about the intricate geometry by which invariants build a network [46]. The beautiful journey of the seer and the seen has been developed over thousands of years, documented in scholar literatures. It offers a unique perspective on the nature of reality and the potential for human transformation by exercising strict mathematical rules of rhythms (Chhanda Shastra).

Here are some of the key features of the Vedic theory of consciousness, where Purusha and Prakriti play out as inseparable entity [44]:

- Consciousness is an essential statistical mathematical aspect of reality it is Guna or property of Purusha.
- Consciousness is a field of awareness of the whole by a part that underlies and permeates all of existence.
- The individual self is not separate from the universal Self, a suitable matrix (aadhaar) is needed to hold consciousness (form Prakriti).
- Consciousness is shared by all conscious beings, not all beings, it is not panpsychism.
- Meditation and other spiritual practices can help to cultivate and expand consciousness of a local entity toward complete consciousness or Purusha.

The Vedic theory of consciousness has been influential in many cultures and traditions, including Hinduism, Buddhism, and Jainism. It continues to be a source of inspiration and guidance for people all across the globe who are seeking to understand the nature of consciousness and to live a more awakened life.

1.1.1.2 Western Philosophy of Consciousness

One of the most influential theories of consciousness in Western philosophy is the dualism of Descartes. Descartes argued that the mind and the body are two distinct substances, the mind being non-physical and the body being physical. He famously claimed that "I think, therefore I am," which means that the fact that I am able to think proves that I exist as a non-physical mind. Descartes' dualism (mind and matter separate, [86]) was challenged by later philosophers, such as David Hume in the eighteenth century and John Locke in the seventeenth century. Hume argued that the mind is simply a collection of ideas and impressions, and that there is no such thing as a separate mental substance. Locke, on the other hand, argued that the mind is a blank slate at birth, and that all of our knowledge is acquired through experience. Most Western philosophers find non-physical part of information, like "redness of color red" as Qualia [36] and continue to isolate it as "hard problem of consciousness".

Three primary theories of consciousness constitute Western philosophies, Integrated Information Theory, IIT [87]; Global Workspace Theory, GWT;

Panpsychism: Panpsychism proposes that consciousness is a fundamental property of the universe and is present in all matter to some degree. In this view, even elementary particles possess some form of consciousness. It seeks to address the "hard problem" of consciousness by positing that consciousness is a fundamental aspect of reality.

- Brain imaging studies have shown that conscious experiences are associated with increased activity in the frontoparietal cortex, which is thought to be involved in the workspace.
- Studies of patients with brain damage have shown that damage to the frontoparietal cortex can lead to impaired consciousness.
- Studies of attention and binding have shown that these processes are disrupted when the workspace is not functioning properly.

Many consciousness theories are often labeled as pseudoscience [87] due to several key reasons:

1. **Non-testability**: Consciousness is inherently subjective, no one tries to make it parametric, a mathematically derivable property. Consequently, many consciousness theories like IIT [87], GWT, HOT, panpsychism rely on speculation rather than empirical evidence. So, there is no prediction. If contested, proponents suggest.
2. **Lack of Falsifiability**: Scientific theories should be falsifiable, meaning they must allow for experiments that could potentially disprove them. However, numerous consciousness theories like IIT, GWT, HOT, panpsychism are so vague and poorly defined that no such experiments can be conceived.
3. **Lack of Parsimony**: Scientific theories should strive for simplicity and clarity, with mathematical derivation of consciousness, irrespective of their definition. Unfortunately, all consciousness theories make it more complex and convoluted, amplifies confusion so that it is difficult to comprehend and test. For example, IIT.
4. **Limited Acceptance**: A significant portion of the scientific community, including neuroscientists and philosophers of mind, does not find existing consciousness theories adequately supported by evidence.
5. **Violating basic premises of Science**: All theories suggest brain is a non-Turing machine but their formulations fall under the purview of a Turing machine and Shannon's information theory.

1.1.2 The Differences Between the Western Philosophy and the Eastern Philosophy of Consciousness

Here are some of the key differences between Vedic consciousness and Western consciousness:

- **Nonduality versus duality**. Since Vedic consciousness has two parts, non-physical or mathematical part called Purusha and physical part that is Prakriti, consciousness is based on the principle of nonduality of Purusha-Prakriti, which means that there is no separation between the individual and the universe. Everything is interconnected and made of the same underlying consciousness. Western consciousness, on the other hand, is based on the principle of duality, which means that there is a distinction between the individual and the world, "body and the mind" even all individuals are separate. In Self-Operating Mathematical Universe, SOMU [78], we have adopted 3D space and 3D time, or 6D space–time as Prakriti and 3D morphogenesis, 3D density of primes or 6D morphogenesis-prime as Purusha. Thus, a unique 12D entity where Purusha and Prakriti are combined as inseparable dynamic entity [79] is better than the Western approach where mind–body separation leads to Qualia as "hard problem of consciousness" which cannot arise in the Vedic school.
- **Unity versus multiplicity**. Vedic consciousness sees the universe as a unified whole, human consciousness is a subset consciousness of the whole, while Western consciousness sees it as a collection of separate objects. Using prime number distribution in the integer space as the code of the universe, SOMU has figured out that if density of primes is the supreme non-physical code, using the engineering of only 15 primes a body could explore 86% of that code [6]. So, the Vedic perspective "I am part of the whole, and I am the whole of my parts", a fractal composition of consciousness is mathematically taken into account in SOMU.
- **Subjectivity versus objectivity**. Vedic consciousness is not exclusively subjective as believed by Western philosophers, meaning that it is experienced from the individual's own point of view, dual point of view and we continue to increase the numbers as triplet of triplet (Tri Netra of Shiva, or three seers, Tri Kala or three-time domains or convergence, Tri-Guna Satva Raja and Tama or generators of 23 elements that makes universe, total triplet of triplet means nine subjects) majority of the whole could mirror its fundamentals in the part of 108 conscious entities or objectiveness. Western consciousness is objective, meaning that it is seen from an outside perspective alone. In SOMU, 108 elements of conscious states have been proposed, and 91 has been detected from human subject experiments (Chapter 5 of this book).
- **Intrinsic versus extrinsic value**. Vedic consciousness sees all things as having intrinsic value, meaning that they are collectively precious in and of themselves, every single entity has a purpose to the universe. Western consciousness sees value as being extrinsic, meaning that it is derived from something else, such as usefulness or beauty, that's why Godel's incompleteness theorem was created in

the west, or we encountered Bertrand Russel's argument on $1 + 1 = 2$, where does the second "1" comes from, it there was only one "1" in the box. The inherent assumption is that something new cannot generate from within, it has to come from outside. Under the SOMU framework, every single object or matter or entity is a fractal network of prime numbers, namely phase prime metric [65]. Thus, it confirms the essential requirement for consciousness, a suitable matrix that can hold consciousness, for SOMU it's a symmetry in the arrangement of frequencies.

- **Unity of knowledge versus specialization**. Vedic consciousness sees all knowledge as being interconnected, and that connection is not linear or nonlinear its fractal, while Western consciousness tends to specialize in specific areas of knowledge, and all discrete pieces need to be connected to understand consciousness. SOMU endorses the fractal connection, and it transforms every single piece of information into space–time-topology-prime (STts) quartet [78–80], so that a human is not required to connect bits and pieces of information, humans could only edit information in its perpetual flow.
- **Enlightenment versus materialism**. Vedic consciousness sees enlightenment as the ultimate goal of life, enlightenment is "becoming the right matrix where universe or the whole could mirror itself", while Western consciousness is often more focused on material success because more "bits" or resources one could lead to more complexity and thus consciousness would be more. SOMU theory of consciousness is all about finding the use of density of primes [61] in the massively complex information structure so that it is reduced to a fractal seed. Thus, SOMU believes all complex information in the universe is originating from a set of primes, and once we find it, we could drastically reduce the major resources as we can invent an engine that if runs, could generate an enormous amount of information. Consequently, consciousness is also all about holding right composition of symmetry of primes, which enables us to create a right matrix or aadhaar that mirrors any information structure of the universe with 86% accuracy.

The idea that the universe is conscious and that there are conscious lives within a conscious life is a central tenet of Vedic consciousness. This is in contrast to the Western view of consciousness, which sees consciousness as being limited to individual beings. The Vedic view of consciousness is also musical and dance move based, seeing the individual as being interconnected with the universe via 108 fundamental forms of dance and music [22, 27].

The Western view of consciousness is based on the idea that information is like a linear or nonlinear tape, i.e., a Turing machine. This means that the amount of information that an object has and the speed at which it can process information determines whether or not it is conscious. This view is more mechanistic and reductionist than the Vedic view, where the tape is fractal, it grows within and above. Every conscious entity is part of the whole, and itself a whole of its constituent parts.

The Vedic view of consciousness is conceptually orthogonal to the Western view of consciousness and one cannot be deduced to another. Thus, Vedic view is still being explored and understood by Western scientists and philosophers. However, it is a growing field of study and there is a growing interest in the potential benefits of incorporating Vedic concepts into Western psychology and medicine.

1.2 The Quest for a Psychological Marker of a Mind and Consciousness: A Mind that Sees Within and Above

1.2.1 Experimental Challenges to Find a Marker for Consciousness

The search for an experimental marker of consciousness [90] has been a long and challenging one. Consciousness is a complex phenomenon that is not fully understood, and there is no single biological marker that can be definitively linked to it. One reason why it is difficult to find an experimental marker of consciousness is that it is not clear what consciousness is, exactly. There are many different theories of consciousness, and each theory has its own set of proposed markers. For example, some theories of consciousness suggest that it is correlated with activity in specific brain regions, while others suggest that it is correlated with the ability to experience emotions or sensations. Another reason why it is difficult to find an experimental marker of consciousness is that it is difficult to measure consciousness in a way that is objective and reliable. Many of the methods that have been used to measure consciousness, such as self-report and behavioral tests, are subjective and can be influenced by factors other than consciousness, such as motivation or attention. Despite the challenges, there is ongoing research into the neural correlates of consciousness (NCCs). NCCs are brain activity patterns that are associated with consciousness. If NCCs can be identified, they could be used to develop experimental markers of consciousness.

Artificial brain research and artificial intelligence are also being used to explore the nature of consciousness [90]. Researchers are using artificial neural networks to model the brain and to simulate consciousness. This research could lead to new insights into the neural basis of consciousness and to the development of new experimental markers. Here are three perspectives. Creating an artificial brain made from time crystals [65] would enable us to fine tune brain signal measurements because we would normalize the data. Time crystals are materials that exhibit periodic behavior in time, and it has been comprehensively argued [6, 71] that they could be used to create artificial brains that are more efficient and less error-prone than traditional artificial brains. The making of hardware always enables a better measurement tool. First, the difficulty of finding an experimental marker of consciousness is one of the reasons why the hard problem of consciousness remains unsolved. Second, the need for a novel approach to the study of consciousness is increasingly recognized. Some promising approaches include the use of neuroimaging techniques, such as functional magnetic resonance imaging (fMRI, to study the neural correlates of consciousness, the development of new experimental paradigms to measure consciousness; and the use of artificial intelligence to model the brain and to simulate consciousness. Third, the development of an experimental marker of consciousness would be a major breakthrough in the study of consciousness. It would allow researchers to study consciousness in a more objective and reliable way, and it could lead to new treatments for disorders of consciousness.

Exploring the brain's essential nature and its involvement in cognition and emotions has piqued researchers' curiosity. While various perspectives, methods, and goals exist, they all converge on the shared aim of gaining deeper insights into human beings. This encompasses both physiological markers and behavioral characteristics. Such investigations have contributed to a more comprehensive understanding of how the brain collaborates across its hemispheres to produce coordinated actions, spanning individual and interpersonal realms [90]. Psychology, renowned for its pursuit of understanding, predicting, and influencing human behavior, has historically lacked a model that incorporates neurobiological mechanisms [57].

1.2.2 *Mind the Fractal Mind: The Within and Above Network of Brain's Emotion, Perception, and Cognition*

1.2.2.1 Functional and Hardware Fractal

The human mind is an intricate amalgamation of various mental states, encompassing emotions (e.g., fear, hatred, love), cognitions (e.g., memory retrieval, future planning, concentration), and perceptions (e.g., facial recognition, color perception, auditory perception). Historically, it was presumed that specific mental states corresponded to distinct psychological faculties, each with its own unique neural correlates. The brain orchestrates its multifarious activities through massive cascades of interacting components, influenced by both innate factors and experiential influences [56]. These neuroelectric processes underlie the essence of the mind's operation. While the precise mechanism by which neural field activities interact with their surroundings to engender cognitive processes remains elusive, emerging perspectives suggest avenues for long-term exploration of these fundamental questions. Dynamic interpretations of brain functions are at the forefront of these explorations. Though deterministic chaos and chaotic bifurcations have not been conclusively established as prerequisites for brain functions or as the source of diverse thoughts and sensations, they offer intriguing, thought-provoking, and empirically investigable theories.

Emotions such as fear, rage, panic, seeking, play, lust, and maternal care traverse specific regions within the diencephalon, particularly the periaqueductal gray (PAG) within the centromedial midbrain. Panic and play are predominantly associated with the medial thalamic and anterior cingulate zones, while other emotional circuits pass through the hypothalamus, projecting to higher brain regions like the amygdala, basal ganglia, and frontal cortical areas. These systems facilitate intricate two-way communication channels connecting lower brain regions governing reflexes, autonomic functions, and hormonal responses to higher cerebral systems responsible for integrating spatiotemporal information, fostering complex learning, planning, and diverse behavioral sequences. Emotional systems significantly impact attentional processes through ascending reticular systems, such as the PAG, in lower brain levels, while also providing a fundamental substrate for core consciousness [55]. Consequently, the intricate interplay between emotional command systems, alongside their enigmatic dynamics and their interaction with higher brain regions responsible for thoughts, perceptions, and cultural constructs, remains a subject of speculative exploration.

It is noteworthy that emotional effects within the brain extend beyond the confines of the integrative "command systems" situated in the core of emotional processing, encompassing a far-reaching dissemination across the brain. This expansive influence challenges traditional concepts of emotional localization, emphasizing the presence of profound evolutionary homologies across mammalian species. Hormones such as oxytocin and prolactin, traditionally associated with childbirth and nursing in mothers, play pivotal roles in modulating maternal behaviors, separation distress, and the formation of social bonds. These complex systems find detailed elucidation in animal brains [55].

1.2.2.2 Fractal of Non-physical Entities: The Saga of Emergent Fractals

Plenty of research articles are published advocating fractals in the brain that grow side by side, for brain and consciousness it does not work that way, find fractals within and above. It means the equations that would represent it would have one or multiple imaginary numbers representing every imaginary world where the system would contribute and accept the contribution from. The most interesting question could be fractal of what? Space, time, morphogenesis, and prime? Four different entities, which have no relation in between, all of them grow their fractals independently. Space and time are physical entities, but not morphogenesis or incidents where an intuitive geometric shape changes in a loop to execute an event almost perpetually. Or in prime numbers in the integer systems. These morphogenesis and primes embedded in the engineers could be astronomical or infinite in number, but all of their variations could be emergent from a few fundamental entities. These are two fundamental differences that we experience, when we study the human psychology or consciousness in the light of Self-Operating Mathematical Universe, SOMU. Non-physical emergent fractals, where three elements fuse together to create missing elements for the new fractal seed to be born and that is what makes fractal brain theory interesting because it's a continued synthesis of new rules leaving an option for the free will that could never be mathematically modeled.

1.2.3 The Role of Thermal Noise in Consciousness

Often we think we get energy for our mind and consciousness from food. We have outlined that energy management in the brain is unique [6, Chap. 1]. Zillions of molecular motors, key to our life and our neurons to live depends on thermal noise, in fact, had nature not arranged for harvesting thermal noise we would not have lived for even a moment. Mental states are believed to be non-physical in the Western world; however, in this book we are trying to provide a mathematical structure of all primary 108 cognitive and conscious experiences in the light of Self-Operating Mathematical Universe, SOMU The study of emotional dynamics in the brain is challenging due to the huge noise found in the brain that surpasses the signal. Nonlinear techniques in electrical neurodynamics measurement offer promise for enhancing this field, but the idiosyncratic dynamics of emotional systems within higher brain regions pose significant challenges. At birth, the brain predominantly operates within the realm of affectivity, with fundamental emotional systems orchestrating responses to the external world. As the child matures, these emotional systems imbue the higher brain regions with values. The question to nature is that, why do you do this? The predominant affective resonances within the cognitive domain of the brain, influenced by genetic predispositions and environmental factors, play a pivotal role in shaping individuals' perceptions of the world.

This affective imprinting profoundly influences an individual's interactions with the world, endowing behaviors and societal affiliations with unique affective qualities. The precise mechanisms underpinning the interaction between neural field activities and their environment, giving rise to the complexity of the human brain, remain enigmatic. However, emerging dynamic perspectives offer avenues for prolonged exploration. The intricate dynamics of emotional systems, their widespread influence across the brain, and their profound impact on individual behaviors and societal dynamics constitute captivating subjects for further inquiry and empirical investigation.

Scale-free fractal structures are found throughout the brain hardware, from the synapses to the neural networks, fractal organic superstructures show fractal truth table, it is a mean to make logic that makes no sense when we look at the information structure globally [24]. These structures have self-similar properties, meaning that they have the same basic pattern repeated at different scales. This self-similarity allows the brain to process information efficiently and robustly, even in the presence of noise. One way that scale-free fractal structures can do noise harvesting is by filtering out the same noise very differently over different scales. Imagine the same source of noise gives different fruitful information in different imaginary worlds that are linked by the fractal hardware. This is possible because noise is typically not self-similar, meaning that it does not have the same basic pattern repeated at different scales. When noise is filtered out over different scales, it is effectively reduced in power but not in form.

Another way that scale-free fractal structures can do noise harvesting is by using the redundancy of the self-similar pattern. This is because the same basic pattern is repeated at different scales, which means that there is a lot of redundancy in the information that is encoded in the structure. This redundancy can be used to correct for noise, by identifying and removing errors that occur at different scales. The ability of scale-free fractal structures to do noise harvesting is essential for the brain to function effectively. Without this ability, the brain would be unable to process information accurately in the presence of noise.

1.3 Quantum Approaches to Consciousness

1.3.1 Self-Operating Mathematical Universe, SOMU: A Fractal Brain Theory Where Prime Numbers Hold Key to Consciousness

1.3.1.1 What Does It Mean that Brain is a Quantum Device?

The quantum mind, also known as quantum consciousness, challenges classical mechanics' ability to explain consciousness. It suggests that quantum processes such as entanglement and superposition may play a role in brain function and potentially elucidate consciousness. Consciousness is intricately linked to brain activity, and quantum theory, being the most fundamental theory of matter, offers a promising avenue for understanding it. Various approaches have emerged in recent decades, falling into three categories: (1) Consciousness arises from quantum processes in the brain, (2) Quantum concepts are used to comprehend consciousness independently of brain activity, and (3) Matter and consciousness are viewed as dual aspects of one underlying reality. In essence, all biology, including the brain, is governed by quantum mechanics due to its atomic composition. Quantum biology delves into phenomena like superposition states, coherence, tunneling, and entanglement, which challenge our classical understanding. Regarding the brain's involvement, it primarily consists of neurons, elongated cells with dendrites, axons, and cell bodies. Neurons transmit information by firing or not firing, influenced by their electrochemical potential. This potential depends on ion movement across the cell membrane, altering its charge. To fire, a nerve cell's resting potential must reach a threshold potential. While the precise mechanism of signal transmission between cells remains unknown, neurotransmitters in the synaptic cleft are believed to bind to receptors on the receiving cell, modifying its electrochemical gradient and triggering neural activation.

Recently, we have shown that the picture depicted to us as neuron is like a balloon, only its membrane is working is wrong. Membrane microtubule junction [70] plays a crucial role in neuron, heterojunctions are important [60] for signal transfer in a fractal structure. The electric potential transmitting through the skin of the neuron cell or membrane is responsible for the cognitive, perceptive, and emotional experiences of a human could not be a proper reasoning. Deep inside the neuron cell, or the balloon model, microtubule nanowires are packed and thousands of helical nanowires of proteins are considered silent. We have proved that this very foundation of neuroscience is wrong. All these nanowires do vibrate 10,000 times faster than membrane, and whole idea of neuroscience that milliseconds time domain cognition originates in the membrane is wrong [2, 25, 70, 72, 78–80]. For human cognition and perception, a millisecond time resolution is required, however, maximum time resolution for membrane is a few milliseconds. Then, how come membrane keeps precise time, unless the microtubules within the neuron assists them with the 10,000 times faster processing. Neuron can fire without microtubules, but cannot regulate precise time without microtubule [70].

1.3.1.2 Microtubule Research Has Argued in Favor of Quantum Processes

Quantum consciousness, a theory linking quantum physics and consciousness, was proposed by Roger Penrose and Stuart Hameroff in the 1990s [29]. The theory, known as "orchestrated objective reduction" (Orch OR), suggests that quantum computations in cellular structures called microtubules influence neuronal firing and, consequently, consciousness. While Orch OR faced criticism due to the seeming incompatibility of quantum effects with the warm, noisy environments of biological systems [85], recent research has suggested that entangled states can persist in such conditions [43]. Moreover, experiments in Japan have provided evidence of vibrational resonances and conductivity features in microtubules, supporting their potential as macroscopic quantum systems [68]. Quantum tunneling could be used to read, write information in such a network of junctions [7]. A fractal network of heterojunctions could process information logically [24], where the perceptions, emotions, and other psychological traits experiences an endless chain of thoughts spread over 12 dimensions [78–80], but all that projects to a singular perception.

This renewed interest in microtubules aligns with the idea that these structures, found in nerve cells, could facilitate coherent energy transfer akin to photosynthesis. Microtubules play vital roles in cell structure, division, and the movement of motor proteins. In summary, quantum consciousness, once met with skepticism, has garnered fresh support through recent studies on microtubules and their quantum potential. The relationship between quantum physics and consciousness continues to spark intrigue and exploration [30].

1.3.1.3 More Than Quantum, Understanding Brain and Consciousness Requires SOMU

The role of quantum effects in the brain is a controversial topic, and there is no scientific consensus on the matter. Theories have suggested that quantum effects could play a role in the brain, as they are capable of processing information faster than classical physics. However, there is no definitive evidence to support this claim. One way to bypass the need for quantum effects in the brain is to use the theory of the SOMU. SOMU proposes that the universe is a mathematical structure that is self-organizing and self-evolving. This means that the universe is constantly creating new information and patterns, and that these patterns are not limited by the laws of classical physics. The theory of the SOMU proposes that the universe is a mathematical structure that is self-organizing and self-evolving. This theory could provide an alternative explanation for the role of quantum effects in the brain. SOMU is still in its early stages, and there is much that we do not know about it. However, it is a promising theory that could revolutionize our understanding of the brain and consciousness. Despite the lack of definitive evidence, the possibility that quantum effects play a role in the brain is an intriguing one. If quantum effects are indeed involved in brain function, then it could have implications for our understanding of consciousness and cognition. If SOMU is correct, then it is possible that the brain does not need to use quantum effects to process information. Instead, the brain could simply tap into the vast reservoir of information that is constantly being created by the universe. This would mean that the brain would be able to respond to queries even before they are made, as it would already be aware of the possible questions that could be asked. This would also explain why the brain is sometimes able to make predictions about the future. The theory of SOMU is still in its early stages, and there is much that we do not know about it. However, it is a promising theory that could revolutionize our understanding of the brain and consciousness.

1.3.2 Polyatomic Time Crystal Model of Human Brain: Consciousness as an Architecture of Time

1.3.2.1 All Discrete Times Used to Make a Clock in the Brain Contains a Clock Inside

The SOMU theory of consciousness is founded on two fundamental pillars. The first is the concept of Fractal Information Theory (FIT; [1]), which introduces us to an enigmatic, self-similar tapestry of information, distinct from the conventional Turing tape. Our journey into FIT commenced in 2014 [23], driven by the resonant harmonies observed in proteins, protofilaments, and neurons—a symphony of fractal resonance bands. Within this intricate interplay, we envisioned the orchestration of a triplet of triplet symmetries, a phenomenon we deemed far from coincidental. By manipulating the geometric phases of these resonant oscillations, we postulated the creation of intricate timekeeping mechanisms—clocks. These clocks, we believed, could elegantly capture cognitive experiences at the millisecond time domain. Leveraging the underlying symmetry, they could then unfurl thoughts across various temporal domains, all while concurrently processing how these symmetries evolved over time. This transformative journey spanned five remarkable years, culminating in 2019 when we embarked on a groundbreaking exploration: the mathematical underpinnings of prime numbers within the human brain's cognitive responses. The Government of

India's Ministry of Human Resource Development (MHRD) generously funded our project, bringing us to the present, where we unveil captivating insights into the unfolding of human cognition, perception, and emotion within the pages of this book.

The second bedrock of SOMU lies in its consideration of prime numbers as the universe's fundamental code—the very essence of the cosmos' construction. The universe, as the entirety of existence, harbors an astounding multitude of prime numbers, with the ultimate boundary yet to be unveiled. However, we have discerned that humans employ a mere 15 prime numbers, evident in the distribution of brain components [6]. Astonishingly, these 15 prime numbers have the capacity to encapsulate a remarkable 86% of all conceivable vibrations within the cosmos. This mathematical serendipity underscores the profound advantage of possessing a code rooted in primes. We can now assert that this code, aptly named phase prime metric (PPM), could potentially serve as a universal language for all conscious entities throughout the universe. If any conscious entity were to conceive novel thoughts or exhibit cognitive, perceptual, or emotional attributes, another system equipped with PPM could emulate these experiences with a striking 86% accuracy.

Hence, the synergy of this new information theory and the universal code, threading through the very essence of consciousness, positions SOMU as a rudimentary yet promising point of departure. Here, we embark on the quantification of the human cognitive, perceptual, and emotional tapestry. As the years progress, our model will undoubtedly evolve, and with each new experiment, we inch closer to unraveling the enigmatic mysteries that shroud the realms of consciousness.

1.3.2.2 Six-Dimensional Clock Architecture Needs Morphogenesis for Human Cognition and Perception

In the tapestry of our existence, a constant stream of spatio-temporal events unfolds, ceaselessly registering in our minds. In the realm of Western scientific thought, the past century witnessed the acceptance of a four-dimensional space–time framework, comprising three dimensions of space and one dimension of time, to comprehend and portray the occurrences within our universe.

However, Hinduism introduces a unique perspective on time, characterizing it as cyclic—a phase cycle completing a full 360-degree rotation to constitute one unit of time. This distinctive viewpoint on time is exceptional globally. Aside from Hindu scholars who have diligently explored the concept of minuscule time increments over millennia, such an approach has not been extensively investigated elsewhere. Thus, we decided to embark on this intriguing journey.

While the scientific community has ventured into the realms of linear and nonlinear time, Hindu scriptures have been replete with contemplative experiments involving discrete time structures. These time loops, which govern the passage of time, are situated on phase spheres. Placing these spheres within a 3D assembly effectively emulates space, resulting in three dimensions of time and three dimensions of space converging to form a 6D space–time structure—a path we've chosen in this book to delve into the intricacies of human cognition, perception, and emotion by leveraging this distinctive approach to time.

Whenever a fundamental shift occurs in our comprehension of information structures, it inevitably carries profound consequences. Over time, the scientific community has gradually adapted to the concepts of linear and nonlinear time, resulting in the development of various valuable tools. However, when we transitioned to the 6D space–time structure, especially one anchored in an unconventional concept of time reminiscent of atomic orbitals where only specific orbitals are permitted, our worldview underwent a profound transformation, albeit with a dearth of pre-existing tools to assist us.

Thus, an imperative task emerged: the creation of the necessary tools to navigate this uncharted territory. As previously mentioned, our chief concern revolved around the notion that if information proliferates within and beyond, each world exists in a state of imagination relative to the others. Consequently, embarking on journeys through these distinct worlds stacked atop one another implies traversing diverse dimensions. Recently, a proposal emerged positing that the human brain processes information across 12 dimensions. In essence, we concur with this perspective, but we diverge in our definition of dimension. For us, transitioning to another dimension entails discovering the enduring structural element that remains unchanged during the shift. Each answer we find in this quest becomes a stepping stone in our journey to higher dimensions.

Now, in the realm of 6D spatio-temporal events, venturing into higher dimensions introduces us to a diverse array of elements. Human thoughts and cognitions are intricate chains that span multiple dimensions. In this unique concept of the mathematical universe, where only elements are invariant, there exists no need for traditional factual or informational storage akin to computer memory. Instead, different imaginative realms house invariants from distinct temporal domains. Consequently, the thread that traverses these varying imaginary worlds embodies our thoughts, representing the intricate tapestry of cognition, perception, and emotion.

1.4 Current Protocols Used by Psychologists for Human Subject Experiments Are Wrong

1.4.1 Beyond Culture, Language, Race, Caste, and Creed: Exploring Deeper Dimensions of Identity

1.4.1.1 Individualism, Societal Harmony and Cultural Variation of Neural Responses

Cultural neuroscience explores how culture influences brain function, perception, and behavior, emphasizing the interplay between genetics and experience. It investigates how cultural values and practices shape cognitive processes, using cutting-edge neuroscience techniques to reveal both behavioral and neural differences across cultures. Recent research has shown that even seemingly universal functions can have underlying neural variations influenced by cultural beliefs. One fundamental aspect studied is self-construal style, with independent cultures emphasizing autonomy and uniqueness, while interdependent cultures prioritize social harmony and conformity. This emerging field aims to unravel the intricate relationship between culture and the human mind, shedding light on the diversity of human cognition.

What we see and how we see are both influenced by culture. Participants in one study were shown a line within a frame and asked to reproduce a line of the same absolute length or proportional length with reference to the printed frame, adapting the classic Rod-and-Frame Test [40]. When asked to attend to their absolute size, American participants were more accurate at reproducing lines of correct length, while Japanese participants were more accurate at reproducing lines of correct length when asked to attend to their relative size. These differences point to cultural differences in visual attention, which could be due to differences in the integration of percepts across cultures. Hedden et al. [32] used fMRI to investigate neural activity during this task, based on these behavioral findings. Different brain regions were recruited to perform the relative and absolute line-judgment tasks in relation to the perceiver's culture, revealing cultural variation in neural responses. When drawing the line that was more incongruent with their cultural patterns, participants recruited frontal and parietal regions associated with attentional control to a greater extent, suggesting that East Asians and Westerners use different neural circuitry for relatively simple perceptual tasks.

1.4.1.2 Identical Neurological Path, yet Very Contradictory Decision-Making

Such culture–brain interactions have clear real-world implications, as they may influence who we choose to affiliate with, both on a personal and broader, political level. Someone that we elect as our leaders, for example, can have a significant impact on our lives and well-being. Researchers made a surprising discovery about leadership preferences across cultures in a recent study using fMRI [67]. Participants from the USA and Japan saw the faces of actual US and Japanese political candidates, formed impressions of them based on their power and warmth, and indicated their voting preferences. Participants in the USA preferred candidates who appeared to be more powerful, while those in Japan preferred candidates who appeared to be warmer, and this preference was reflected in their actual voting decisions. Nevertheless, fMRI scans showed that both American and Japanese participants have used the exact same area of their brains in this task—the amygdala. Members from both cultures followed the same neurological path in this case, but the differences in behavioral outcomes were due to cultural preferences. As a result, researchers observed that emotional signals from in-group members have a stronger neural response than emotional signals from outgroup members.

Our brains actively absorb the regularities in our cultural environments. The brain's representations, in turn, influence how we interact with and shape our surroundings. The way the brain is wired and activated is influenced by culture, but the brain is malleable. Much like the changing tide can erode a footprint in the sand, so too can changing experience over time reshape these brain activations. In this way, the brain can be thought of as a "cultural sponge," absorbing the regularities of our physical and social surroundings.

1.4.2 The Reason for Redesigning Human Classification Experiments in Psychology

1.4.2.1 Pattern of Thoughts: Are We Truly Capturing the Pattern?

Pattern thinking has a strong influence on cognition and understanding human behavior [45]. Inferences from pattern thinking are however based on observation of behavior rendering it subjective. However, studies by physiological psychologists have helped reduce this subjectivity by exploring the brain behavior relationship. The human brain has mainly been explored through localization, lateralization, and regionalization studies. Further, with the advancement of neuroimaging techniques, understanding, analyzing, and predicting human behavior became more scientific by reducing object observation. Individual differences replaced generalization of human behavior. Same versus distinctiveness has always plagued the mind of researchers. For example, perception, every individual has a unique way of connecting the dots to perceive the same stimulus. This uniqueness is based on their past experience, cognition, emotion, and other factors. How much of the stimulus is to be perceived for further interpretation also varies from individual to individual. Thus, this made for an interesting study to understand the boundaries of the brain that creates them. The study of the brain and behavior relationship had to happen with the convergence of knowledge and tools available across disciplines.

1.4.2.2 Have We Mapped Properly Even Introverts and Extroverts?

The field of psychology is dedicated toward understanding, controlling, and predicting human behavior. But this proves difficult at times when behavior does not follow a predicted pattern and are performed beyond conscious awareness of the individual. Individual differences defy universality of behavior, reducing psychologist prediction to probabilities. Another reason why predicting behavior becomes difficult is the presence of a host of factors that intervene between the stimulus and behavior. These factors are both internal (physiological) and external (situational). This is a reason why understanding human behavior has always been challenging for the psychologists, to evaluate the experimental data carried out on the mind and its expressions. Most expressions of a human are masked by emotional states and preconditioning is not sufficient. A human mind wanders continuously and it cannot be restricted to a particular state so that scientifically designed experiments could continue on a static mental state. Therefore, discovering all possible invariants that would enable credibility to the data acquisition and analysis are welcome. There have been several efforts on the front of data analysis, increasing the number of datasets does not help, statistical tools are applied only to minimize random fluctuations by a certain individual, it does not take into account generic features of a human mind.

For example, in our day-to-day life we encounter introverts and extroverts. Have we truly charted the intricate landscapes of introverts and extroverts, unraveling the fascinating depths of their inner worlds and the vibrant tapestries of their social interactions? An introvert's orientation is more toward self while an extrovert is oriented more toward the outside. An introvert enjoys solitude, both physical and emotional and an extrovert seeks the attention of others and is more energized with others. So introverts and extroverts usually exhibit opposite behavior Often we think that there are two kinds of people in the society when we carry out the psychological tests. However, we all know that there are not two but many categories made of mixed mental attitudes between introverts and extroverts. We simply do not have the tool to measure and classify attitudes and expressions in a set of parameters that could be used to filter the experimental studies on the mental states.

1.4.2.3 Preconditioning of Human Subjects?

Another problem in the psychological experiments is regarding the preconditioning of the subjects, that is widely used by psychologists to make sure during the experiments, the subjects are all in a singular fixed mental state. However, all psychologists are well aware that preconditioning does not work. There are a wide variety of human subjects whose responses are well beyond the normal scales at all times, at all states, they are neglected during the tests. However, there are people who deliver normal responses, like other subjects, but the performance is not according to the examiners desire. Therefore, two major problems are there which all psychologists are aware of, yet there is hardly any solution.

1.4.2.4 Why is It Absolutely Essential to Classify Human Based on Their Emotion?

In our case we surveyed the literature and found that emotions, perceptions, and intelligence are grossly studied considering that all humans are identical. On one hand we have theories to explain human behavior and on the other we believe that there are individual differences. Primarily one school of thought negates the other. We felt this is fundamentally wrong. In our day-to-day life, we encounter similar classes of humans and in our social interactions we often suggest that my known person A is close to person B and those people, C, D and E have similar emotional outbursts. This helps us to set boundaries to within and outside group; us vs they, differentiating one from another. So, classifications of humans based on emotions we do every now and then, however in literature the examples are scarce. But such differentiation is based on observation and are naive. The need of the hour is to develop categorization based on scientific judgments, for example, classifying emotions. Primarily there are six emotions-happy, sad, fear, anger, surprise and disgust. The popular tools used in Psychology to measure these emotions has been with the use of questionnaires. Results of questionnaires have always been plagued by the respondent giving socially desirable responses. So, with the passage of time measurement of emotions shifted from the use of questionnaires to the use of coding system.

One another initiative of measuring emotion was the use of thermal cameras- this marked the beginning of the use of infrared cameras to measure human emotional expressions. It was a scientific physiological-based measure that allowed contactless measuring of emotions. But, there are plenty of literature where for a particular mental state one kind of facial thermal changes has been assigned. As if all humans in the world would have a similar thermal signature. Take another example. Mirror drawing tasks are being performed since the 1920s. The underlying assumption of the mirror drawing task is that you train one hemisphere with your non-preferred hand and then a few trials using your preferred hand which helps in training the other hemisphere. And then when you try performing the task back with your non-preferred hand you find that the error and time taken have both decreased suggesting a transfer of learning that has facilitated the learning of the non-preferred hand. Till date, all tasks consider that all subjects develop motor nerve training skills similarly. For that reason, there has been no theory or experimental proposal ever that left brain and right brain could communicate differently in different humans. The training would transform the skill differently but not randomly. A particular class of people may train similarly. The question is how many classes are there in the world? We don't know. We wanted to learn as many classes as possible.

1.4.2.5 The Textbook Teaching of Cognitive Processing Has no Scientific Foundation:

Let's move to the third example, cognitive task where most psychological tests are performed to understand the learning ability and development of perception in a human being. Cognitive tasks require either left brain or right brain and there is a subtle need for the exchange of energy between the two parts of the brain. However, does it happen similarly as we see it in case of mirror drawing? While mirror drawing involves motor nerves and exchange energy between the left and right brain to build a cyclic energy transmission cycle, cognitive tasks involve different sensory parts of the brain to build a loop of energy transfer. We were excited whether left to right brain communication is similar for both the cases? Also, another important question that haunts us, if we classify humans into different categories based on their mode of left and right brain energy transfer, should we find conclusive evidence that categories of human subjects are universal if classifications are done based on their left–right brain energy exchanges?

The final most radical question that arises is that if humans are classified in terms of typical emotions and learning processes, then how do they communicate? Can communication be only verbal or does the human mind have the capacity to connect silently? Though hard to believe but communication can be done without the use of words and in complete silence. What happens in a real society, is it information content that makes a decision or let society run on that view or it is classification of different categories of peoples who are participating in the decision-making process. To understand this question, we certainly need to make the subjects silent and let them communicate without signs only looking at each other and try to map their brain and verify later if they were exchanging really meaningful features. This is the most important factor in decision-making, most decisions made by humans in today's world are collective and the primary factor if not information alone, but the categories of people participating then that would change the way we study human psychology forever.

1.5 Fractal Brain Theory: Information Self-Similarity or Hardware Self-Similarity?

1.5.1 If Brain Hardware is a Fractal Structure, Why Cannot Be Thoughts?

1.5.1.1 The Magical Use of Prime Numbers in Constructing the Brain

Electromagnetic resonance of biomaterials is a widely studied subject, but, we are the only group who noticed that depending on the structural symmetry, during electromagnetic resonance electric and magnetic parts get separated on the biomaterials and they occupy different regions. These isolated regions collectively emerge unique resonance spectrum [77]. The brain is a complex organ with a vast number of interconnected neurons. Recent research suggests that the brain may be organized in a fractal way, meaning that it contains self-similar structures at different scales [48, 58, 59, 62]. This has led to the development of the fractal brain hypothesis, which proposes that the brain's fractal structure is essential for consciousness [21, 72, 81]. We are not interested in the self-similarity features of the fractal, but the junction of these self-similar regions, where the geometric phase is born due to adiabatic shielding of inner layer of resonators from the outer layer of resonators. Consequently, membrane-microtubule junction plays a crucial role in neuron's time modulation of nerve spikes that is fundamental to brain's cognition, thus, heterojunctions are important [60] for signal transfer in a fractal structure.

Our group has been instrumental in designing experiments where neuron circuits were mapped at different times using a newly invented machine called Scanning Dielectric Microscope, SDM. This machine helped us, seeing circuits operating in the microseconds time domain, 10,000 times faster than neurons [2, 25, 78–80]. It means at the cellular level, we have already observed EEG like pattern, but in different time domains using a new instrument called Scanning Dielectric Microscope, SDM. While SDM could live measure neural circuit interactions, it was essential for us to invent an analogous machine to check human behavior and its cognitive responses. Here in this book, we have explored the new invention Dodecanogram, DDG that is analogous to SDM, in Chap. 5 we have shown how this instrument measures twelve time domain signals from the brain.

One of the key features of fractals is that they are scale-invariant, meaning that they look the same at different magnifications [8, 49]. This suggests that the brain's fractal structure could allow it to process information at different levels of abstraction, from the smallest details to the most general concepts [58, 59]. Another important feature of fractals is that they are complex and unpredictable [34, 76]. This could explain why the brain is so good at generating new ideas and solving problems. The fractal brain hypothesis is still a new and controversial idea, but it has the potential to revolutionize our understanding of consciousness [64]. By understanding how the brain's fractal structure works, we may be able to develop new treatments for neurological disorders and even create artificial intelligence that is truly conscious.

Self-similar structures are abundant in the brain. A few symmetries or order parameters have been used to develop structures, repeatedly used for constructing several brain components. For example, hexagonal close packing of the cortical columns. A 2D sheet of cortical columns form the cortex and a 2D sheet of cortical columns fold uniquely to form the hippocampus. There are roughly 23 different geometries of neurons repeatedly building a massive 80 billion neural networks that constitute the brain. Similarly, 23 types of glial cells which spirally wrap around neurons generate a massive statistic of 8000 billion replicas. Therefore, structural self-similarities are fundamental to brain construction. In the 1920s, Brodmann wanted to find how thought processes differentiate in self-similar brain structures. He found roughly 47 distinct regions of the brain where neurons are uniquely arranged. All cognitive processes thought, perception, and intelligent decisions are finally made in these 47 regions [6].

1.5.1.2 The Many Proposals of Fractal Brain and the Limitations of Fractal Brain Theory

Observing an abundance of self-similarity in the ordered arrangement of the brain, several researchers proposed that the brain is a fractal machine. It means only a few key pieces of information repeat to generate all kinds of thoughts at the perception level. Neurogenesis, or correlating the brain thoughts with the structure of the brain hardware is still a distant dream, however, significant efforts are made to simplify the process of decoding the thoughts. Fractal shapes are one kind of approach to the understanding of the brain and fractal information theory is another. In fractal information theory, researchers do not argue about what the components look like. Rather, they talk about a few pieces of information repeating and overlapping to frame a thought. Here, on this point of argument there are extensive debates. It is generally considered that due to self-awareness all thoughts generated in the brain are new. We never see an apple twice similarly.

Fractal brain proposals have several other problems. First of all, assigning a fundamental component that generates the fractal signal. For example, if we consider it to be a neuron, and self-similar pulse streams made of neuron spikes are responsible for the cognition and perception that eventually leads to consciousness. The problem remains the same, how thoughts are encoded, what is the real-world physical significance of a pulse. If there is none, at which level of fractal integration does thought acquire a conscious experience?

1.5.1.3 The Difference Between Fractal Analysis of Signal and Fractal Brain Theories

Whenever there is astronomical data, one generic tendency among researchers is to fight the self-similarity index of the data and these fractal coefficients are often used to find key governing rules for the information generation. However, in practice the origin of self-similarity could have multiple contradictory probabilities. Often fractal analysis are dead values with no physical significance attached to it. We find application of fractal analysis in large-scale financial, astrophysical genetic proteomic healthcare data, mostly delivering accidental repeating factor. Historically, in the 1940s it was suggested that most events happening in nature are repeating themselves. Even our brain registers repeating events as key to building its intelligence. By the 1970s, complex mathematical theories appeared to find periodic events.

Surprisingly, the historical journey of fractal analysis and exploration of periodic events by hardcore complex mathematical modeling of the data had very different routes and they never merged. One of the fundamental differences between the two approaches is that in case of periodic event hunting inside a big unknown data, the physical significance of the repeating event is very well known and all fundamental associated parameters related to that event are well organized and understood. Therefore, while evaluating the fractal brain theory it is utmost important to cover the periodic events, because it leads to the discussion of invariants, key to brain's learning. When a kid learns it finds a geometric invariant that never changes with the training subjects. Mostly, the invariants are hidden in the abstract geometric features of the information content.

1.5.1.4 The Quest for Invariants to Understand the Mechanism of Human Thoughts

The question that researchers always wanted to find an answer to was, if it is possible to have a few geometric transformations that would cover all possible transformations in the universe? For example, we see the transformation of spirals and vortices all around the universe. Therefore, searching for spirals and vortices in anything we see and feel, would not be a futile process. To some extent we will be successful. Similarly, teardrop to ellipsoidal transmission, that is also, we see everywhere. It is also nice to see that spirals are electromagnetic fields while teardrop to ellipsoid is gravity-like. Therefore, if we can build a grammar of how 10–15 geometric transformations are all possible to occur in nature, we can really build a new kind of computer. That computer would search for 15 geometric transformations everywhere, thus, identify the clock in any input data. We simply assume that whatever be the form and format of the information, we do not need to know, but they would fall into the category of these 15 geometric transformations. Therefore, we can get input as a code of 15 geometric transformations, then convert them into 15 geometric shapes embedded into clocks holding geometric shapes, then convert that as a composition of metrics, running in a loop. The infinite series-like clock architecture is converted back into the input image, using 15 geometric shape transformations. The interface between the computer and nature is the 15 geometric shapes.

1.5.1.5 Geometric Self-Similarity is also a Fractal: Is Confusion a Door to the Brain's Cognitive Power?

Let's explore the intriguing insights provided by ancient Indian philosophers regarding the concept of confusion. In Hinduism, two distinct types of confusion have been delineated.

The first type comprises a set of confusions that can be interlinked in a time loop. These confusions emerge sequentially and follow a logically circular pattern.

The second type involves temporal loops of confusions nested within one another. Within this framework, we encounter confusion structures consisting of elementary confusions, and each confusion within the loop serves as an elementary or constituent element for another, higher-level confusion.

These two categories of confusion naturally emerge as we delve into inquiries guided by the universal symmetries of the cosmos, as advocated by Maharshi Kapil in his Sankhya Darshan. He proposed that "contexts and contents are naturally created in the universe to regenerate the symmetry [17]."

The idea of creating confusion within confusion aligns with one of the fundamental principles in Hinduism, which posits that there must be similarities in everything we observe. By grouping these similarities into a time loop and replicating such loops, we can construct an intricate network of universes both within and beyond our immediate reality.

Over a span of 3000 years, Hindu religious scholars ardently adhered to this protocol [17, 27]. They applied it diligently to various aspects of existence, endeavoring to uncover the right confusion and traverse through the different layers of confusion by embarking on journeys through diverse realms of imagination and understanding.

1.5.2 *Fractal Thoughts, Fractal Perception, and Fractal Decision-Making?*

1.5.2.1 Shared Values at All Levels Without Amplifying Resources Exponentially

Fractal organization is all about shared values. It leads to integrity of the organization and generates a perception of reality. A **fractal organization** can be formed either by multiple inter-(ter-) dependent organizations which share a common (primary) driver, or by multiple branches. Self-similarity or the fractal feature is found in the governing parameters that hold the organizations. Having multiple constituents within a common structure helps in alignment, sharing of learning, increase in wisdom, skills, and decision-making. It also helps in adapting quickly to a changing situation. While fractal management is a methodology for observing and changing scalability attunement in organizations. Exploring the field of management from a fractal perspective is in its nascent stage and so open to ideation. Fractals primarily refer to systems made up of systems; however, there should be a level of self-similarity. When people interact in organization, following common drivers' patterns emerge from a common feature. During the pattern evolution, complex decisions are made and often it becomes difficult to fathom the intricate details. It is interesting to see and understand these patterns. Complexity theory helps understand how these interactions unfold.

1.5.2.2 Interhemispheric Transfer of Information and Fractal Organization

In many aspects, an institution running by fractal organization actually depends on the individual making decisions by interaction with tasks, training and other aspects. All kinds of learning processes require exchange of cognitive processes between the left brain and the right brain. In this book we have emphasized on the Mirror drawing task, where we could see live how different patterns emerge when signals transfer from one part of the brain to another before transferring the lobe. Now, learning process affects the evolution of organization.

1.5.2.3 Role of Interference in Cognitive Tasks: ASMR, Random Noise, and Periodic Beats

In the Vedic wisdom, sound is considered supreme, the geometry of sound has been a subject of exploration for thousands of years. In the pre-Vedic era, it was prohibited to write pure wisdom. The 3D field distribution of mechanical energy created by pronunciation contains an intricate 3D geometric shape, and the listener takes a bath in this 3D field distribution. This has been compared to taking dips in the holy rivers. Thus, sensing the 3D geometry of sounds has been an integral part for the development of Sanskrit language, Indian classical dance and raga-based music. C V Raman's work on anharmonic overtone was not an accident, Indian drums and musical instruments are tuned to generate these anharmonic overtones. Thus, 3D geometry of sounds that appears as noise to most ears, would be the coded language for a Vedic practitioner. ASMR is often a 3D geometry of sound, which means, imagine a cube, whose eight corners flow eight periodic sounds, now if the cube rotates, it would appear as if a random noise, but it is not. It has always been a common argument that brain signals operate under noise which has magnitude greater than the processing signal [45]. ASMR or random sounds of objects with no order in it affects cognitive tasks in the positive direction. Several ASMR research has established that it assists in faster sleep, greater attention and computation with greater efficiency. On the other hand, periodic beats which are supposed to generate an ordered mental behavior, affects very negatively in attention and consequently cognitive tasks. While the argument would continue on how noise and ASMR would be different as far as geometric arrangement of ordered signal is considered.

Brain under noise behaves in a very organized manner. The reason for its greater efficiencies under random noise is unknown. There exists no theory or plausible explanation on its origin. However, one speculation is that the brain is a fractal device and noisy signals access a greater number of communication channels and activate more modes in the sensory system. That enhances memory elements in a greater bandwidth. Moreover, geometric information is more stable than linear stream of signals. Therefore, we have advocated for reshaping any composition of information into a geometric structure, whose corners activate simultaneously, or at a time gap (geometric musical language, GML; [1, 6, 65]).

Here in this book, we have carried out an extensive analysis of noise and ASMR effects on the left and right brain communication in Chap. 4 and thereby enhancing the efficiency of a human subject. But the question is why should the brain be so happy in receiving the geometric shapes? Brain gets happy because it brings rich significance and a set of events that builds a new story, a sequence of thoughts, which makes any brain happy. To the best of our knowledge, there have not been any report for the fact that periodic signals in reality acts as a negative catalyst to the thought process.

1.5.2.4 Fractals in Decision-Making from Self-Organized Criticality

The nature that we see around us is perpetually out of balance, it is not random but organized in a poised state as far as its global superstructure is concerned. The critical state where an organized phenomenon could emerge from randomness falls within well-defined statistical laws. The science of self-organized criticality [3] explore the fundamental question of why Nature is complex, whether it is a composition of key governing laws, or emergence of new laws continuously from a random concept set or infinite series generated by apparent self-similarity that changes its form in the long run. Self-organized criticality tries to discover some ubiquitous self-similar patterns in nature, instead of parameter-based specific mathematical constructs.

One important aspect is that any conscious being always considers methodologies so that decisions are made as a predictable sequence of events. Dynamic decision-making often requires making a series of decisions. In that series the constituent decisions are not independent, internally connected and an environment that apparently does not have any role to play, decides the grammar of decision changes. When the state of the world changes continuously due to the decision made by the system itself or autonomously due to accidental match between environment and the system and finally, as a consequence of the decision-maker's actions to change the world, we find systems as dynamic and decisions should be made as a sequence. Often the decisions are to be made in real time, even under the scenario that the world is changing, then two dynamic processes, the decision made by the user and the changing environment work together modifying each other.

Dulov [14] addressed three principle questions applicable to any particular decision-making theory: (1) How many criteria do we need to reach a decision? (2) How well they are defined/described from the information database from which the decision is arrived at? (3) Are there any relations between the key parameters governing the decision, or can we consider them to be independent entities?

(1) How many criteria do we need to reach a decision?
(2) How well they are defined/described from the information database from which the decision is arrived at?
(3) Are there any relations between the key parameters governing the decision, or we can consider them to be independent entities?

1.6 Meta-analysis of Functional Brain Imaging Datasets

1.6.1 Global Databases of Brain Scans and Automated Brain Image Analysis

1.6.1.1 Involuntary Human Behaviors and Connecting Hardware

Spontaneously operating neural circuits that runs various biological rhythms, enzyme cycles, hormone cycles and cell, protein regeneration cycles are considered silent; however, they play a crucial role in setting up a body that acts like trillions of clocks forming a matrix. The cognitive experiences would only have to choose a few and assembly to fit local environment. A large database of brain signals shows variations of the internal brain activations for specific cognitive functions. The brain-to-brain variations for the singular cognitive function reveals us that there is practically no absolute marker for human emotion, cognition, and emotion. At the same time, certain signatures are always there that we could intuitively suggest as a marker for a particular brain signal. Automated brain image analysis has emerged as a crucial research area, simplifying the complexity of brain signaling by distilling essential information from vast datasets [41]. Functional Magnetic Resonance Imaging (fMRI) images of the brain provide insights into the activation of different functional regions when a person engages in thought, expresses emotions, or processes perceptions.

To facilitate behavioral analysis, researchers extract regions of interest (ROIs) from these functional images, anatomical scans, or brain atlases. This extraction process relies on spatial normalization to standards such as the MNI or Talairach coordinate systems. Researchers primarily categorize brain functions into five behavioral domains: Action, Cognition, Emotion, Interoception, and Perception. These primary behaviors further branch into 51 subcategories ([20], see Table 1). We are emphasizing these 51 subcategories because eventually we want to create mathematical representatives of all cognitive processes, we have listed 91 cognitive traits of a human mind retrieved experimentally using DDG in Chap. 5. We have made one to one correspondence between our method of psychology studies and the conventional database in Chap. 2, to set a ground for the fMRI and EEG link, which has been a primary concern for many researchers for over three decades.

A comprehensive list of these subcategories and corresponding brain images is meticulously organized in the Brain Mapping Database, accessible at http://www.brainmap.org/taxonomy/behaviors.html. This resource serves as a valuable repository for researchers exploring the intricate landscape of brain activity and behavior.

The table available at http://www.brainmap.org/taxonomy/behaviors.htm provides a comprehensive overview of various behavioral domains explored within the realm of brain mapping. These domains encompass:

1. **Sensory-Motor Behaviors**: This category delves into interactions between the body and the environment. It encompasses faculties like vision, hearing, touch, movement, and balance. While conscious thoughts are generally considered to

be the primary driver of behavior, there is evidence to suggest that unconscious influences can also play a role [53]. Unconscious influences on decision-making have been found to have limited and narrow effects, but they still contribute to shaping behavior

2. **Cognitive Behaviors**: Cognitive behaviors pertain to the processing of information. They encompass facets such as attention, memory, learning, and decision-making. Furthermore, involuntary attention, which is influenced by trait impulsivity, has been shown to affect various psychological processes such as decision-making, social success, and psychopathology [42]. This highlights the wide-ranging effects that involuntary behaviors can have on an individual's psychological well-being
3. **Emotional Behaviors**: Emotional behaviors encompass both the expression and experience of emotions. This domain covers emotions like happiness, sadness, anger, and fear. In addition, research has shown that voluntary actions and conscious awareness are closely linked [29]. Involuntary movements caused by brain stimulation can reverse the effects of voluntary actions, indicating the intricate relationship between voluntary and involuntary behaviors.
4. **Social Behaviors**: Social behaviors involve interactions with other individuals. They encompass aspects like communication, cooperation, and competition. Involuntary human behaviors can have a significant impact on human psychology. These behaviors are often conditioned and regulated by environmental stimuli [5]. For example, involuntary admission to psychiatric institutions, which is a coercive action, has been a topic of international concern due to its potential long-term impact on a person's well-being and human rights [82]. This suggests that involuntary behaviors can have profound psychological consequences.
5. **Autonomic Behaviors**: Autonomic behaviors are involuntary actions that regulate bodily functions, including heart rate, breathing, and digestion. The extent to which unconscious information can influence behavior has been a topic of debate in psychology [11]. While there is evidence to suggest that unconscious influences can have an impact, the magnitude and scope of these effects are still under investigation.

The table further provides insights into the specific brain regions associated with each behavioral domain. For instance, the visual cortex plays a pivotal role in vision, the auditory cortex is crucial for hearing, and the motor cortex governs movement. Overall, involuntary human behaviors can have a profound impact on human psychology. These behaviors are influenced by environmental stimuli, trait impulsivity, and unconscious influences. Understanding the psychological consequences of involuntary behaviors is crucial for promoting well-being and addressing potential negative outcomes.

This table serves as a valuable resource for gaining an understanding of various behavioral domains and the corresponding brain regions implicated in them. Researchers can leverage this information for planning studies focused on brain function and behavior, while clinicians can employ it for the diagnosis and treatment of conditions affecting behavior.

1.6.1.2 Unveiling the Power of Invariants in Science

In science, the pursuit of understanding and explaining the natural world often involves grappling with complexity and change. Nature's intricate systems and phenomena can challenge researchers to find underlying patterns and principles that govern the dynamic and ever-evolving universe. One invaluable tool in this quest is the concept of invariants. Invariants serve as the bedrock upon which many scientific disciplines stand, enabling researchers to uncover hidden truths, make predictions, and unravel the mysteries of the cosmos. Thus far, mind researchers, psychologists, and specially consciousness researchers have not delved into this domain [87]. However, since SOMU is based on architecture of invariants using protein and its complexes retrieved language, namely GML.

The Essence of Invariants

At its core, an invariant is something that remains unchanged under specific transformations or operations. In science, invariants are essential because they provide stability and predictability amid the constant flux of variables and conditions. These unchanging properties or principles allow scientists to simplify complex problems, uncover hidden relationships, and make meaningful predictions. Invariants are mathematical properties or relationships that remain unchanged despite variations in the values of variables. They provide a way to link a large number of variables by identifying common patterns or constraints that hold across different instances or conditions [89].

In computer science, invariants play a crucial role in software testing and debugging. Daikon, a dynamic detection system, identifies likely invariants by analyzing program executions. These invariants can involve multiple variables and help identify patterns or relationships that hold true during program execution [18].

Overall, invariants serve as powerful tools for linking a large number of variables across different domains. They provide insights into common patterns, constraints, or relationships that remain unchanged despite variations in the values of variables. By identifying and analyzing these invariants, researchers can gain a deeper understanding of the underlying structure or dynamics of complex systems.

Types of Invariants

Invariants come in various forms, each tailored to the specific field of science where it is employed. Here, we explore some common types of invariants and their applications in different scientific disciplines:

1. **Symmetry Invariants**
    - **In Physics**: Symmetry invariants play a fundamental role in physics, particularly in the realm of particle physics and quantum mechanics. Noether's theorem, formulated by mathematician Emmy Noether, states that for every continuous symmetry in a physical system, there is a corresponding conserved quantity or invariant. For example, the conservation of angular momentum is a result of rotational symmetry in a system. Measurement invariance is a

concept commonly used in organizational research. It refers to the consistency of the measurement properties of a scale or instrument across different groups or time points. Tests of measurement invariance examine the associations between observed scores and latent variables, which can help establish the link between a large number of variables [89].

- **In Chemistry**: In molecular chemistry, symmetry operations help identify the symmetry elements and classify molecules into specific point groups. These symmetry invariants are crucial for predicting a molecule's spectroscopic properties and understanding its reactivity.

2. **Invariant Manifolds**

   - **In Astronomy and Celestial Mechanics**: Invariant manifolds are used to describe the long-term behavior of celestial bodies within the framework of the three-body problem. These manifolds help predict stable regions in the solar system, such as Lagrange points, where spacecraft can remain relatively stationary with respect to multiple celestial bodies.

3. **Topological Invariants**

   - **In Mathematics and Materials Science**: Topological invariants, such as the Euler characteristic or topological charge, are used to classify and study shapes and structures. In materials science, they are applied to characterize the electronic properties of topological insulators, materials that exhibit unique conducting properties on their surfaces. In mathematics, invariants are widely studied in various fields, such as algebraic topology and differential equations. For example, the Alexander polynomial is a famous link invariant that captures properties of knots and links in three-dimensional space [51]. Differential invariants are used to integrate ordinary differential equations and classify transformations in mathematical analysis [88]. Laplace invariants are fundamental semi-invariants for scalar linear hyperbolic partial differential equations [46]. These mathematical invariants provide a way to link a large number of variables by capturing essential properties or relationships.

4. **Invariant Networks [46]**

   - **In Neuroscience**: Invariant networks in the brain refer to patterns of neural activity that remain constant across various cognitive tasks or conditions. Understanding these invariant networks helps researchers uncover the neural basis of cognitive functions and brain disorders.

5. **Invariant Theory**

   - **In Algebra and Algebraic Geometry**: Invariant theory deals with polynomial functions that remain unchanged under transformations. It has applications in various mathematical and scientific fields, including coding theory, robotics, and computer graphics.

The Power of Invariants [46]: Invariants serve as invaluable tools for scientists across disciplines, offering a consistent framework for understanding and predicting natural phenomena. Here's why invariants are indispensable in the scientific endeavor: Most of the current research is limited to using neural network to find invariants. Here we provide a strong literature survey to compile the contemporary works. In Chap. 5, we would describe finding topological and knot invariants for human cognition, emotion, and perception.

1. **Simplicity Amid Complexity**: Invariants distill complex systems or problems into simpler, more manageable forms by identifying the unchanging aspects. This simplification allows researchers to focus on the essential elements of a problem.
2. **Prediction and Control**: Invariants often enable scientists to make predictions about the behavior of a system or phenomenon. By identifying what remains constant, researchers can gain a better understanding of how to control or manipulate these systems.
3. **Discovery of Fundamental Laws**: In physics and other natural sciences, the discovery of invariants has often led to the formulation of fundamental laws and principles. These laws, like the conservation of energy or angular momentum, have profound implications for our understanding of the universe.
4. **Bias and Attribution-Free Decision-Making Using Knots**: The concept of invariants transcends individual scientific disciplines. Since mathematics has no bias, it is very useful to find knots and their symmetries, counting knots is called knot invariant [35], and it provides a good transitions from morphogenesis of geometric shapes to the phase prime metric, PPM of SOMU. There are abundance of literatures on integrating a large number of knot invariants [12].
5. **Robustness and Resilience**: Invariants represent stability and resilience in dynamic systems. Understanding these aspects can lead to the development of more robust technologies and strategies.

Invariants are Silently Running the Universe [46]: Invariants are the silent heroes of the scientific world, providing stability, predictability, and simplicity in the face of complexity and change. From the realm of particle physics to materials science, from celestial mechanics to neuroscience, invariants play a crucial role in unlocking the secrets of the universe. As science continues to advance, the power of invariants remains a guiding light, illuminating the path toward greater understanding and discovery.

So far there has been no attempt to include invariants in cognition and attempts to realize neurogenesis. In this book, we have experimentally reviewed a new theory of consciousness, namely Self-Operating Mathematical Universe, SOMU.

1.7 Conclusions and Malpractices in the Psychology Research: Science or Pseudoscience

1.7.1 Universally Accepted Challenges of Psychology Research

Psychology research faces several challenges and criticisms, some of which can be considered fundamental problems. These problems, along with psychologists' reluctance to adopt more experimental approaches, contribute to the perception of psychology research as pseudoscientific in certain contexts. The credibility crisis in psychology is not limited to the field itself but also extends to related disciplines, such as information systems [19]. The lack of adequate research on credibility and its impact on organizational reputation is a gap that needs to be addressed [37]. Trustworthiness of the scientific literature in industrial and organizational psychology has also been questioned [39].

Here are 10 fundamental problems and reasons for reluctance:

1. **Replication Crisis**: Many psychological studies cannot be reliably replicated, because of the variations among human subjects even when similar conditions are implemented, casting doubt on the validity of their findings. This crisis of replicating a particular human subject variation undermines the credibility of psychology as a science. Human subject classification is a primary concern that we have addressed in this book in every single chapter from Chaps. 2 to 9.
2. **Publication Bias**: Societies promote views of its creators, most journals run by a few foundational editors, and the organization accepts only works that follow their line of thoughts. There is a tendency to publish studies with positive results while neglecting those with null or negative findings, leading to an overrepresentation of significant results in the literature.
3. **Questionable Research Practices**: Some researchers engage in practices like p-hacking, selective reporting, or hypothesizing after results are known (HARKing), which can inflate the likelihood of obtaining significant results. Fear to collaborate and lack of resources to measure human subjects using state-of-the-art instrument add to acquiring a credible data.
4. **Small Sample Sizes**: Psychological studies often use small sample sizes, making it difficult to generalize findings to larger populations and increasing the risk of Type I and Type II errors. There is no grammar or established protocol on how to choose subjects, color bias is very much prominent, since majority of research journals and dominating scientists are from the west. Psychological traits dramatically differs among different cultures. The credibility crisis in psychology research has been a topic of concern in recent years [75]. One of the main issues is the reliance on a narrow sample of participants, known as WEIRD subjects (Western, Educated, Industrialized, Rich, and Democratic) [33]. This disproportionate reliance on WEIRD subjects raises questions about the generalizability of findings to the broader population [33]. The replication crisis in psychology has further cast doubt on the credibility of researchers [75].

Concerns about fraud, replication, publication bias, and false positive results have contributed to the crisis [27].

5. **Lack of Standardization**: Psychology research lacks consistent standards for methodologies, measurement, and reporting, making it challenging to compare or combine studies. Even those groups who undergoes a large human datasets, certainly ignores various brain hardware factors that we have outlined here in this book.
6. **Heterogeneity of Subjects**: Human subjects in psychology research are highly diverse, leading to substantial individual variability that can obscure underlying patterns or effects. This book uncovers the grammar of human subject variations by classifying human subjects on the basis of time management in Chap. 2, on the basis of blood vessel arrangement for thermal expression of emotion in Chap. 3, noise harvesting and periodic signal tolerance ability in Chap. 4, group synchronization ability in Chap. 5, silent communication ability in Chap. 6, response ability to unknown problems in Chap. 8, sensitivity to universal primordial love in Chap. 8, switching ability between two compelling attractions, awe and sexual tension in Chap. 9. Thus, this book has made an unprecedented effort to classify humans based on their hardware and subconscious mind structure.
7. **Complexity of Human Behavior**: Critical psychology offers an alternative perspective on the study of human subjects. It emphasizes the importance of understanding the subjective experiences and conflictuality of individuals in relation to society [73]. By exploring the contradictions between individuals being subject to their conditions and their potential to influence and change those conditions, critical psychology aims to provide a more comprehensive understanding of human subjectivity [73]. Human behavior is influenced by myriad factors, circadian rhythms affect humans differently, sleep cycle runs in different humans differently, making it challenging to isolate and manipulate variables in controlled experiments. All these factors are never taken into account in psychology research, thus, majority of research outputs are preset intuitively in the mind of the researcher and they look for supportive data. Thus, majority of psychological studies databases could never be integrated for betterment of humanity. Here we have addressed this concern and all 289 subjects who were primarily studied in IIM Ranchi were thoroughly investigated for all kinds of rhythms that periodically shifts the mind of a single human subject.
8. **Ethical Constraints**: Ethical concerns limit the types of experiments that can be conducted, restricting researchers from manipulating certain variables or employing hardcore experimental tools.
9. **Inaccessible Biological Mechanisms**: Unlike fields such as physics or chemistry, psychology often deals with phenomena that are difficult to observe or measure directly at the molecular or cellular level. Thus, environmental, societal, cultural background of human subjects regulate their free will, thus, we remain regulated by behavioral responses. For the first time, using DDG described in Chap. 5 we are going to change it forever.
10. **Interdisciplinary Nature**: Psychology research often draws on concepts from various disciplines, making it challenging to fit into a traditional experimental

framework. In the current project, we assembled electronics engineer, software engineer to write codes, physicists, chemists, mathematicians, psychologists of various specialties, and we put all experts in one room, doing human subject studies.

Psychologists may be reluctant to use more experimental tools for several reasons:

1. **Ethical Considerations**: Many experiments that might be considered "hardcore" could raise ethical concerns, such as conducting invasive procedures on human subjects.
2. **Resource Limitations**: Hardcore experimental tools can be expensive, require specialized training, or involve sophisticated equipment, limiting their accessibility to many researchers.
3. **Tradition and Inertia**: Some psychologists may be entrenched in traditional research methods and may resist adopting more experimental approaches due to inertia or skepticism.
4. **Complexity of Human Behavior**: Human behavior is not always amenable to reductionist experimental approaches, and some researchers argue that more holistic or qualitative methods are better suited to understanding it. One of the primary challenges is to prepare the human subjects for the experiments. Often, the human subject gets nervous and want to quit, sometimes they take it as a fun, and sometimes they ignore it.

The perception of psychology as a pseudoscience can arise from some of these challenges and limitations:

1. **Lack of Falsifiability**: Some psychological theories and hypotheses may be difficult to falsify or test empirically, leading to concerns about their scientific validity. Psychology includes subjective linguistics, which cannot be parameterized. In this context, for the first time we have made a bold move in this book in Chap. 5. Using a newly invented machine, DDG, where one could map 12 active time domains from seconds to picoseconds, we have mapped invariant geometric shapes representing 91 cognitive, perceptive and emotional traits which are mathematical structures that could be falsified. This is not an absolute finding, but such practices would help psychologists and neuroscientists to at least present psychological states in terms of parametric abstracts, not undefined subjective terms.
2. **Unsubstantiated Claims**: In some cases, psychology research may make claims or interpretations that lack robust empirical support. Even robust empirical supports are intuitive and predetermined conclusions. Since majority of terminologies used in psychology are subjective, interpretations for the same experimental observations could be dramatically different. Thus, making the psychological research a pseudoscience, where readers are free to conclude the findings in their own way, and human bias for fundamental drives are already wild. Here, in this book we have stressed to invent new machines, look into Hindu perspective of consciousness to bring light to the Western perspective, so that very orthogonal views to the Western intellectual discourse receives a formidable challenge. It could generate conflicting claims leading to fruitful debates.

3. **Overreliance on Surveys**: The prevalence of surveys carried out by researchers with a very different mindset, and analyzed by an expert of statistical analysis with a very different perspective and self-report measures, which can be prone to biases and inaccuracies, can contribute to the perception of psychology as pseudoscientific. Under the framework of SOMU, the new consciousness theory that we are experimentally verifying in this book contains a code of density of primes, namely, phase prime metric, PPM or Purusha for those who hates pure mathematics, and wants to stick to philosophy and subjective matters only. This PPM is a pure mathematical structure that includes all possibilities in strict mathematical form, not probability that leaves an abundance of free will. The statistical dominance we debated in the argumentative discourse of psychology could now be a discussion on quantitative terms, possibly reducing the wild interpretation of the surveys.
4. **Media Sensationalism**: Misinterpretation or oversimplification of psychological research by the media can lead to sensationalized claims or pseudoscientific portrayals. Moreover, it is triggered by the drive of researchers to be viral, helping them gaining attention to funding and reputation. When media controls reputation and recognition to research, and no control study is feasible, the subject is bound to become a pseudoscience.

It's important to note that many psychologists are actively working to address these challenges and improve the rigor and validity of their research. The field of psychology continues to evolve, and there is ongoing debate and reform aimed at enhancing its scientific credibility.

1.7.2 Ten Ways This Book Wants to Change the Research on Psychology Forever

To address the credibility crisis, several recommendations have been proposed. These include improving the rigor of research methods, increasing transparency in reporting, promoting replication studies, and addressing publication bias [9]. Efforts to reflect upon and increase generalizability in hiring and promotion decisions in psychology are also important for maintaining credibility and quality in psychological science [83]. In this book, we aim to reshape the field of brain science by drawing insights from behavioral psychology seen through the lenses of brain signals in different time domains. We identify ten critical areas in psychology where a paradigm shift is overdue:

1. **Philosophical Arguments Over Instrumental Measurements**: Much of psychology remains mired in philosophical discussions, often avoiding instrumental measurements. Many papers lack scientific rigor, with subjective studies leading to non-scientific outcomes. Current psychology often falls short of the scientific practice of making falsifiable predictions, contributing to its classification as a pseudoscience.

2. **Mathematical Definitions for Cognitive and Emotional Traits**: Psychology needs precise mathematical definitions for every cognitive and emotional trait, yet many psychologists shy away from mathematical modeling. We've taken steps, as detailed in Chap. 5, to define 91 cognitive features using geometric and spatio-temporal representations, thanks to the groundwork laid by Singh P. et al.
3. **Neglecting Biological Rhythms**: Psychology often overlooks the significant role of biological rhythms in cognitive, perceptive, and emotional responses, focusing primarily on biological drives.
4. **Lack of Brain or Consciousness Models**: Current psychology lacks comprehensive brain or consciousness models; it often treats the entire neural network as a black box. Our experimental approach, including the creation of the DDG instrument for measuring subconscious traits, introduces the Self-Operating Mathematical Universe (SOMU) model, offering an archive for all possible decisions without requiring human intervention.
5. **Ignoring Individual Variations**: Psychology frequently treats all human subjects equally, neglecting the need for classification into multiple categories. Our work in Chaps. 2 to 9 illustrates the importance of such classifications through eight diverse experiments.
6. **Neglecting Time Management**: Time management within the brain, involving various time scales and rhythmic processes, significantly impacts perception, cognition, and emotion. Current psychology lacks interest in these aspects, often treating the mind as a simple software. We've demonstrated the presence of four parallel time management systems in the brain through eight experiments.
7. **Lack of Correlated Maps for Consciousness**: Psychology rarely attempts to build correlated maps connecting the subconscious and conscious mind, overlooking the various layers and hierarchies of awareness that influence cognitive responses and decision-making.
8. **Disregard for Self-similarity**: Psychologists often avoid self-similarity or fractal structures, favoring linear or nonlinear models. However, self-similarity plays a crucial role in understanding the brain's inner workings, with its many layers and integrated architecture.
9. **The Need to Embrace Confusion**: Current psychology insists on factual, logical explanations, but the brain's behavior can often be illogical, circular, and intuitive. We introduce the "architecture of confusion" model in Chap. 5 to better understand decision-making and cognitive processes.
10. **Ignoring Symmetry and Symmetry Breaking**: Symmetry and symmetry breaking are fundamental in the universe, yet psychology rarely considers them in understanding brain behavior. We propose using patterns of primes to model brain resonant oscillations and gain insights into cognitive and perceptive behavior.

In conclusion, the credibility crisis in psychology research is a complex issue that involves concerns about the generalizability of findings, replication, publication bias, and trustworthiness of the scientific literature. Efforts to address this crisis include improving research methods, promoting transparency and replication studies, and considering alternative perspectives such as critical psychology. These measures are crucial for maintaining the credibility and quality of psychological science. By addressing these ten challenges, in this book we would aim to revolutionize the field of psychology and usher in a new era of understanding the human mind.

References

1. Agrawal L, Chhajed R, Ghosh S, Ghosh B, Ray K, Sahu S, Fujita D (2018) Fractal information theory (FIT)-derived geometric musical language (GML) for brain-inspired hypercomputing. Soft Comput Theories Appl 548:343–372
2. Agrawal L, Sahu S, Ghosh S, Shiga T, Fujita D, Bandyopadhyay A (2016) Inventing atomic resolution scanning dielectric microscopy to see a single protein complex operation live at resonance in a neuron without touching or adulterating the cell. J Integr Neurosci 15(04):435–462
3. Bak P (1996) How nature works. Springer, New York
4. Balasubramanian SV (2018) Genealogy and the Vedic timescale. viXra
5. Bandura A (1999) Social cognitive theory: an agentic perspective. Asian J Soc Psychol 2(1):21–41. https://doi.org/10.1111/1467-839x.00024
6. Bandyopadhyay A (2020) Nanobrain: the making of an artificial brain from a time crystal. Boca Raton CRC Press, Taylor and Francis. ISBN: 9781439875490. https://doi.org/10.1201/9780429107771
7. Bandyopadhyay A, Miki K, Wakayama Y (2006) Writing and erasing information in multilevel logic systems of a single molecule using scanning tunneling microscope. Appl Phys Lett 89(24)
8. Barnsley MF (1988) Fractals everywhere. Academic Press
9. Boeck P, Jeon M (2018) Perceived crisis and reforms: issues, explanations, and remedies. Psychol Bull 144(7):757–777. https://doi.org/10.1037/bul0000154
10. Brehmer B (1992) Dynamic decision making: human control of complex systems. Acta Physiol (Oxf) 81:211–241
11. Bussche E, Noortgate W, Reynvoet B (2009) Mechanisms of masked priming: a meta-analysis. Psychol Bull 135(3):452–477. https://doi.org/10.1037/a0015329
12. Craven J, Hughes M, Jejjala V, Kar A (2022) Illuminating new and known relations between knot invariants. Mathematics
13. Diaz S (2003) Bharata Natyam: classical Indian dance: a Hindu fractal. Int J Human Peace 19:83
14. Dulov E (2020) Evaluation of decision-making chains and their fractal dimensions. Integr Psych Behav.https://doi.org/10.1007/s12124-020-09566-9
15. Duquette J, Ramasubramanian K (2017) Śrīharṣa on the indefinability of time. https://doi.org/10.1007/978-3-319-44418-5_4. Corpus ID: 171149707
16. Durant W (1961) The story of philosophy. Washington Square Press Publication of Pocket Books, New York
17. Dutta AK, Sriram M (2016) Mathematics and astronomy in India before 300 BCE
18. Ernst M, Perkins J, Guo P, McCamant S, Pacheco C, Tschantz M, Xiao C (2007) The daikon system for dynamic detection of likely invariants. Sci Comput Program 69(1–3):35–45. https://doi.org/10.1016/j.scico.2007.01.015
19. Firth D, King J, Koch H, Looney C, Pavlou P, Trauth E (2011) Addressing the credibility crisis in is. Commun Asso Inform Syst 28. https://doi.org/10.17705/1cais.02813
20. Fox PT, Laird AR, Fox SP, Fox M, Uecker AM, Crank M, Loenig SF, Lancaster JL (2005) BrainMap taxonomy of experimental description and evaluation. Hum Brain Mapp 25:185–198
21. Freeman WJ (1999) The physiology of perception. Scientific American Library
22. Ghosh M (1967) Bharatmuni Natya Shastra; The Natyasastra: a treatise on ancient Indian [Hindu] dramaturgy and histrionics) ascribed to Bharata-muni. Asiatic society, Calcutta

23. Ghosh S, Sahu D, Fujita A (2014) Bandyopadhyay; Design and operation of a brain like computer: a new class of frequency-fractal computing using wireless communication in a supramolecular organic, inorganic systems. Information 5:28–99
24. Ghosh S, Fujita D, Bandyopadhyay A (2015) An organic jelly made fractal logic gate with an infinite truth table. Sci Rep 5(1):11265
25. Ghosh S, Sahu S, Agrawal L, Shiga T, Bandyopadhyay A (2016) Inventing a co-axial atomic resolution patch clamp to study a single resonating protein complex and ultra-low power communication deep inside a living neuron cell. J Integr Neurosci 15(04):403–433
26. Giner-Sorolla R (2012) Science or art? how aesthetic standards grease the way through the publication bottleneck but undermine science. Perspect Psychol Sci 7(6):562–571. https://doi.org/10.1177/1745691612457576
27. González-Reimann L (2023) Cosmic cycles, cosmology, and cosmography. In: Jacobsen KA, Basu H, Malinar A, Narayanan V (Editor-in-Chief Associate eds) Brill's encyclopedia of Hinduism online. Consulted online on 02 October 2023. 10.1163/2212-5019_BEH_COM_1020020. First published online: 2016
28. Haggard P, Clark S, Kalogeras J (2002) Voluntary action and conscious awareness. Nat Neurosci 5(4):382–385. https://doi.org/10.1038/nn827
29. Hameroff S, Penrose RJM (1996) Orchestrated reduction of quantum coherence in brain microtubules: a model for consciousness. Math Comput Simul 40:453–480
30. Hameroff S, Penrose RJ (2014) Consciousness in the universe: a review of the 'Orch OR' theory. Phys Life Rev 11:39–78
31. Hansotia PJC (2003) A neurologist looks at mind and brain: "The enchanted loom." Clin Med Res 1:327–332
32. Hedden T, Ketay S, Aron A, Markus HR, Gabrieli JD (2008) Cultural influences on neural substrates of attentional control. Psychol Sci. 19(1):12–7. https://doi.org/10.1111/j.1467-9280.2008.02038.x. PMID 18181784
33. Henrich J, Heine S, Norenzayan A (2010) The weirdest people in the world? Behav Brain Sci 33(2–3):61–83. https://doi.org/10.1017/s0140525x0999152x
34. Holland JH (1995) Hidden order: how adaptation builds complexity. Basic Books
35. Hughes MC (2016) A neural network approach to predicting and computing knot invariants. J Knot Theory Its Ramifications
36. Jackson F (1982) Epiphenomenal Qualia. Philos Q 32(127):127–136. https://doi.org/10.2307/2960077.JSTOR2960077
37. Jamal J, Bakar H (2017) Revisiting organizational credibility and organizational reputation—a situational crisis communication approach. SHS Web Conf 33:00083. https://doi.org/10.1051/shsconf/20173300083
38. Kar M, Pathak A (2013) Vedic Mathematics—a compendium of Sutras. Mathematical Theory Model 3:146–147
39. Kepes S, McDaniel M (2013) How trustworthy is the scientific literature in industrial and organizational psychology? Ind Organ Psychol 6(3):252–268. https://doi.org/10.1111/iops.12045
40. Kitayama S, Duffy S, Kawamura T, Larsen JT (2003) Perceiving an object and its context in different cultures: a cultural look at new look.https://doi.org/10.1111/1467-9280.02432
41. Lancaster JL, Laird AR, Eickhoff SB, Martinez MJ, Fox PM, Fox PT (2012) Automated regional behavioral analysis for human brain images. Front Neuroinform 6:23. https://doi.org/10.3389/fninf.2012.00023
42. Landau A, Elwan D, Holtz S, Prinzmetal W (2012) Voluntary and involuntary attention vary as a function of impulsivity. Psychon Bull Rev 19(3):405–411. https://doi.org/10.3758/s13423-012-0240-z
43. Li J, Paraoanu GJ (2009) Generation and propagation of entanglement in driven coupled-qubit systems. New J Phys 11:113020
44. Lochtefeld JG (2001) The illustrated encyclopedia of Hinduism. A-M, Rosen Publishing, pp 224, 265, 520. ISBN 978-0823931798
45. Logan RK, Tandoc M (2018) Thinking in patterns and the pattern of human thought as contrasted with AI data processing. Information 9(4):83. https://doi.org/10.3390/info9040083

46. Mahomed F, Qadir A, Ramnarain A (2011) Laplace-type semi-invariants for a system of two linear hyperbolic equations by complex methods. Math Probl Eng 2011:1–15. https://doi.org/10.1155/2011/202973
47. Malinar A (2007) Interconnecting parallel times: notions of time in the Caitanya tradition
48. Mandelbrot BB (1983) The fractal geometry of nature. W. H. Freeman
49. Mandelbrot BB (1980) Fractal aspects of the human brain. Phys Scr 21(4):467–474
50. Muniandy R, Viknarasah RG (2017) Multiple states of consciousness: ancient Hindu spiritual perspectives
51. Murakami J (1993) A state model for the multivariable alexander polynomial. Pac J Math 157(1):109–135. https://doi.org/10.2140/pjm.1993.157.109
52. Narayanan V (1999) Y51K and still counting: some Hindu views of time. J Hindu-Christian Stud 12:7
53. Newell B, Shanks D (2014) Unconscious influences on decision making: a critical review. Behav Brain Sci 37(1):1–19. https://doi.org/10.1017/s0140525x12003214
54. Nuland SB (ed) Doctors: the biography of medicine, 2nd ed. Vintage Books, New York, p 13
55. Panksepp J (1998) Affective neuroscience: the foundations of human and animal emotions. Oxford University Press, New York
56. Panksepp J, Knutson B, Pruitt D (1998) Toward a neuroscience of emotion: The epigenetic foundations of emotional development. In: Mascolo MF, Griffin S (eds) What develops in emotional development? Plenum, New York, pp 53–84
57. Parada FJ, Rossi A (2018) If neuroscience needs behavior, what does psychology need? Front Psychol 9:433. https://doi.org/10.3389/fpsyg.2018.00433
58. Peitgen HO, Jürgens H, Saupe D (2004) Chaos and fractals: new frontiers of science. Springer Science & Business Media
59. Peitgen HO, Jürgens H, Saupe D (2004) Fractals for the classroom: an introduction to fractal geometry with exercises. Springer Science & Business Media
60. Pradhan B, Bandyopadhyay A, Pal AJ (2004) Molecular level control of donor/acceptor heterostructures in organic photovoltaic devices. Appl Phys Lett 85(4):663–665
61. Pramanik S, Singh P, Sahoo P, Ray K, Bandyopadhyay A (2023) 1D to 20D tensors like dodecanions and icosanions to model human cognition as morphogenesis in the density of primes. In: Kaiser MS, Waheed S, Bandyopadhyay A, Mahmud M, Ray K (eds) Proceedings of the fourth international conference on trends in computational and cognitive engineering. Lecture notes in networks and systems, vol 618. Springer, Singapore. https://doi.org/10.1007/978-981-19-9483-8_38
62. Pribram KH (1991) Fractals in brain function, perception, and cognition. In Fractals in physics. Springer, Berlin, Heidelberg, pp 263–290
63. Radhakrishnan Sarvepalli, Moore CA (1957) A Source Book in Indian Philosophy, Princeton, New Jersey: Princeton University Press, ISBN: 0-691-01958-4
64. Ramachandran VS (2011) The tell-tale brain: a neuroscientist's quest for what makes us human. W. W. Norton & Company
65. Reddy S, Sonker D, Singh P, Saxena K, Singh S, Chhajed R et al (2018) A brain-like computer made of time crystal: could a metric of prime alone replace a user and alleviate programming forever? Soft Comput Appl:1–43
66. Rian IM, Park J, Ahn H, Chang D (2007) Fractal geometry as the synthesis of Hindu cosmology in Kandariya Mahadev temple, Khajuraho. Build Environ 42:4093–4107
67. Rule NO, Moran JM, Freeman JB, Whitfield-Gabrieli S, Gabrieli JD, Ambady N (2011) Face value: amygdala response reflects the validity of first impressions. Neuroimage 54(1):734–41. https://doi.org/10.1016/j.neuroimage.2010.07.007. Epub 2010 Jul 12. PMID 20633663
68. Sahu S, Ghosh S, Ghosh B, Aswani K, Hirata K, Fujita D, Bandyopadhyay AJ (2013) Atomic water channel controlling remarkable properties of a single brain microtubule: correlating single protein to its supramolecular assembly. Biosens Bioelectron 47:141–148
69. Satyanarayana B, Eye K (2012) Some mathematical concepts in ancient Sanskrit works (survey article)
70. Saxena K, Singh P, Sahoo S et al (2023) Self-survival of quantum vibrations of a tubulin protein and microtubule: quantum conductance and quantum capacitance. In: Kaiser MS, Waheed S,

Bandyopadhyay A, Mahmud M, Ray K (eds) Proceedings of the fourth international conference on trends in computational and cognitive engineering. Lecture notes in networks and systems, vol 618. Springer, Singapore. https://doi.org/10.1007/978-981-19-9483-8_43

71. Saxena K, Singh P, Sahoo P, Ghosh S, Krishnanda D, Ray K, Fujita D, Bandyopadhyay A (2022) All basics that are wrong with the current concept of time crystal: learning from the polyatomic time crystals of protein, microtubule, and neuron. Proc Trends Electron Health Inform 376:243–254
72. Saxena K, Singh P, Sahoo P, Sahu S, Ghosh S, Ray K, Fujita D, Bandyopadhyay A (2020) Fractal, scale free electromagnetic resonance of a single brain extracted microtubule nanowire, a single tubulin protein and a single neuron. Fractal Fract 4(2):11
73. Schraube E (2015) Why theory matters: analytical strategies of critical psychology. Estudos De Psicologia (Campinas) 32(3):533–545. https://doi.org/10.1590/0103-166x2015000300018
74. Sen Gupta, Anima (1986), The evolution of the Samkhya school of thought, New Delhi: South Asia Books, ISBN: 81-215-0019-2
75. Sim J (2022) Holding the door open for the practitioner community. Ind Organ Psychol 15(4):542–545. https://doi.org/10.1017/iop.2022.71
76. Simon HA (1996) The sciences of the artificial, 3rd ed. MIT Press
77. Singh P, Ray K, Bandyopadhyay A (2022) Biological antenna to the humanoid bot: electromagnetic resonances in biomaterials. In: Studies in rhythm engineering (SRE). Springer Singapore. https://doi.org/10.1007/978-981-16-9677-0
78. Singh P, Sahoo P, Saxena K, Ghosh S, Sahu S, Ray K, Fujita D, Bandyopadhyay A (2021a) Quaternion, octonion to dodecanion manifold: stereographic projections from infinity lead to a self-operating mathematical universe. Proc Int Conf Trends Comput Cogn Eng 1169:55–77
79. Singh P, Sahoo P, Saxena K, Ghosh S, Sahu S, Ray K, Fujita D, Bandyopadhyay A (2021b) A space-time-topology-prime, stTS metric for a self-operating mathematical universe uses Dodecanion geometric algebra of 2–20 D complex vectors. Proc Int Conf Data Sci Appl 148:1–31
80. Singh P, Sahoo P, Saxena K, Manna JS, Ray K, Ghosh S, Fujita D (2021c) Cytoskeletal filaments deep inside a neuron are not silent: they regulate the precise timing of nerve spikes using a pair of vortices. Symmetry 13(5):821
81. Singh P, Saxena K, Singhania A, Sahoo P, Ghosh S, Chhajed R, Ray K, Fujita D, Bandyopadhyay A (2020) A self-operating time crystal model of the human brain: can we replace entire brain hardware with a 3D fractal architecture of clocks alone? Information 11(5):238
82. Smyth S, Casey D, Cooney A, Higgins A, McGuinness D, Bainbridge E, Murphy K (2016) Qualitative exploration of stakeholders' perspectives of involuntary admission under the mental health act 2001 in Ireland. Int J Ment Health Nurs 26(6):554–569. https://doi.org/10.1111/inm.12270
83. Stengelin R, Bohn M, Sánchez-Amaro A, Haun D, Thiele M, Daum M, Schuhmacher N et al (2023) Responsible research is also concerned with generalizability: recognizing efforts to reflect upon and increase generalizability in hiring and promotion decisions in psychology. https://doi.org/10.31234/osf.io/ykh9x
84. Talairach J, Tournoux P (1988) Co-planar stereotaxic atlas of the human brain. Thieme, New York
85. Tegmark MJ (2000) Importance of quantum decoherence in brain processes. Phy Rev 61:4194
86. Thibaut F (2018) The mind-body Cartesian dualism and psychiatry. Dialogues Clin Neurosci. 20(1):3. https://doi.org/10.31887/DCNS.2018.20.1/fthibaut. PMID 29946205; PMCID PMC6016047
87. Tononi G (2 Nov 2004) An information integration theory of consciousness. BMC Neuroscience 5(1):42. https://doi.org/10.1186/1471-2202-5-42. ISSN 1471-2202. PMC 54347
88. Tsaousi C, Sophocleous C (2010) Differential invariants for systems of linear hyperbolic equations. J Math Anal Appl 363(1):238–248. https://doi.org/10.1016/j.jmaa.2009.08.023
89. Vandenberg R, Lance C (2000) A review and synthesis of the measurement invariance literature: suggestions, practices, and recommendations for organizational research. Organ Res Methods 3(1):4–70. https://doi.org/10.1177/109442810031002
90. Wager TD (2006) Do we need to study the brain to understand the mind. APS member article

Chapter 2
Quantized Interhemispheric Energy Transfer: Learning Motorized Tasks

2.1 Introduction

2.1.1 Motor Learning and Interhemispheric Energy Transfer

2.1.1.1 Procedural Learning: Basic Concepts of False-Positive Psychology

Studying human behavior and psychology has been quite challenging, primarily because there have been instances where our findings turned out to be incorrect. This is called false-positive psychology. False-positive psychology is when we mistakenly think something is true when it's not [13]. One interesting experiment that helps us understand the brain's inner workings is called mirror drawing (MD) [99]. This test shows us how our brain's two halves communicate with each other, which is vital for many tasks we perform. Think of it like teamwork between user brain's left and right sides. Now, when we talk about this communication, it's not as straightforward as we might think. It's not like a super-fast text message, it's more like a relay race where information is passed from one side to the other, where time flow is not linear, sometimes its fast and sometimes it is slow [53]. And interestingly, it seems like this relay race isn't always equal in both directions, there's some time difference, called interhemispheric information transfer time (IHTT) [67]. This time difference tells us that our brain's left and right sides have their specialties, like being right or left-handed [63]. It's like having a dominant hand when we're playing a game. But what's even more fascinating is that when us practice tasks like MD, our brain starts connecting the dots between what our eyes see and what our hands do, it is called procedural learning [16]. It's like learning to dance, at first, our feet are all over the place, but with practice, they move smoothly to the rhythm. This process is related to something called procedural memory, which is like muscle memory for our brain [19]. It helps us get better at tasks through practice. Now, the communication between

T. Dutta and A. Bandyopadhyay, *Emotion, Cognition and Silent Communication: Unsolved Mysteries*, Studies in Rhythm Engineering,
https://doi.org/10.1007/978-981-99-9334-5_2

our brain's halves isn't always the same, it can vary [21]. Imagine it like different ways of talking, sometimes us use words, sometimes gestures, but the conversation never stops. So, in a nutshell, mirror drawing helps us unlock the mysteries of how our brain's two sides work together, how they have their unique strengths, and how practice can make us better at things. It's like a window into the fascinating world of the human mind!

2.1.1.2 Time is not Linear, Nonlinear, or Continuous: What Does that Mean?

Understanding the perception of time and its implications for human cognition has been a focal point of psychological research for decades. While the prevailing view suggests that time is perceived as a continuous and unbroken flow this traditional perspective may not provide a comprehensive understanding of the complex interplay of temporal dynamics within the brain, particularly in the context of motor learning and skill acquisition. The prevailing body of psychological research suggests that human perception of time unfolds as a continuous phenomenon, yet this perspective falls short of a comprehensive understanding of the intricate processes governing interhemispheric energy transfer within the brain. In this chapter, we delve into experimental evidence that presents a compelling argument against the notion that all interactive time domains equally enhance motor learning efficiency, drawing insights from a comprehensive meta-analysis of brain activity databases [47]. The study by Karolis et al. [47] stands as a significant milestone in this endeavor, offering valuable insights into the temporal characteristics of brain function during various cognitive tasks, including motor learning.

The conventional view in psychology has long portrayed human perception of time as continuous. However, recent experimental evidence challenges this notion, suggesting that interactive time domains do not uniformly enhance motor learning efficiency, as supported by data from the brain's meta-analysis database [47]. In mirror drawing (MD) tasks, subjects attempt to trace shapes within the confines of polygonal double lines (e.g., stars, diamonds, squares, or triangles), relying solely on the inverted reflection of their hand in a mirror [46]. Although there are claims that MD establishes new associations between visually rotated tasks and the motor nerves controlling arm movements [66], EEG and brain hardware studies have yet to confirm its fundamental principles. With repetition of a task, the brain forms associations, and subjects experience reduced difficulty on subsequent attempts (measured by a familiarity index) due to enhanced bidirectional visual-motor spatial awareness [21]. Such reduced inter-subject variability can lead to erroneous conclusions during statistical analyses [3]. This study revisits meta-analysis [47] to uncover evidence of energy quantization in hemispheric transfer [17], introducing the novel concept of favored and restricted time zones. Surprisingly, despite frequent use of EEG in observing motor skill development tasks [102], no prior studies have investigated the existence of these time zones.

While engaging in the task of tracing geometric shapes, subjects are challenged to remain within the confines of double lines outlining polygons such as stars, diamonds, squares, or triangles [46]. Notably, this task is conducted with subjects observing an inverted reflection of their own hand movements in a mirror placed before them during the tracing process. While it has been postulated that mirror drawing (MD) tasks establish novel associations between visually perceived tasks, rotated by 180°, and the intricate neural pathways governing arm movements [66], in-depth investigations involving EEG and neuroimaging techniques have not conclusively confirmed the fundamental rules underlying these processes. Furthermore, once a task is executed, the brain actively constructs associations, leading to markedly reduced difficulty during subsequent attempts, as evident in the concept of the familiarity index [21]. This bidirectional visual-motor awareness of spatial dimensions contributes to a convergence of individual subject performance differences [9], often to a degree where statistical analyses may yield misleading conclusions [3]. In this chapter, we revisit the meta-analysis conducted by Karolis et al. [47] to seek substantial support for the notion of quantization in hemispheric energy transfer. Notably, our exploration extends to the uncharted territory of identifying favored and restricted temporal zones within the context of motor skill development tasks, an aspect that has thus far eluded scientific scrutiny, despite the frequent utilization of EEG-based assessments [102].

2.1.2 *Methodologies Adopted for Mirror Drawing Test and Analysis Protocols*

In our study, we delve into the intricate temporal dynamics of interhemispheric energy transfer during the process of motor learning. To construct our insights, we draw upon a comprehensive body of experimental evidence gleaned from previous research endeavors. A central focus of our investigation centers on mirror drawing (MD) tasks. In these tasks, participants are assigned the challenge of tracing geometric shapes while observing their actions through a mirror, resulting in a mirrored reflection of their hand movements [43].

One pivotal facet of MD tasks lies in their potential to forge novel connections between visual perceptions and the intricate pathways controlling motor functions. While this hypothesis has generated considerable interest within the scientific community, the precise mechanisms underpinning these associations have remained somewhat elusive. Additionally, we probe into the concept of the familiarity index, as proposed by Chein and Schneider in [20]. This index reflects the reduction in task difficulty that occurs with repeated attempts. This reduction is attributed to the brain's remarkable ability to construct associations and enhance bidirectional visual-motor awareness, particularly in relation to spatial dimensions.

Our analysis extends to a reexamination of a meta-analysis conducted by Karolis et al. in [47], with a particular focus on its implications for quantifying energy in interhemispheric transfer. Furthermore, we embark on an exploration of the concept of favored and restricted time zones within the temporal dynamics of tasks related to the development of motor skills. The scrutiny of the experimental evidence surrounding mirror drawing tasks offers captivating insights into the potential formation of associations between visual tasks and the intricate pathways controlling motor functions. Despite the proposal of such associations, neuroimaging techniques like EEG have not yet definitively confirmed the underlying mechanisms.

Moreover, the concept of the familiarity index sheds light on the remarkable adaptability of the brain, enabling it to optimize motor performance through repeated task execution. This fascinating phenomenon, which converges individual differences in subject performance, poses intriguing challenges for statistical analysis. Our review of the meta-analysis by Karolis et al. [46] underscores the pressing need for further exploration into the temporal dynamics governing interhemispheric energy transfer during the development of motor skills. Our exploration extends to the relatively uncharted territory of favored and restricted time zones, a subject that remains underexplored within the realm of cognitive neuroscience, despite the frequent utilization of EEG-based assessments.

2.2 Is Lateralization Real? Does It Really Happen?

2.2.1 Critical Challenges Associated with Lateralization

The lateralization of brain function, often referred to as brain laterality, represents a captivating facet of neuroscience research. It pertains to the specialization of cognitive and motor processes in the left and right hemispheres of the brain, a phenomenon that has been a subject of extensive study for decades. While significant progress has been made in elucidating the lateralization phenomenon, numerous challenges persist, hindering a comprehensive understanding of this complex phenomenon.

2.2.1.1 The Multifaceted Nature of Lateralization

One of the foremost challenges in understanding brain lateralization lies in its multifaceted nature. The brain's specialization is not limited to a simple division of labor between the left and right hemispheres. Instead, it encompasses a spectrum of functions, including language processing, visuospatial skills, emotional regulation, and motor control. The intricate interplay of these functions complicates efforts to decipher the precise neural mechanisms governing lateralization [69]. One could clearly notice that several brain regions are activated and the pathway of activation from left brain to the right brain changes dramatically.

2.2.1.2 Methodological Limitations

Methodological limitations pose a significant obstacle in the study of brain lateralization. Traditional neuroimaging techniques, such as functional magnetic resonance imaging (fMRI) and positron emission tomography (PET), offer valuable insights but are constrained by factors like spatial resolution and the need for controlled experimental conditions. Furthermore, these techniques may not capture the dynamic and distributed nature of lateralization. In this book we have made an effort to introduce a new method of looking into the structure of time to understand a human mind and its cognitive ability. Pixels from an image of brain response do not reveal a cognitive mind, pieces of clocks do.

2.2.1.3 Individual Variability

Individual variability represents another formidable challenge. While general trends in lateralization exist, considerable variation exists among individuals. Genetic factors, developmental influences, and life experiences all contribute to this variability. Consequently, designing experiments that account for this heterogeneity is a complex endeavor. We contend that this book represents a pioneering endeavor, distinct from any prior efforts, to chart the spectrum of human diversity, transcending considerations of race, caste, and creed. While it may be regarded as rudimentary, it stands as the inaugural initiative to categorize human individuals based on the essential intricacies of their brain's internal hardware.

2.2.1.4 Dynamic Nature of Lateralization

Lateralization is not static but instead exhibits a dynamic quality that can change over time. Neuroplasticity, the brain's ability to reorganize itself in response to experiences, can influence lateralization patterns. Understanding how and why these changes occur remains a critical question in the field. May be there is no lateralization but a vertical processing. The hemispheric dominance observed at any time is the sum result of the dynamic interaction between bottom-up and top-down processing tendencies [41].

2.2.2 Embracing Interdisciplinary Approaches: Familiarity Index

To surmount these challenges, researchers must adopt interdisciplinary approaches that incorporate insights from genetics, psychology, cognitive science, and neurobiology. Leveraging advanced neuroimaging techniques with enhanced spatial and temporal resolution, such as magnetoencephalography (MEG) and electroencephalography (EEG), can provide a more nuanced understanding of dynamic lateralization.

The lateralization of brain function is a captivating and multifaceted phenomenon that continues to challenge our understanding of cognitive and motor processes. Addressing the key challenges associated with lateralization requires a holistic and interdisciplinary approach. As technology advances and our methodologies become more refined, we inch closer to unlocking the secrets of lateralization, paving the way for a deeper appreciation of the intricate workings of the human brain. In doing so, we may uncover not only how the brain is lateralized but also why, shedding light on the fundamental processes that underpin human cognition and behavior. The findings presented in this chapter underscore the complexity of temporal dynamics in interhemispheric energy transfer during motor learning. While the prevailing view in psychology supports the notion of continuous time perception, our analysis suggests that this perspective may not fully capture the intricacies of brain function in the context of skill acquisition. Mirror drawing tasks, which offer a unique window into the relationship between visual perception and motor control, have generated intriguing hypotheses regarding the formation of associations. However, the absence of conclusive neuroimaging evidence regarding these mechanisms calls for further research. The concept of the familiarity index highlights the brain's remarkable adaptability and its capacity to reduce task difficulty with repeated practice. This phenomenon poses challenges for statistical analysis and warrants consideration in future studies. In conclusion, our exploration of the meta-analysis by Karolis et al. [47] has revealed promising avenues for research into the quantization of energy in hemispheric transfer. Additionally, the concept of favored and restricted time zones within the temporal dynamics of motor skill development tasks represents a novel and uncharted area of investigation that holds great potential for advancing our understanding of cognitive neuroscience.

In the context of motor skill learning tasks, the initial performance serves as a window into the controlled processes [62, 93] and offers valuable insights into the genuine state of the brain [54]. In our novel approach to data analysis, we place a significant emphasis on the initial data and its variability, positing that it can unveil the fundamental workings of the human brain when presented in a generalized manner without omitting any seemingly extraneous data points [81].

During mirror drawing (MD) tasks, EEG data was simultaneously recorded [105], revealing three fundamental dynamics: synchronous (S), silent (S), and randomness (R), collectively forming the SSR protocol. Plotting this EEG data in a two-dimensional representation unveiled six distinct brain dynamics associated with procedural learning [51]. This study involved 20 management students selected from a total of 89 subjects within the age group of 21–37 years.

What's particularly noteworthy is the consistent emergence of these six distinct SSR features at regular intervals, effectively quantifying the visual-motor learning process [46]. This implies that the duration an individual takes to complete a task doesn't necessarily correlate with their procedural learning ability [100, 101]. Rather, it hinges on how much time is expended during the initial phases of the task, providing a predictive indicator of whether an individual will demonstrate rapid improvement or evolve as a slower learner.

2.3 A Critical Experiment that Challenges the Very Idea of Lateralization

2.3.1 Experimental Procedure

Impact of Variations in Human Subjects in Psychology Research and Our Effort

Variations in human subjects can significantly impact the reproducibility of psychology research [4, 11, 12, 27, 42, 55, 72, 84, 103, 106]. Researchers often assume that there is little variation across human populations or that the "standard subjects" used in studies are representative of the entire species [43]. However, the evidence suggests that there is a reproducibility crisis in psychology and other life sciences, with low statistical power, researcher's degrees of freedom, and an emphasis on publishing surprising positive results contributing to the problem [11]. The complexity and context dependency of human behavior also pose challenges to reproducibility [12]. Additionally, the lack of standardized experimental procedures and the failure to address methodological factors further hinder reproducibility [84, 106]. It is crucial for researchers to address these issues and adopt practices that promote reproducibility in psychological research [55].

In our experimental setup, we meticulously selected a total of 289 subjects, ranging in age from 21 to 37 years. It's important to note that all of these individuals were in good health, with body temperatures hovering around 97–98 °F, initial pulse rates falling within the range of 78–132, which were later adjusted to a consistent 80–85 beats per minute. Their blood pressure levels ranged from 85 to 130, and their blood oxygen levels were consistently maintained between 95 and 97%.

To ensure that our subjects' circadian rhythms were aligned, all experiments were conducted during the evening, following the conclusion of their daily classes. We took great care to exclude individuals with high fever, neurological disorders, a history of major diseases, or any signs of abnormal responses. Additionally, subjects who exhibited reduced concentration had been diagnosed with ADHD, displayed nervousness during testing, excessive excitement, or failed to meet consistent medical conditions were excluded from the study. It's worth mentioning that all participants in our study were right-handed.

Our selection process employed three rigorous criteria to whittle down the initial pool of over a hundred applicants to a final group of 289 subjects who underwent testing. Here for the purpose of clarity we have used only 20 subjects, characterized by substantial similarities in their mental, physical, and attitudinal attributes, were deemed suitable for inclusion in our study. As we discussed in the Sect. 1.7 about selection of the human subjects, as one could see here, that we have made a sincere effort to find nearly similar human subjects.

Procedure: Figure 2.1a–c provide a visual representation of EEG patterns associated with three distinct forms of information processing within the brain: random, silent, and synchronous. These patterns were observed during a mirror drawing (MD) task, specifically involving the tracing of a star-shaped figure outlined by double lines, with a 3 cm gap between these lines (refer to Fig. 2.1d for visual reference). Figure 2.1a–c provide an overview of the experimental setup employed in our combined EEG and mirror drawing study. In Fig. 2.1, one can observe a subject engaging in the act of drawing along a star-shaped path while concurrently monitoring the EEG pattern of their brain [105]. To ensure participant comfort and prevent any discomfort during the experiment, we utilized wireless 16-probe EPOC EEG machines, eliminating the need for cumbersome cables. In Fig. 2.1a, we capture a snapshot of a typical random state of EEG. It's worth noting that different frequency domains exhibit dissimilar potential distribution patterns. Occasionally, the EEG activities dip below a threshold, resulting in a predominance of blue regions covering the EEG landscape, as illustrated in Fig. 2.1b.

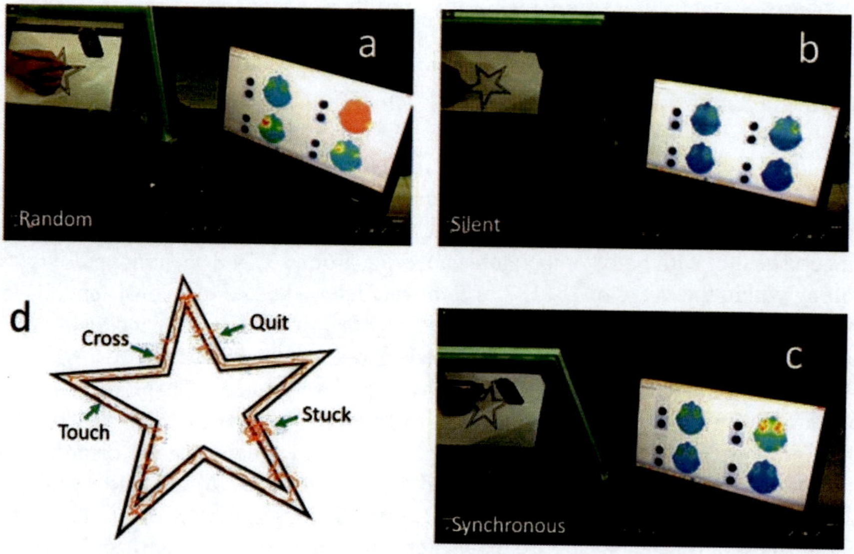

Fig. 2.1 In this illustration, we observe the mirror drawing (MD) task in progress, where an additional eye guard has been strategically placed to obstruct any potential visual access to the subject's hand movements. The EEG data is divided into four distinct quadrants, each corresponding to specific frequency ranges: the top left quadrant representing delta waves (3–4 Hz), the top right quadrant capturing theta waves (4–7 Hz), the bottom left quadrant displaying alpha waves (7–13 Hz), and the bottom right quadrant exhibiting gamma waves (13–39 Hz). The depicted mental states include a snapshot of random motion (**a**), a snapshot of silent dynamics (**b**), synchronous activity across both hemispheres (**c**), and the representation of a star-shaped geometry with parallel lines (**d**). These visuals provide insight into the diverse cognitive processes at play during the MD task

We've used arrows to indicate instances where a subject becomes fixated on a specific region, essentially "stuck" in one place, moving aimlessly within that area. Eventually, frustration sets in, leading the subject to abandon that spot, which we refer to as "quit." Some participants may attempt to trace the borderline and lack the confidence to venture into the middle region of the path, a behavior we term as "touch," as depicted in Fig. 2.1d. On occasion, a subject may cross the borderline as they lose their sense of direction and then return to the central path; we refer to this as "cross."

Furthermore, it's important to note that across all four frequency domains (alpha, beta, gamma, and delta) of EEG, we did not observe synchrony between the left and right sides of the brain [78], as demonstrated in Fig. 2.1a. However, a notable similarity exists in the activity patterns on both sides of the hemisphere across all frequency domains of EEG, as shown in Fig. 2.1c. Therefore, when we reference synchrony in this paper, we consistently allude to activity patterns where both sides of the brain exhibit similar behavior.

The experimental setting took place in a dimly lit room, where participants were comfortably seated in front of a table. Positioned on the table was a response button aligned with their body's midline. Each participant's instructions were clear: they were to trace the star pattern without lifting the pen, avoid touching the borderline, and follow the middle region of the gap while completing the star's path. The MD task was performed as follows: each subject initiated with their non-dominant hand (L), followed by three consecutive trials using their dominant hand (RRR). Subsequently, the subject returned to using their non-dominant hand. This protocol was denoted as L-RRR-L. The participants faced a mirror placed at a distance of 80 cm. EEG measurements were acquired using a wireless EPOC 16 probe EEG machine. Prior to the experiment, all device probes were prepared by applying a conductive gel or saline solution to ensure optimal conductivity.

Throughout the experiment, each subject underwent a total of five trials: the initial trial with their non-dominant hand, followed by three consecutive trials with their dominant hand, and concluding with the last trial using their non-dominant hand once more. Data from the experiments were meticulously recorded for future reference. Additionally, the time taken to complete each MD trial was measured and averaged. The recorded times ranged from a maximum of 6.12 to a minimum of 0.674 s. The mean time across all trials for all participants was 2.21345 s, with a median time of 1.975 s and a standard deviation (S.D.) of 1.381756 s.

2.3.2 Data Analysis Protocols

Recent research challenges the traditional perspective that a single locus governs bimanual coordination in the brain. Instead, emerging evidence points to a distributed neural network responsible for neural synchronization and desynchronization, crucial for various coordinated functions [95]. Observations of disrupted interlimb coordination in movement disorders further support this distributed network concept [95].

Studies investigating the impact of left-hand training on interlimb coordination have yielded intriguing findings, demonstrating performance improvements not only in learned tasks but also in transfer tasks [56].

Moreover, investigations into frontal EEG spectra have provided valuable insights into motor skill acquisition, revealing significant modulations in frontal EEG activities during skill acquisition [65]. The transfer of learning between hemispheres in interlimb coordination research has been characterized by symmetrical brain activation patterns [56]. Additionally, the complexity of the task and training duration have been identified as factors influencing bilateral transfer [49]. However, our recent research introduces a novel perspective, suggesting that the time required for learning transfer is primarily influenced by an individual's chosen strategy. These strategies encompass a wide range, including "winners" and "quitters," and intriguingly, each strategy is associated with distinct brain activation patterns, adding a new dimension to our understanding of interlimb coordination and learning transfer.

In the context of mirror drawing, our research addresses a fundamental question: What scientific parameter might be compromised when interpreting the correlation between visual cognition and motor nerve function? One avenue of exploration centers on EEG correlates of haptic feedback within a visuomotor tracking task [59]. Through the examination of these parameters, we aspire to illuminate the intricate relationship between visual cognition and motor control, ultimately advancing our comprehension of interlimb coordination and learning transfer within the realm of neurobiology.

2.3.2.1 EEG Generic Data Analysis

Individual differences in SSR protocol were significant, so we took only those subjects with similar SSR responses when they talked and behaved normally. We made an effort to acquire a consistent response. Frontal parietal lobes blink periodically between high-intensity signal (red) and low-intensity signal (blue), then a person performs mirror drawing task fast (1–2 min finishes the task). Those subjects, whose one side of the parietal lobe blinks average, take a little longer (4–8 min). Finally, some subjects have nonperiodic, occasional blinks. One side of the parietal lobe is active in EEG. Most of the time, the activity of the entire brain region is null (blue). Then, it takes a very long time to finish the mirror drawing task. Practice increases periodicity in the blinking. It improves the time taken to complete the task during practice. The increase in efficiency has a one-to-one correspondence with the periodicity of blinking in the brain. Those who are best performers occasionally show activity in the Alpha and Delta frequency region. This effect is silent for those who take a long time. One subject's left brain is more active, hence less movement. Movement means periodic blinking with a similar pattern between both sides of the brain. If both sides of the brain are not symmetrically active in the frontal region, the subject gets very confused and reveals frustration by body gestures. Confusion is directly proportional to the asymmetry between the left and right brain.

2.3.2.2 Data Management for "Quit" and "Procrastination"

The "quit" who procrastinate activates all frequency bands in EEG, and there is no distinct interhemispheric transfer. In contrast, the winners show activity only at one (beta) frequency bands and a good left–right interhemispheric transfer. One of the winners' striking features is that the time taken for completing the task is the same for all attempts, be it the first test, training test, or the return confirmation test. The detailed analysis of the EEG of losers shows that they activate multiple functional domains of both brains as if multitasking is going on in the brain, just before the quit drawing, express frustration, and move to another point.

2.3.2.3 Time Resolution Used During Data Analysis

Those who are best performers occasionally show activity in the Alpha and Delta frequency region, totally silent for those who take a long time. We measured the duration of EEG response for each action with 1/100th of a second-time resolution. Movement means periodic blinking with symmetry between both sides of the brain. If both sides of the brain are not symmetrically active in the frontal region, the subject gets very confused. Confusion is directly proportional to the asymmetry between the left and right brain patterns in EEG. The changes in such responses were calculated part by part, freezing the EEG video and slowing the frame transition rate. The time ranking for Fig. 2.2 is made of the total time taken by subjects. That is why horizontal axis data varies from 1 to 36 (the maximum time taken by a subject is 36 min). For Figs. 2.3 and 2.5, the time ranking is a simple numbering of subjects according to their time taken to complete the MD task.

2.3.2.4 Declaration on Human Subjects

The local ethics committee of the institute approved the study. It was performed following the ethical standards laid down in the Declaration of Helsinki (last modified, 2004). All 89 subjects were above 18 years, and written informed consent was taken from all of them.

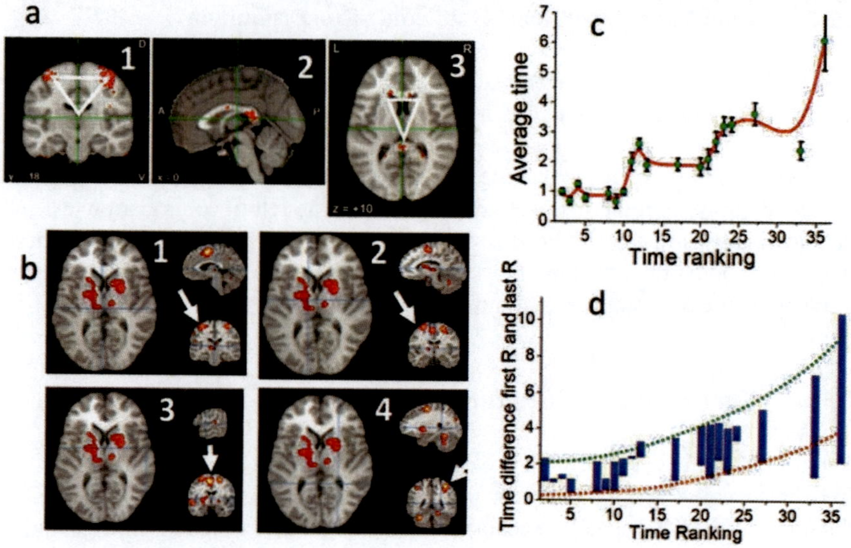

Fig. 2.2 **a** This figure displays the load of the "interhemispheric" meta-analysis, comprising data from 81 research papers extracted from the Neurosynth brain scan database. The "association" function was applied to generate the metadata (refer to the methods section for detailed information). **b** The figure also illustrates data related to action motor learning from the BrainMap database [10], consisting of a compilation of 27 relevant publications, encompassing 114 experimental contrasts and 1080 reported coordinates. **c** and **d** In both panels (c) and (d), the time ranking is depicted, representing the total time taken for five mirror drawing (MD) tests for each subject. The time is scaled in minutes. In panel (a), the graph showcases the total time for performing five consecutive MD tasks (LRRRL), where the error bar represents the sum of errors across all five measurements. The total time for five tests is averaged for each of the 20 subjects and plotted along the horizontal axis in ascending order, denoted as time ranking. In panel (b), the figure illustrates the difference between two times: the time taken for the first MD test and the time taken for the last MD test after training. Similar to panel (a), the time ranking is calculated in the same manner to facilitate comparison

2.3.3 Metadata Analysis from "Brainmap's" Behavioral Taxonomy of "Axon Motor Learning" and "Neurosynth's" Interhemispheric Energy Transfer

2.3.3.1 Online Databased Sources Used for Meta-analysis

In this research, we adopted a meta-analytic approach, drawing upon data from 81 distinct studies as detailed by Yarkoni et al. [104] (http://Neurosynth.org). To access the relevant interhemispheric transfer data, we retrieved the Neurosynth database (https://neurosynth.org/analyses/terms/interhemispheric/). These files were processed and visualized using two web-based platforms SOCR [94] https://socr.umich.edu/HTML5/BrainViewer/ and Papaya [70] https://rii-mango.github.io/Papaya/. Furthermore, we applied the Talairach labels atlas before employing the

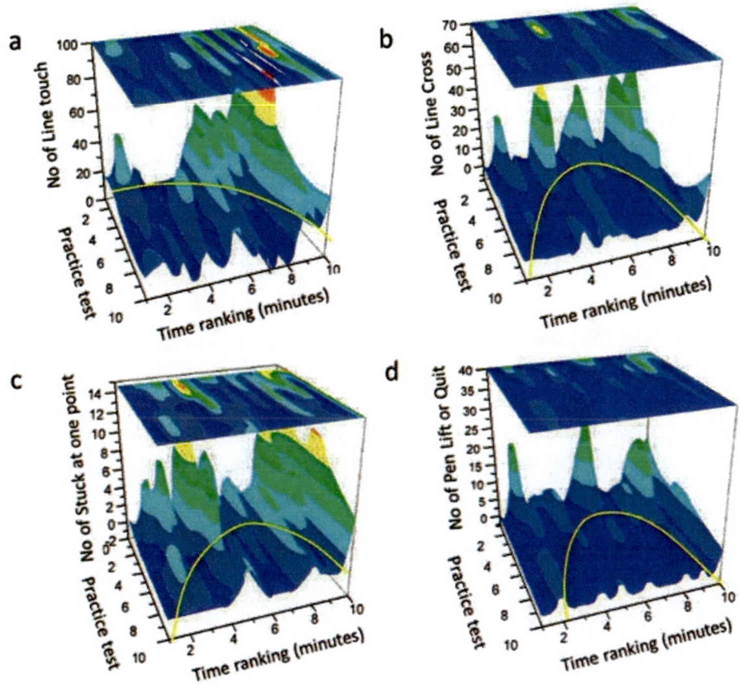

Fig. 2.3 3D plots of line touch (panel **a**), line cross (panel **b**), stuck at one point (panel **c**), and lift or quit (panel **d**) are plotted vertically against the surface made of practice test LRRRL (five experiments divided into ten by 2 × 2 matrix operation) and time ranking (minutes). Here "time ranking" includes the ranks for 20 subjects. The value 20 is reduced to 10 for processing, and each value depicts two subjects

behavioral analysis plugin in the metadata analysis software available at http://www.brainmap.org/ale/ [57].

Specifically focusing on action motor learning, we compiled data from 27 relevant publications, which can be accessed at http://www.brainmap.org/taxonomy/behaviors/Action.Motor_Learning.html. For additional guidance and tutorials related to our methodology, please refer to http://rii.uthscsa.edu/mango//videos.html.

2.3.3.2 fMRI Support for the EEG Data

In order to find the fMRI support for the EEG results obtained here, we performed meta-analysis of the term "interhemispheric" automatically identifying all reports in the Neurosynth database (see methods), finding the load to identify the brain regions consistently or preferentially noted in the reported tables of the selected papers in that database. The search had a false discovery rate (FDR) of 0.01. The association test map for "interhemispheric" displays voxels reported in the main texts of the 81 articles more often than in their abstracts alone. In Fig. 2.2a of the EEG images, we have put the fMRI images obtained from the "interhemispheric" meta-analysis. Three images show that two triangular domains work simultaneously in two perpendicular planes in the brain (see images 1 and 3). They together govern the energy transfer between the left and right hemispheres. However, this particular meta-analysis does not reveal the energy transfer process. So, we carried out behavioral analysis from another brain database called BrainMap (see methods). We selected metadata of motor control learning that closely associates to MD task, the sum of 27 associated publications presented in Fig. 2.2b shows an unprecedented three-point energy transfer (see 3). Three bright spots blink in sequence, two spots are identified in the two lobes of the brain, but the central one is a common energy transfer domain that regulates the lateralization process. Both meta-analyses of panels a and b suggest that energy transfer happens by jumping through space by spending specified time, leading to spatio-temporal quantization of lateralization. We wanted to find whether such quantization in the hardware-based behavioral analysis has a real effect on human subject responses.

2.3.4 *Quantization of the Time Taken to Complete the Task*

Figure 2.2a, b reveal two significant insights into the workings of the human brain. In Fig. 2.2a, we explore the total time taken by all 20 subjects, plotting it on the horizontal axis, while the average time for each test is represented on the vertical axis [97]. Typically, one might expect a linear increase, but instead, we observe a quantized jump followed by a flat region, an unprecedented finding. This indicates that visual image processing and motor nerve coordination do not exhibit a linear relationship [50]. Notably, subjects within the flat regions of Fig. 2.2a share a consistent average time, defying the expectation that they should take progressively longer to complete all five tests.

Figure 2.2b delves into the experiences of procedural learning [38] for a specific subject. Following the LRRRL protocol, we calculate the time difference between the first L test and the last L test in the MD task, assuming that the practice gained from three right-hand tests (RRR) influences the second left-hand test post-training. As expected, test completion time decreases post-training [3, 21, 60, 77]. However, the time gap plotted in Fig. 2.2b displays a unique pattern. Subjects who take longer to complete the initial L test experience the maximum impact on procedural learning. There is also a discernible pattern in the reduction of task time among the twenty subjects, which we've connected with dotted lines.

In essence, Fig. 2.2a, b provide key observations: the presence of quantized jumps in Fig. 2.2a suggests that visual image processing and motor nerve coordination do not follow a linear trajectory. Subjects within the flat regions share the same average time for the MD task, regardless of the number of tests. Additionally, our investigation reveals that during training (RRR), subjects in the plateau region exhibit a compensatory response between the left and right brain. If the left brain responds faster, the right brain is slower, effectively maintaining a consistent average time. However, this adjustment has its limits, and beyond a threshold, there is a noticeable jump in average time.

Similarly, Fig. 2.2b illustrates that the impact of fixed training on procedural learning adheres to a particular rule, suggesting that there is no universal hardware limit for procedural learning. We observe a grouping of subjects and cases where training has no discernible impact between the two extreme regions. These findings provide valuable insights into the clustering and isolation of specific time domains within the context of our study.

2.3.5 *Touch, Cross, Stuck, and Quit, Four Fundamental States of a Human Psychology*

2.3.5.1 Unlocking the Brain's Temporal Patterns: Insights from Mirror Drawing and Procedural Learning

In our quest to understand the intricacies of the human brain, we often encounter questions about how our minds perceive and interact with time. Figure 2.3, a collection of four 3D plots derived from EEG data, unveils fascinating insights into the human brain's response to time during the mirror drawing (MD) task and procedural learning [58]. These visual representations offer a window into the unique temporal preferences of individuals, sparking a profound inquiry: Is time truly universal, or do our brains carve out distinct time zones? This article delves into the significance of these findings, examining how four fundamental mental states—touch, cross, stuck, and quit—correlate with specific timeframes, and whether these patterns extend beyond the confines of this study.

The human brain's relationship with time has long fascinated scientists and researchers. In Fig. 2.3, we explore this intricate connection by analyzing the temporal dynamics observed during the MD task and procedural learning. These insights raise essential questions about the universality of time perception and who or what shapes these restricted time zones within the brain's response. Figure 2.3 consists of four 3D plots generated from EEG data. In these plots, the horizontal 2D base comprises two parameters: time ranking and the test protocol (LR...RL), while the vertical axis represents different states: "touch" for Fig. 2.3a, "cross" for Fig. 2.3b, "stuck" for Fig. 2.3c, and "quit" for Fig. 2.3d, as previously described in Fig. 2.1d. Each of these mental states is quantified as the frequency or number of times the event occurred during the entire mirror drawing (MD) test sessions conducted with 20 carefully selected subjects from a pool of 89. The selection criteria involved rigorous medical assessments, including blood pressure fluctuations, body temperature variations, oxygen flow rates in the blood, left–right handedness, among other factors (see methods section).

In all four 3D plots, we have included guidelines to highlight quantization and an optimized time domain where the brain's procedural training is most efficient. Notably, a dip in the central zone is observed in all four plots, indicating that subjects who complete the first MD task in approximately 3–4 min undergo the most efficient procedural learning process. Subjects who either finish the task very quickly or very slowly require more extensive training steps to improve their efficiency in MD [16, 25].

Quantization of Time

The central theme that emerges from Fig. 2.3 is the concept of quantization—a phenomenon where time appears to be discretized into distinct intervals rather than flowing continuously. Figure 2.3a reveals that the inclination to follow the borderline or exhibit insecurity increases among subjects who take a significantly long time to complete the task. The probability of adopting an insecure path grows with higher time rankings. This quantization is most evident in Fig. 2.3b, where subjects display clear instances of crossing the star border at regular intervals. The presence of these quantized time segments prompts us to question whether our brains inherently prefer certain timeframes for specific actions.

Optimized Time Domain

Another striking observation is the existence of an optimized time domain within the brain's procedural training process. Subjects who complete the first MD task in approximately 3–4 min undergo the most efficient procedural learning. This begs the question: What drives this optimization, and can it be harnessed to enhance learning and skill acquisition across various domains?

Mental States and Time

Figure 2.3 further demonstrates how distinct mental states—touch, cross, stuck, and quit—correlate with specific timeframes. For instance, subjects taking significantly longer to complete the task in Fig. 2.3a are more inclined to follow the borderline or exhibit insecurity. This raises intriguing questions about the relationship between psychological states and temporal preferences. In Fig. 2.3b, quantization is particularly pronounced among the four mental states. Instances of crossing the star border are observed at regular intervals, with distinct dips in the data. These dips signify that subjects falling into specific temporal categories do not require extensive training.

The Challenge of "Stuck"

Among these mental states, overcoming the tendency to get "stuck" at a particular location in the star shape, as depicted in Fig. 2.3c, proves to be the most challenging. This insight prompts us to explore the neural mechanisms responsible for resisting change in temporal patterns. Figure 2.3c focuses on the mental state of getting "stuck" at a particular point. Here, quantization and the generalized dip are most prominent. Among all four mental states, overcoming the tendency to get "stuck" at a particular location in the star shape proves to be the most challenging.

The Ease of "Quit"

Conversely, Fig. 2.3d illustrates that the probability of quitting can be more easily eliminated through training. This finding suggests that our brains may possess adaptable mechanisms for shifting temporal preferences. Finally, Fig. 2.3d illustrates the "quit" mental state, where quantization is observed. Notably, among all four mental states, the probability of quitting can be more easily eliminated through training. These observations provide valuable insights into the brain's response to procedural learning during the MD task.

2.3.5.2 Universal Versus Individual Time

One of the central questions raised by these observations is whether time perception is a universal trait shared by all humans or if individual differences play a significant role. The diverse temporal patterns depicted in Fig. 2.3 highlight the possibility that each person's brain may have its unique temporal signature. In conclusion, the insights gleaned from Fig. 2.3 challenge our understanding of time perception and its relationship with human behavior. The presence of quantization, optimized time domains, and the influence of mental states open exciting avenues for future research. Ultimately, this exploration of the brain's temporal dynamics invites us to ponder the age-old question: Is time a universal constant, or does it bend to the whims of individual minds? Further investigations may shed light on the profound interplay between the human brain and the enigmatic realm of time.

2.3.6 Spontaneous, Random, Asymmetric Activity Between Both Hemispheres and Localized Responses

2.3.6.1 Quantized Brain Dynamics: Decoding Procedural Learning

In our exploration of the human brain's intricate response to the mirror drawing (MD) task through EEG studies, we unveil the presence of quantization within the brain's hardware. Figure 2.4 presents a fascinating depiction of three distinct brain events during the MD process—synchronous activity, asymmetric responses, and the silent domain. These events, when analyzed across 20 subjects, reveal a remarkable quantization of brain dynamics into six distinct groups. Each group exhibits unique characteristics, from U-turn dynamics to convergent and oscillatory patterns. Additionally, Fig. 2.5 offers insights into how these quantized dynamics manifest before and after practice, shedding light on the brain's capacity to adapt and optimize its response. These findings challenge our understanding of procedural learning and highlight the role of quantization in shaping our cognitive processes.

The human brain's response to the MD task has long intrigued researchers, prompting a deeper exploration of its temporal dynamics. Figures 2.4 and 2.5 offer

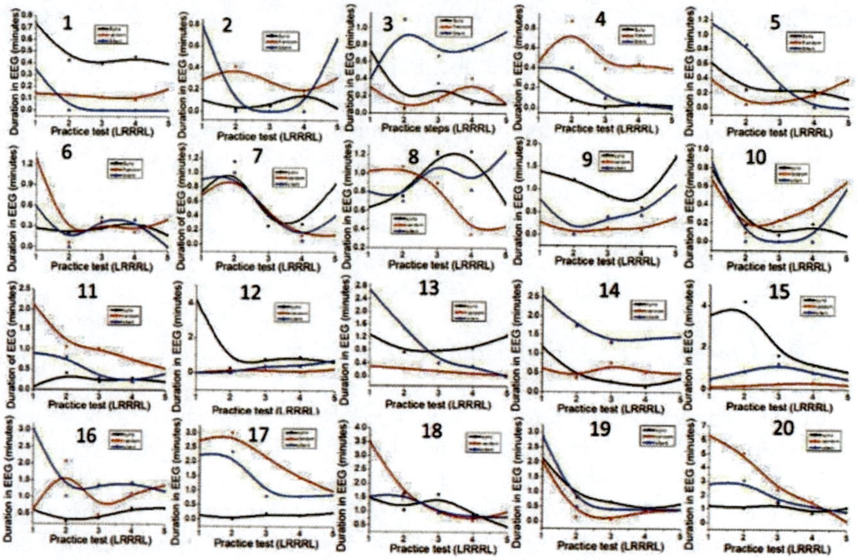

Fig. 2.4 Tracking mirror drawing practicing using EEG for 20 subjects. The horizontal axis shows a practicing left hand, L, and right-hand R sequence, LRRRL, depicted as 1 to 5. The vertical axis depicts the duration for which one could observe sync (black), random (red), and silent (blue) responses in the EEG. Twenty plots for 20 subjects were ranked by the time taken for completing the MD task

a glimpse into the quantization of these dynamics, raising questions about how the brain's hardware governs our procedural learning.

Quantization in Brain Dynamics

Figure 2.4 presents three key brain events during MD processes: synchronous activity, asymmetric responses, and the silent domain. Strikingly, when we analyze these events across 20 subjects, we observe quantization inherent to the brain's hardware. This quantization manifests as six distinct groups, each characterized by unique dynamics.

Decoding the Quantized Groups

1. *U-Turn Group (Subjects 2 and 10)*: In this group, subjects exhibit maximum asymmetry between their left and right hemispheres. This finding challenges conventional notions of brain symmetry [47]. The asymmetry between left and right hemispheres [74] is the maximum [47]
2. *Convergent Groups (Subjects 11, 12, 13, 18, 19, 20)*: Minimum training leads to highly efficient procedural learning in these subjects, emphasizing the adaptability of the human brain [52]. Minimum training builds an extremely efficient procedural learning in the brain [52].
3. *Oscillatory Groups (Subjects 6, 7, 8, 15, 16, 17)*: Here, both hemispheres interact in an oscillatory manner, revealing dynamic exchanges between synchronous,

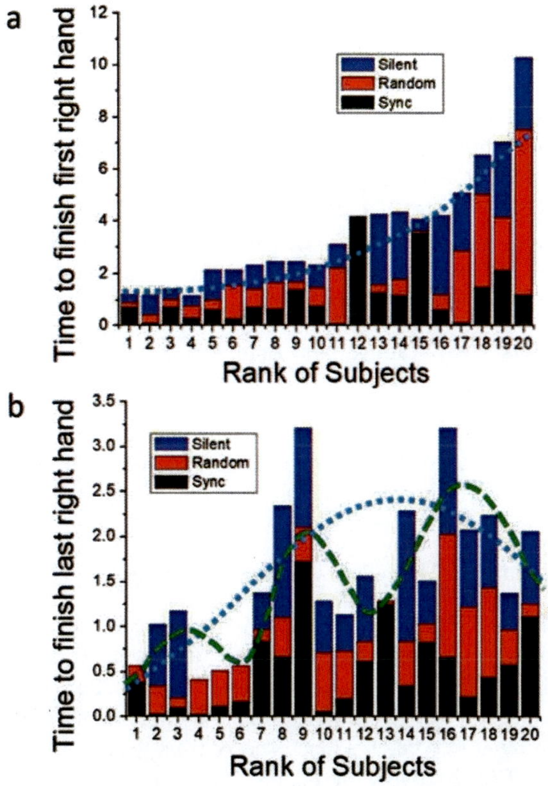

Fig. 2.5 Contribution of silent (blue), random (red), and sync (black) dynamics in EEG during the entire MD task. The top plot is for the L test before training (panel **a**), and the bottom plot is for the L test after training (panel **b**)

random, and silent features during MD testing, both the hemispheres interact in an oscillatory manner [96].

4. *Silence Group (Subjects 3 and 14)*: Dominated by the lowest frequency bands of EEG, this group represents a unique state of brain activity during MD. The silent domain where EEG activity reduces to below threshold [48] is depicted in blue.
5. *Synchronization Group (Subjects 1 and 9)*: Subjects in this group exhibit nearly identical brain activity patterns in both hemispheres, challenging notions of hemisphere independence [80]. The pattern of activities for both the hemispheres is identical [80]
6. *Random Dynamics Group (Subjects 4, 11, 17, 20)*: Governed by random dynamics, this group showcases a distinct pattern of brain activity during MD.

2.3.6.2 Temporal Management Before and After Practice

Quantized Brain Dynamics: A Technical Exploration of Procedural Learning: This technical analysis delves into the intricate nature of quantized brain dynamics as revealed in Figs. 2.4 and 2.5. These findings challenge conventional notions about brain symmetry, independence, and adaptability, prompting a closer examination of how the brain manages time and responds to complex tasks. The study of procedural learning has taken a significant step forward with the discovery of quantized brain dynamics. These dynamics, evident in Figs. 2.4 and 2.5, open up a new realm of possibilities for understanding how our brains adapt to various temporal demands.

Figure 2.4 illustrates three critical brain events during the mirror drawing (MD) task: synchronous activity, asymmetric responses, and the silent domain. Intriguingly, the analysis of these events across 20 subjects uncovers inherent quantization within the brain's hardware. This quantization categorizes brain dynamics into six distinct groups, each characterized by unique features and responses.

Temporal Management Before and After Practice

Figure 2.5 offers a detailed examination of the temporal aspects of these quantized dynamics, contrasting intervals before and after practice. Figure 2.5a reveals a continuous increment in brain dynamics before practice, while Fig. 2.5b exposes quantized management post-practice. This transition underscores the adaptability of the human brain, as quantization replaces continuous dynamics.

Complex Time Zones in Brain Response

The oscillatory patterns observed among various subjects indicate that those falling within favorable and unfavorable time domains face oscillatory time responses. This implies that quantization, characterized by sharp boundaries in restricted time zones for brain responses, transforms into a gradient of allowed and favorable time zones when observed on a macroscopic time scale.

The discovery of quantized brain dynamics within procedural learning has profound implications for our understanding of human cognition and learning processes. These findings challenge traditional beliefs and prompt further investigation into the mechanisms that govern the brain's temporal management. As we explore this fascinating territory, we aim to unlock the mysteries of human cognition and shed light on the role of quantization in shaping our cognitive processes on both microscopic and macroscopic time scales.

2.3.7 *Unlocking the Enigma of Quantized Procedural Learning*

We now enter into the fascinating realm of quantized procedural learning, a phenomenon uncovered through a meticulous meta-analysis of motor learning and interhemispheric energy transfer. This discovery challenges conventional paradigms, sparking inquiries into the spatio-temporal quantization of energy transfer and its impact on human subjects. Leveraging mirror drawing (MD) tasks as our experimental conduit, wherein subjects replicate geometric shapes by gazing into a mirror, we unveil a profound connection between visual data and motor nerve movements—forged in a quantized fashion within the human mind. We have explored quantized procedural learning [23, 52] in the time domain.

Traditionally, MD task analyses have neglected to explore temporal correlations among diverse subjects. However, this study, facilitated by the integration of EEG with MD tasks, pioneers the exploration of quantized procedural learning within the temporal domain. This endeavor marks the inaugural classification of human minds based on their unique responses to procedural learning. While some subjects exhibit oscillatory and quantized responses tailored to their individual learning experiences, a central revelation lies in the correlation between subjects and their distinct temporal domains of response—a discovery that beckons an anatomical investigation.

Drawing on the distinct brain dynamics observed in Fig. 2.4 during procedural learning, it can be posited that the composition of these dynamics dictates the resulting MD responses. Intriguingly, the brain's temporal dynamics cyclically privilege certain patterns over others (as evidenced in [22]), affording these patterns periodic prominence and ultimately culminating in the quantization of procedural learning.

The temporal aspect of procedural learning unraveled herein suggests that each brain may necessitate varying durations to master such tasks. While our study primarily resides within the realm of behavioral psychology, we have employed diverse methodologies to validate quantization phenomena. Recent assertions, notably [30] claim of quantized visual cognition, align with our findings. Furthermore, the data's temporal reduction quantities present limits, hinting at a consistent reduction rate governing procedural learning within the human mind—a tantalizing implication that invites deeper exploration into the cognitive mechanisms governing human learning.

Quantized procedural learning stands as a groundbreaking revelation that challenges traditional notions of cognitive processes. Through the prism of MD tasks and EEG analysis, we unravel the complex temporal dynamics at play in this phenomenon. The correlation between subjects' responses to distinct temporal domains beckons further anatomical scrutiny. The unique brain dynamics governing procedural learning and the periodic privileging of specific patterns offer profound insights into the quantization of cognitive processes. As we ponder these findings, we contemplate the profound implications they hold for our understanding of human cognition and learning.

2.4 Geometrodynamics of the Brain: Redefining Unit of Information and Integration Theory

2.4.1 *Identification of Polyatomic Time Crystal (PTC): Fractal Integration of PTC*

Helmholtz suggested that the brain minimizes energy to make decisions, and later, the concept of free energy was introduced to mathematically represent this idea. Free energy minimizes the probability differences between environmental variables [7] and the network's configuration induced by arbitrary contributors through changes in network geometry [32, 33]. A comprehensive review [14] begins by providing an overview of time crystals and their properties [76], followed by discussions on potential applications in artificial intelligence and computing, proposing that time crystals could revolutionize the efficiency and reliability of such systems. We have developed an integrated clock architecture by measuring resonant oscillation frequencies of neurons, microtubules, etc. [1], and discovering a fractal network of clocks. There are many hidden circuits in the brain, each operating in a particular time domain [89, 90]. Based on triplet of triplet fractal scale-free resonant bands connecting the microtubule, protein, and the neuron [85], we have already created a SOMU model for understanding the human brain [91, 93], eventually developing a complete time crystal model of the human brain [86].

It is now well established that the brain has two coupled hierarchically organized structures, one anatomical and the other functional. The multilayered integration of brain hardware, such as the Rich club organization of the connectome and feedback signal transmission loops in the visual cortex, links the dense network of cortico-cortical axon paths with human behaviors. This hallmark of neurogenesis unfolds these networks from the spatial and topological centrality of the core in the cortex [34], suggesting a scale-free energy transmission [31, 37, 64, 98].

Integrating different energetically active brain parts with their specific behavioral and functional features into a singular correlated hierarchical architecture has been a challenge [24]. Linking active ROI sites in fMRI, EEG, or PET scans has required a mathematical model to integrate information from all evaluation and attention systems to generate perceptual cognitions simultaneously using short-term and long-term memory. Modeling the causal relationship between active hardware parts and behavioral responses remains difficult.

2.4.2 *Looking Beyond Neurons: Integrating All Operations Using Only Discrete Clocks*

Optical imaging of brain's intrinsic signals (OIS) is used for mapping the functional topography of the human cortex [34, 75]. To model spatial scale hierarchy,

Simon et al. have outlined the limitations of current models and described the recent trend of imagining neurobiological atoms or functional nodes that act as a basic unit, self-assembling with different multimodal functional features in the brain [28]. Our time crystal-based brain model extends this multimodal approach to 537 modes (one mode = one clock system) in the brain, integrating 12 layers of fractal hardware from proteins to the entire body's neural network [86]. We imagine a hypothetical neuron based on the global workspace of cognitive responses [26].

One of the novel ideas contributing to the understanding of the brain's emotional structures is the ordering of the phase of transmitted signals. The distribution of phase across the brain, whether ordered or disordered, allows for entropy calculations to estimate information quantity and quality [61]. Since the direction of information flow depends on the signal's frequency [39], our time crystal-based model of the entire human brain, integrating 537 classes of clocks to estimate the whole brain's spatio-temporal network, holds true [86].

A wide spectrum of cortical functional features arises from structural constraints, leading to invariants of functional properties [40]. In this work, we have made an effort to map these invariants in the complete integrated emotional and functional maps of the brain [86].

Automated brain image analysis employs keywords to mine several papers and identify the complexity of brain signaling by filtering essential information from large free datasets [57]. Emotion and perception-driven thoughts and expressions have been correlated with bright active regions in fMRI, EEG, and PET scans across extensive databases. However, identifying a particular emotional state with a distinct region of interest (ROI) is challenging due to the significant overlap of most functional domains.

2.5 Meta-analysis of Brain Database: Confirming the Geometric Invariants

2.5.1 A Tutorial on the Brain-Behavior-Hardware Database

Here we mention a large database of brain images for action, emotional, cognitive, interoception, and perceptual events. One could learn the number of papers involved in creating the metadata, the number of datasets that took part in creating the metadata and the number of datapoints plotted together in the combined image that has been created. The metadata analysis of brain images provides us a global aspect of various brains from studies carried out globally from different laboratories for humans of different races and creeds. Therefore, in these data the localization of bright signals means that many papers suggest that for a particular behavioral expression, a selected part of the brain is activated. z-score of the images and sleuth workspace could be downloaded from the database http://www.brainmap.org/taxonomy/behaviors/Action.Motor_Learning.html. Both the files are in the *.nii format.

What is a z-score: A behavioral expression is often defined as a normalized distribution of activated signals in the brain mapping? The normalized distribution has a mean value at a location where probability of activating a particular behavioral expression is the most. Thus, we get a map of region of interest, ROI. This map is saved as a reference and compared with a new brain image where particular regions are activated. The difference between mean of a reference point with the activated region of a testing brain image is *z*-score. A high *z*-score means the data point is multiple standard deviations separated from the mean. This could happen as a matter of course with heavy/long tailed distributions, or could signify outliers. To avoid false-positive analysis a 3D histogram with multiple variables density distribution is analyzed carefully before a *z*-score is taken seriously.

The Searching Protocol with Sleuth: Foundation of Creating Metadata: Sleuth is a search mechanism to build metadata from published research articles. http://www.brainmap.org/sleuth/ Sleuth is the BrainMap application to search papers based on keywords, read their corresponding metadata, and plot their results as coordinates on a standard brain map converted to Talairach space. However, often the keywords are misleading. A paper could discuss a whole different topic, analyzing a whole new data, yet use statistically dominating keywords for different purposes. A machine algorithm cannot differentiate variable reasons used by authors to justify their argument in brain imaging literature. Using sleuth (http://www.brainmap.org/sleuth/) we find complete information on each paper and each experiment used to build the metadata. The workspace contents and vital statistics are shown in the top portion of the panel. On the right, one can choose what information we wish to view about each experiment, including Citation, Submitter, Prose Description, Subjects, Conditions, Brain Template, Experiments, and Results Synopsis. One could download the Sleuth workspace as *.nii data. For example happiness, a positive emotional state. http://www.brainmap.org/taxonomy/behaviors/Emotion.Positive.Happiness.html

There are several neuroimaging databases available. Neuroimaging databases providing access to metadata from functional human brain research can help make more concise interpretations of behavior (Neurosynth—http://neurosynth.org/, Brede—http://neuro.imm.dtu.dk/services/jerne/brede/, PubBrain—http://www.pubbrain.org/, and BrainMap—http://www.brainmap.org/).

Neurosynth Brain Database: One of the largest brain databases (https://neurosynth.org/). We have used this particular database in our studies for interhemispheric data transfer (https://neurosynth.org/analyses/terms/interhemispheric/#faq) [68]. One could check various forms of brain events and analyze signal generated in the metadata analysis. Lateralization of left and right brain is fundamental to the learning process [47]. During lateralization how energy transfers from one brain to another is a debatable topic, and we argue that brain learning has preferred time window. In this book we would address this particular temporal favoring of human brain significantly.

How to Use Mango Software for Meta-analysis of Brain Images? We have to download papaya [70] for offline analysis or work in the browser version namely papaya https://rii-mango.github.io/Papaya/ http://rii.uthscsa.edu/mango/mango.html We want to carry out behavioral analysis of the brain images as described above. fMRI basic images have active regions. There are several tutorials on youtube, to cut brain slice and see inside https://www.youtube.com/watch?v=fIBLuX5Vyas and 3D reconstruction https://www.youtube.com/watch?v=qWPUmMXJ7B0 Mango has its own tutorials which describes several brain operations http://rii.uthscsa.edu/mango//videos.html Different buttons of the GUI are described here http://rii.uthscsa.edu/mango/mango_guide_image.html#createwebpage

Behavioral Analysis Using Mango software: The Behavior Analysis plugin performs regional behavior analysis based on a user selected brain region of interest (ROI). The image where the ROI is defined must be in the Talairach brain space before processing. You should verify that the origin is correctly positioned (press 'o' key in Mango), since analysis is coordinate based. Talairach spatial normalization can be done using Mango's SN plugin for anatomical images. If your image was aligned using the MNI brain template you can use the MNI-to-Tal transform, which is an option in the transform menu (Mango v2.6 and later).

Results of behavior analysis are presented for BrainMap's five Behavioral Domains (Action, Cognition, Emotion, Interoception, and Perception) and sixty sub-domains (figure). Only z-scores ≥ 3.0 are considered significant ($p \leq 0.05$ with Bonferroni correction for multiple comparisons). A behavior profile chart and z-score ranked table (figure) are provided to assist with interpretation. http://rii.uthscsa.edu/mango//plugin_behavioralanalysis.html

Talairach leveling: The 1988 Talairach Atlas brain (http://www.talairach.org/) has been considered as the ad hoc standard for reporting the locations of activation foci in the functional brain mapping studies in part to its detailed anatomical labeling: Talairach coordinates, also known as Talairach space, is a three-dimensional coordinate system (known as an 'atlas') of the human brain, which is used to map the location of brain structures independent from individual differences in the size and overall shape of the brain. http://rii.uthscsa.edu/mango//plugin_sn.html

Brain image is aligned to the Talairach standard through a three-step process:

- *Mid-sagittal alignment (1)*—During the first step, the manual transform editing tool is used to align the brain image mid-sagittal plain with the displayed mid-sagittal markers.
- *AC–PC and Origin alignment (2)*—During the second step, four anatomical locations are identified—ventral aspect of the anterior Corpus Callosum (CC), ventral aspect of the Thalamus (TN), Superior Colliculus (SC), and superior aspect of the Cerebellum (CB)—which are used to align the image to the AC–PC line. Then the origin is identified as the Anterior Commissure.
- *Scaling (3)*—Finally, in the third step, the overall size of the brain is determined by selecting each side of a bounding box cube. An option is provided to automatically set the bounding box.

Talairach labels atlas (atlas is also another format https://fsl.fmrib.ox.ac.uk/fsl/fslwiki/Atlases-Reference) is applied on input image before behavior analysis plugin is run. MNI means classic *.nii image files obtained from the brain are transformed using Talairach leveling. If our brain image is aligned using the MNI brain template we can use the MNI-to-Tal transform, which is an option in the image menu, scroll down to transform menu, the window pops up and we have to select whether we want to convert MNI to Talairach or Talairach to MNI (Mango v2.6 and later).

There is an alternate method to do Talairach transformation. Download the Talairach plugin http://rii.uthscsa.edu/mango//plugin_sn.html talairach.jar folder. To add it to the mango platform. The process to do it is to click options, plugin manager, add, plugin jar file and select the jar file. It will be loaded you can see above. Now when you load any image *.nii file then a new window opens where we could see the brain images. In that window, you would see plugins, click plugins. We could see the added plugins to the mango platform. Click SN spatial normalization. Scroll down to transform menu, the window pops up and we have to select whether we want to convert MNI to Talairach or Talairach to MNI http://rii.uthscsa.edu/mango//atlas_spec.html

GingerALE Version 3.0.2 GingerALE is the BrainMap application that is used to perform an ALE meta-analysis on coordinates in Talairach or MNI space. GingerALE can also be used to convert coordinates between MNI and Talairach spaces using the icbm2tal transform. http://www.brainmap.org/ale/

Human Connectome Project: Structural connectome data were derived from the diffusion-weighted imaging dataset of 163 participants acquired at 7T by the Human Connectome Project Team (http://www.humanconnectome.org/study/hcp-young-adult/) of humanoid brain project [47].

Meta-analysis of fMRI for Emotional and Cognitive States Shows Hierarchical Invariant Optimization in Brain

The brain's cognitive operation for emotion and perception is captured by fMRI images, which activates or deactivates different functional regions in synchronization with the human thoughts and expressions of emotional states. These synchronized pairs of emotional state and images of activated brain region of interest (ROI) are called functional images. These images are not useful until we couple the brain's anatomical map or brain atlases with the ROI images. The coupling of two maps is called normalization, here we used both MNI and Talairach standards. Then, we investigated five ROI domains of behavioral response shifts, e.g., Action, Cognition, Emotion, Interoception, and Perception to find spatial jumps, periodic jumps between spaces or multiple ROIs to find invariant geometric shapes. Each brain function has a specific set of geometric shapes that remain invariant in a 3D orientation, invariants are subject independent, correlates brain behavior and functions with comparative geometric shapes. Our finding paves the way to integrate spatio-temporal dynamics of hierarchically connected behavioral responses.

Source Links of the Brain Database

In this study, we employed a meta-analytical approach to analyze functional MRI data from 81 distinct research studies, as originally described by Yarkoni et al. [104]. To gather the necessary data, we accessed the Neurosynth database [68], which contains valuable information related to interhemispheric transfer (Neurosynth n.d.). You can find the interhemispheric transfer details we used for our analysis at the following URL: https://neurosynth.org/analyses/terms/interhemispheric/.

Once we obtained the relevant data, we proceeded to visualize it using specialized tools. We utilized two web-based platforms for this purpose:

1. **SOCR Brain Viewer**: You can visualize the downloaded files from the Neurosynth database [68] using the SOCR Brain Viewer (SOCR n.d. [94]) at https://socr.umich.edu/HTML5/BrainViewer/.
2. **Papaya Viewer**: Another visualization tool we employed is the Papaya Viewer (Papaya n.d. [70]), which you can access at https://rii-mango.github.io/Papaya/ or http://rii.uthscsa.edu/mango/mango.html.

To facilitate our behavioral analysis, we applied the Talairach labels atlas to the input images before running the behavioral analysis plugin. This crucial step was performed using metadata analysis software available at http://www.brainmap.org/ale/ [79]. Our focus in this meta-analysis was on action motor learning, and we compiled data from 27 relevant publications. You can explore the specific taxonomy for action motor learning within the BrainMap database (BrainMap n.d.) at the following URL: http://www.brainmap.org/taxonomy/behaviors/Action.Motor_Learning.html. If you require additional guidance on using these tools and performing similar analyses, you may find the tutorials [79] at http://rii.uthscsa.edu/mango//videos.html helpful.

2.5.2 *Understanding the Invariants from EEG or FMRI*

2.5.2.1 Methods for Determining the Invariant Map

In Fig. 2.6a, we illustrate the analysis of fMRI data obtained from the BrainMap database project. This data reveals activated brain regions corresponding to different functions, interconnected by loops that exhibit variations. It's important to note that our data is static, unlike dynamic regions observed in online videos, which often manifest as spatial loops. As a result, we treat the statistical database as a collection of various activated loops, each representing a clock. We then assemble these 3D clocks to account for phase relations between them. We have been making consistent effort to use already available database in addition to our lab generated database to find invariants [73].

Once the clock structures are constructed, we initiate the process of identifying differential clocks, as outlined in Fig. 2.6b. By treating the differential clocks as individual points, we observe that elementary 1D, 2D, and 3D geometric shapes remain consistent across all the papers used to compile this brain metadata. Consequently, we can confidently conclude that these shapes are invariants in our analysis.

In Fig. 2.6c, we present a typical feature observed in EEG patterns and FMRI images. We note that 1D, 2D, and 3D patterns of excited regions combine to form differential 3D clocks, which we refer to as 4D. Within these differential clocks, some vibrate within a plane, leading us to examine the 2D shapes, and we aptly term this as 5D. Additionally, certain planes exhibit consistency across different cases (for

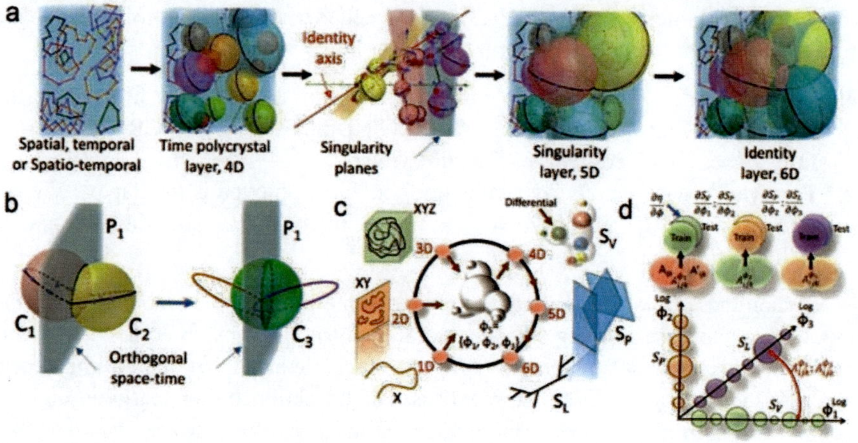

Fig. 2.6 **a** Processing spatio-temporal data from fMRI, EEG, or PET scans follows a step-by-step procedure. Loops of similar intensity values are connected to form clocks. Clocks synchronized in a plane vibrate together, and planes vibrating along a line are assigned a new clock in the second step. This clock integration continues until a singular clock is obtained at the highest level. **b** Hierarchical clock integration is achieved differentially. When two clocks, C1 and C2, overlap, a new clock is assigned to the volume of the overlapping sphere, creating an invariant. This concept underlies the development of hierarchical clock integration. **c** Summary of the procedures outlined in panels (a) and (b). 1D, 2D, and 3D datasets are used to construct the invariant network. The brain's information is encoded within this invariant network, and the brain integrates invariants to make decisions. **d** Mathematical representation of panel (c). Three types of invariants—volumetric, planar, and linear arrangements of differential clocks—are placed along three orthogonal axes. The resulting invariant pattern represents the decisions made by the brain

a list of papers used in the database, please refer to the online resources), allowing us to establish 1D invariants, which constitute the 1D invariant database. The journey from the molecular to the behavioral time scale serves as a testament to the profound interconnectivity of the universe [44]. It highlights the intricate relationship between the brain's temporal dynamics and our daily lives. The brain's temporal architecture influences our behaviors, perceptions, and interactions, shaping our experiences [29].

Remarkably, we extract three types of invariants from both EEG and fMRI databases. Significantly, the geometric shapes identified as invariants, whether on the brain's surface or within its structure, remain consistent for specific brain functions across various instances.

2.5.3 *Interhemispheric Transfer, Language, Speech, Memory, and Positive/Negative Emotions*

In Fig. 2.7a, three images illustrate the simultaneous operation of two triangular domains in two perpendicular brain planes (as seen in images 1 and 3). Together, these domains regulate energy transfer between the left and right hemispheres, as depicted by the invariant geometry in the figure. However, this particular meta-analysis doesn't provide insights into the energy transfer process itself. To delve deeper, we conducted behavioral analysis using data from another brain database called BrainMap. We specifically focused on metadata related to motor control learning, closely associated with EEG tasks for five emotional states. Figure 2.7b presents the results of this analysis, featuring an unprecedented three-point energy transfer (as shown in 3). These three bright spots blink sequentially, with two located in different brain lobes and a central one serving as a common energy transfer domain that regulates lateralization. Both meta-analyses in panels (a) and (b) suggest that energy transfer occurs by leaping through space over specified time intervals, resulting in the spatio-temporal quantization of lateralization. This prompted us to investigate whether such quantization observed in the behavioral analysis has real effects on human subject responses. This exploration of temporal structures and their impact on human psychology is part of a grand symphony of science [83]. It represents humanity's enduring quest to unravel the mysteries of time itself [5]. The significance of this journey lies not only in its theoretical implications but also in its potential to captivate and inspire new discoveries [45].

In Fig. 2.7c, we explore three cognitive functions: language, speech (represented by a pentagon), attention (depicted as a circle with a V shape on the top part of its perimeter in Fig. 2.7d), and explicit memory (illustrated by a U shape with an ellipsoid on top in Fig. 2.7e). These invariant patterns are readily identifiable and understandable without extensive analysis. This initial exploration into finding brain invariants focuses on apparent solutions for several reasons. Firstly, we adhere to a hierarchical network of clocks based on fractal information theory (FIT; [2]) and view the brain as a time polycrystal [87]. Secondly, database creation typically notes only the apparent area and consolidates contributions from numerous independently conducted experiments. Consequently, conducting a comprehensive analysis to validate our brain model becomes challenging. Most information captured from the brain often overlooks how differential loops behave and instead combines them crudely. In the future, we aim to create our own brain signal database to address these issues. The exploration of polyatomic time crystals adds another layer of complexity to our understanding of temporal phenomena [6, 87]. If proven to exist, these enigmatic entities challenge our conventional understanding of time and have implications ranging from quantum mechanics to neurobiology [6].

In Fig. 2.8a–f, we delve into the investigation of positive and negative emotions to identify invariant geometric shapes. Figure 2.8a–c correspond to positive emotions, while Fig. 2.8d–f represent negative emotions. For positive emotions, the invariant geometries manifest as lines with distinct 3D orientations when observing transverse cross-sectional views of the images. In stark contrast, when examining longitudinal cross-sections, loops emerge for negative emotions. In the case of negative emotions, the transverse view reveals an X shape for anger, two connected squares for fear, and a single line for sadness. Thus, negative emotions give rise to two orthogonal invariant patterns. Activation associated with happiness takes the form of a discrete S shape in the transverse cross section and a semicircular loop in the longitudinal cross section.

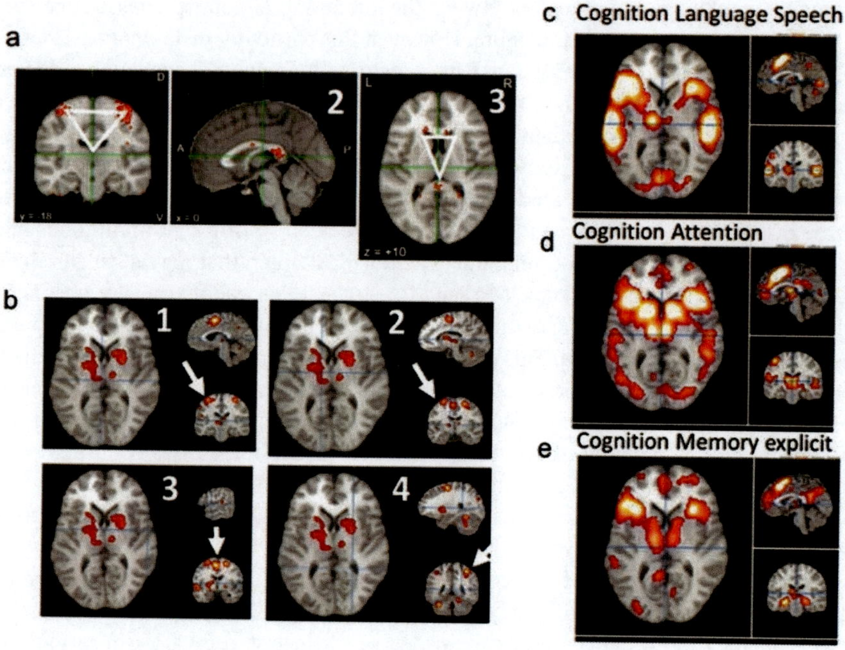

Fig. 2.7 **a** Load of "interhemispheric" meta-analysis, in 81 papers from Neurosynth brain scan database [68]. "association" function has been applied for generating the metadata (see methods for details). **b** Action motor learning of BrainMap database. Here metadata is a compilation of 27 relevant publications; 114 experimental contrasts; 1080 coordinates reported. **c** Cognition language speech, 228 relevant publications, 1028 experimental contrasts, 8977 coordinates reported. **d** Cognition attention, 664 relevant publications, 2748 experimental contrasts, 21,797 coordinates reported. **e** Cognition memory explicit, 262 relevant publications, 1110 experimental contrasts, 8403 coordinates reported

Humor, on the other hand, appears to be a complex emotional state, with distant and discrete activations that do not exhibit any inherent geometric shape. Valence, representing the most conservative emotional state, originates from a small looped activation in the frontal lobe of the human brain.

In Fig. 2.9a–d, we present four interception-related brain states: hunger, sexuality, sleep, and thirst. Meanwhile, Fig. 2.9e–h illustrate four perception-related brain states: vision for color, motion, shape, and somesthesis pain. Hunger predominantly affects a very localized area in the midbrain, while sexuality forms an open loop, visible in the orthogonal cross-section. Sexuality is unique in that it appears to be the only brain state that creates an incomplete loop. Sleeping and thirst, on the other hand, originate as highly localized effects stemming from isolated and discrete memory locations. In terms of perception, the activation for color resembles a V shape, while motion presents as a flowerpot-like shape with a nearly closed neck. Shape-related activation takes the form of a hexagon, while pain is characterized by a dumbbell-shaped activation pattern.

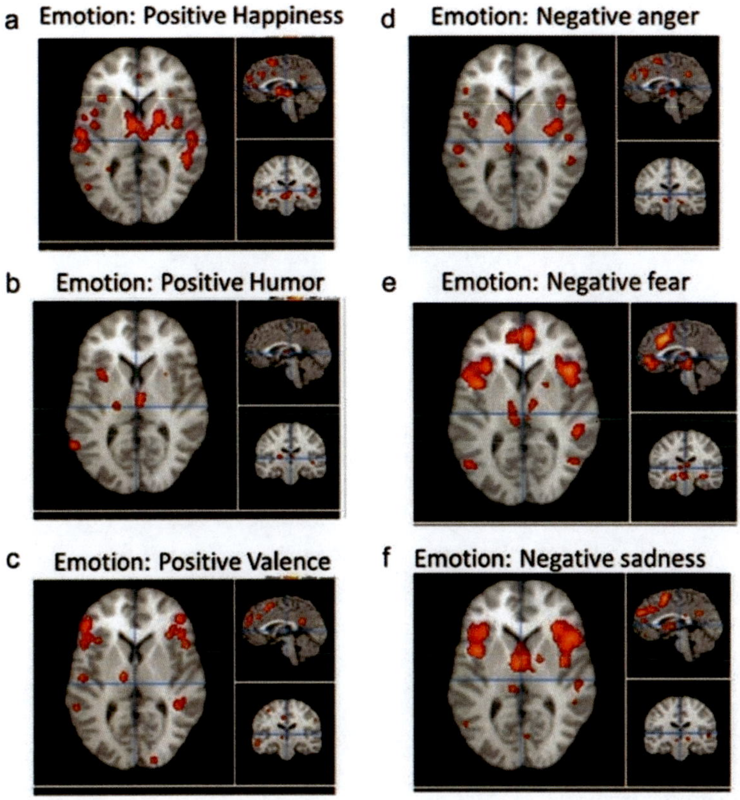

Fig. 2.8 Illustrates various meta-analyses related to emotional states and valences. **a** presents a meta-analysis on positive happiness, involving 117 relevant publications, 411 experimental contrasts, and 2458 reported coordinates. Moving on to **b**, we have a meta-analysis for positive emotion humor, comprising 7 relevant publications, 34 experimental contrasts, and 202 reported coordinates. In **c**, the focus is on emotion with positive valence (intrinsic goodness), involving 43 relevant publications, 150 experimental contrasts, and 1119 reported coordinates. Shifting to negative emotions, **d** presents a meta-analysis on anger with 64 relevant publications, 275 experimental contrasts, and 1496 reported coordinates. **e** delves into emotion with negative valence, specifically fear, with data drawn from 133 relevant publications, 531 experimental contrasts, and 3089 reported coordinates. Finally, **f** explores emotion with negative valence, sadness, utilizing data from 92 relevant publications, 350 experimental contrasts, and 2399 reported coordinates. These insights have been extracted from the Neurosynth brain scan database [68], employing the "association" function for metadata generation, and provide a comprehensive view of emotions and valences in neuroimaging research

As we delve into the intricate nature of temporal structures, our understanding of time's essence deepens [5]. The fusion of temporal structure analysis with brain databases opens up the possibility of exploring 3D temporal architectures and polyatomic time crystals [45]. These advancements push the boundaries of our knowledge and have both theoretical implications and real-world applications [18].

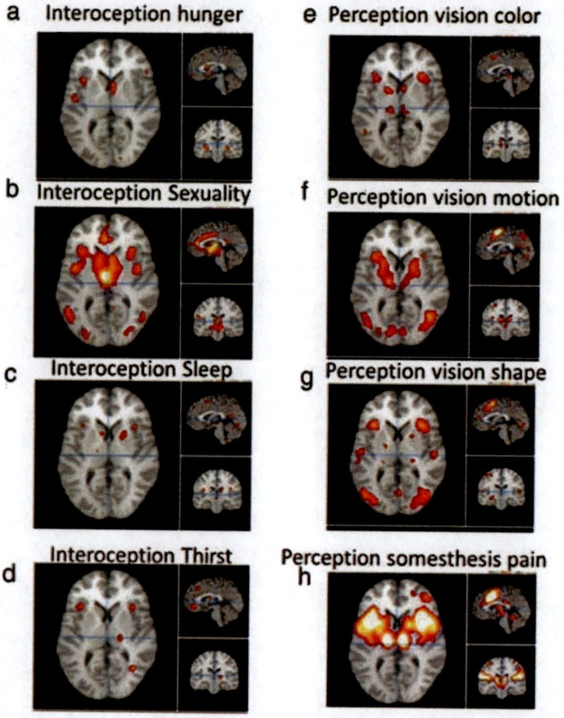

Fig. 2.9 Load of "emotion" meta-analysis, **a** Interoception hunger; 23 relevant publications, 90 experimental contrasts, 505 coordinates reported. **b** Interoception sexuality, 73 relevant publications, 426 experimental contrasts, 3486 coordinates reported, **c** Interoception sleep, 12 relevant publications, 45 experimental contrasts, 278 coordinates reported. **d** Interoception thirst, 6 relevant publications, 28 experimental contrasts, 246 coordinates reported. **e** Perception vision color, 40 relevant publications, 117 experimental contrasts, 1112 coordinates reported. **f** Perception vision motion, 94 relevant publications, 318 experimental contrasts, 3038 coordinates reported. **g** Perception vision shape, 165 relevant publications, 606 experimental contrasts, 4483 coordinates reported. **h** Perception somesthesis pain, 147 relevant publications, 553 experimental contrasts, 5354 coordinates reported. Collected from Neurosynth brain scan database [68]. "association" function has been applied for generating the metadata

2.6 Conclusions and Mathematical Model for Cognitive Engineering

2.6.1 Time is not Linear in the Brain: Different Time Domains Are Favored Differently

In a novel approach, we employed EEG in conjunction with MD to investigate four mental states: insecure, boundary-crossing, stuck, and quitting. We analyzed three explicit EEG features—the duration of random or localized brain activity, the duration of silence where no EEG frequency domain dominates, and the duration of synchronous activity between hemispheres. Intriguingly, we found that subjects completing MD tasks within specific time domains exhibit an innate aptitude for procedural learning through repeated MD tasks. Conversely, more rigorous training is required for others. This study revealed the presence of six quantized time domains across all tasks and mental states [82]. This discovery, that the procedural learning of the human mind is quantized into six restricted and favored time zones, not only holds potential for efficient brain training but also sheds light on the subjectivity challenges in psychology.

In recent years, advancements in neuroscience and the availability of large-scale brain activity databases have opened up new avenues for investigating the temporal aspects of interhemispheric energy transfer. In our quest to comprehend the intricate tapestry of temporal dynamics, we have ventured into uncharted territory. By harmonizing a meticulous temporal structure analysis with the wealth of knowledge contained within brain databases, we have embarked on a journey that may unveil the enigmatic 3D architecture of temporal clocks or the existence of polyatomic time crystals. The polyatomic time crystal model of the brain is another effort [76, 86] that provides a tool to look into the whole brain as integrated clock architecture, not two distinct systems.

The experiments disclosed in this chapter bear an immense significance, as they bridge the divide between what we've observed at the molecular scale, within single proteins and protein-made nanowires, and what now manifests on the grand canvas of the behavioral time scale. The captivating narrative of brain dynamics, which initially unfurls at the molecular level, unfalteringly continues its mesmerizing trajectory into the realm of behavioral phenomena.

The role of the membrane-microtubule junction in neurons is crucial, with heterojunctions being of significant importance for signal transfer in a fractal structure [71]. This fractal, scale-free electromagnetic resonance is observed in a single brain-extracted microtubule nanowire, a single tubulin protein, and a single neuron. It suggests that a triplet of triplet resonance bands, or nine frequency or time domains in discrete groups, operate together [85]. Electromagnetic resonance of biomaterials could couple distantly located parts in a biological system [88]. Quantum tunneling has the potential to be utilized for reading and writing information in such a network [8]. Furthermore, a fractal network could logically process information [35], it is not undefined throughout.

A simultaneous measurement of neuron firing at the neuron membrane and the microtubular bundle inside a neuron reveals that microtubules process firing signals 10,000 times faster than the membrane [1], demonstrating that they are not silent players in the brain's information processing [90]. Interestingly, even neurons not connected by synaptic junctions can exchange energy, as observed in neural circuit imaging using a scanning dielectric microscope (SDM) [36]. This particular discovery of electromagnetic communication among neural and subneural elements raises serious concern on the current interpretations of MD tasks. We need to figure out how much energy of the interhemispheric transfer is coming from electromagnetic energy transfer.

2.6.1.1 Large Database of Brain Dynamics and EEG Analysis Confirms Discrete Times or Crystals of Time

The understanding of temporal dynamics in interhemispheric energy transfer is a complex and evolving field, with implications for various aspects of human cognition, particularly in the context of motor learning. While mainstream psychology has traditionally viewed time perception as continuous, our analysis suggests that this perspective may not fully capture the intricate processes occurring within the brain. Mirror drawing tasks offer a unique opportunity to explore the relationship between visual perception and motor control, although definitive evidence regarding the formation of associations remains elusive. The concept of the familiarity index highlights the brain's ability to adapt and optimize performance over time, posing challenges for statistical analysis. Our review of the meta-analysis by Karolis et al. [47] has opened new avenues for research, indicating the need for further exploration of the quantization of energy in interhemispheric transfer. Additionally, the concept of favored and restricted time zones within motor skill development tasks presents an exciting area for future investigation, with the potential to advance our understanding of cognitive neuroscience.

2.6.1.2 The Winner and the Looser

In the study of left and right hemispheric transfer using mirror drawing, EEG spectra were examined to understand how subjects approached the task. Participants completing the task in approximately 30 min tended to frequently quit, procrastinate, and navigate the parallel pathways with caution. On the other hand, those who completed the task swiftly within 40–50 s, referred to as "winners," demonstrated improvisation, minimal procrastination, and were less meticulous when crossing barriers. The EEG spectra analysis revealed a marked contrast between the "winners" and the participants who procrastinated ("losers"). The "losers" displayed activation across various frequency bands, with no clear interhemispheric transfer evident. In contrast, the "winners" exhibited activity primarily in the beta frequency band, indicating effective left–right interhemispheric transfer. A noteworthy aspect of the "winners" was their consistent completion time across all attempts, whether it was the initial test, training sessions, or return confirmation tests. In contrast, the EEG analysis of the "losers" suggested that their brains activated multiple functional domains, resembling multitasking behavior shortly before quitting, often expressing frustration before moving to another point.

The experiments chronicled in this chapter are not confined to the laboratory; they possess far-reaching implications for our comprehension of the human experience. What transpires at the molecular scale, within the intricate dance of single proteins and nanowires, is now vividly manifested on the stage of behavioral time scale. This transition represents a transformative moment, where the abstract becomes tangible, and the microscopic finds resonance in the macroscopic. The brain, with its astonishing complexity, emerges as the protagonist of this narrative. Its dynamics, ignited at the molecular level, traverse the vast expanse of neural networks, ultimately influencing our behavior, thoughts, and perceptions. This journey offers a profound insight into the unity of science, where the infinitesimal and the monumental are intricately intertwined.

2.6.2 Brain's Ability Should Be Measured as Ability to Rearrange the Geometric Shapes

2.6.2.1 Nineteen Cognitive States of the Brain

At the molecular scale, the brain's intricate dance of dynamics shapes our behaviors, perceptions, and interactions [29]. The dynamics of the brain's temporal architecture play a crucial role in understanding time perception and synchronization within the brain [18]. The existence of standalone 3D temporal architectures, similar to time clocks, may hold the cognitive key to unraveling the fundamental mechanics of time perception [45].

We investigated five Regions of Interest or ROI domains of behavioral response shifts, e.g., Action, Cognition, Emotion, Interoception, and Perception to find spatial jumps, periodic jumps between spaces or multiple ROIs to find invariant geometric shapes. The above discussion accounts for nineteen cognitive states of the brain which could be modeled using dodecanions or 12D tensors [91, 92] and applying them in the space–time-topology-prime metric. All nineteen brain states have a typical geometric shape of activation in the human brain. Therefore, our geometric language of time crystals could integrate periodic vibrations in different organs occurring at different time domains into a unified structure that represents resonant oscillations of the brain components as well as the brain functions, behavioral responses.

2.6.2.2 The Future of Perception, Cognition, and Emotional Studies Using Polyatomic Time Crystal

As we navigate the labyrinthine corridors of temporal structures, our understanding of time's essence deepens. The fusion of temporal structure analysis with brain databases unveils the tantalizing possibility of 3D temporal architectures and polyatomic time crystals, pushing the boundaries of our knowledge. However, the true significance of this exploration lies not only in its theoretical implications but also in its real-world applications. What unfolds at the molecular scale now finds resonance in our daily lives, as the brain's intricate dance of dynamics shapes our behaviors, perceptions, and interactions. In this grand symphony of science, the journey from the molecular to the behavioral time scale serves as a testament to the profound interconnectivity of the universe. It is a testament to the enduring quest of humanity to unravel the mysteries of time itself, a journey that continues to captivate and inspire.

Our exploration commences with a profound analysis of temporal structures, an endeavor that promises to illuminate the very nature of time itself. Through the fusion of data-driven insights and theoretical foundations, we embark on a quest to unearth the existence of standalone 3D temporal architectures. These structures, akin to time clocks, may hold the key to understanding the fundamental mechanics of time perception and synchronization within the brain. Moreover, our journey takes a fascinating detour into the realm of polyatomic time crystals. These enigmatic entities, if proven to exist, challenge our conventional understanding of temporal phenomena. Their discovery would signify a paradigm shift, ushering in a new era of temporal physics, with implications ranging from quantum mechanics to neurobiology.

References

1. Agrawal L et al (2016) Inventing atomic resolution scanning dielectric microscopy to see a single protein complex operation live at resonance in a neuron without touching or adulterating the cell. J Integrative Neurosci 15(4):435–462
2. Agrawal L, Chhajed R, Ghosh S, Ghosh B, Ray K, Sahu S, Fujita D (2018) Fractal information theory (FIT)-derived geometric musical language (GML) for brain-inspired hypercomputing. Soft Comput Theories Appl 548:343–372
3. Adi-Japha E, Karni A, Parnes A, Loewenschuss I, Vakil E (2008) A shift in task routines during the learning of a motor skill: group-averaged data may mask critical phases in the individuals' acquisition of skilled performance. J Exp Psychol Learn Memory Cogn 34(6):1544–1551. https://doi.org/10.1037/a0013217
4. Aarts A, Anderson J, Anderson C, Attridge P, Attwood A, Axt J, Zuni K et al (2015) Estimating the reproducibility of psychological science. Science 349(6251). https://doi.org/10.1126/science.aac4716
5. Allman M, Teki S, Griffiths T, Meck W (2014) Properties of the internal clock: first- and second-order principles of subjective time. Annu Rev Psychol 65(1):743–771. https://doi.org/10.1146/annurev-psych-010213-115117
6. Aoki K, Giovangigli V (2021) Kinetic theory of chemical reactions on crystal surfaces. Physica A 565:125573. https://doi.org/10.1016/j.physa.2020.125573
7. Bandura A (1999) Social cognitive theory: an agentic perspective. Asian J Soc Psychol 2(1):21–41. https://doi.org/10.1111/1467-839x.00024
8. Bandyopadhyay A, Miki K, Wakayama Y (2006) Writing and erasing information in multi-level logic systems of a single molecule using scanning tunneling microscope. Appl Phys Lett 89(24)
9. Bussche E, Noortgate W, Reynvoet B (2009) Mechanisms of masked priming: a meta-analysis. Psychol Bull 135(3):452–477. https://doi.org/10.1037/a0015329
10. BrainMap (n.d.) Activation likelihood estimation (ALE). Retrieved from http://www.brainmap.org/ale/
11. Bavel J, Mende-Siedlecki P, Brady W, Reinero D (2016) Contextual sensitivity in scientific reproducibility. Proc Natl Acad Sci 113(23):6454–6459. https://doi.org/10.1073/pnas.1521897113
12. Boeck P, Jeon M (2018) Perceived crisis and reforms: issues, explanations, and remedies. Psychol Bull 144(7):757–777. https://doi.org/10.1037/bul0000154
13. Bach DR, Melinščak F, Fleming SM, Voelkle MC (2020) Calibrating the experimental measurement of psychological attributes. Nat Hum Behav 4(12):1229–1235. https://doi.org/10.1038/s41562-020-00976-8

14. Bandyopadhyay A (2020) Nanobrain: the making of an artificial brain from a time crystal. Boca Raton CRC Press, Taylor and Francis. ISBN: 9781439875490. https://doi.org/10.1201/9780429107771
15. Barakat M, Carrier J, Debas K, Lungu O, Fogel S, Vandewalle G, Doyon J et al (2013) Sleep spindles predict neural and behavioral changes in motor sequence consolidation. Human Brain Mapping 34(11):2918–2928. https://doi.org/10.1002/hbm.22116
16. Berlucchi G, Aglioti S, Marzi CA, Tassinari G (1995) Corpus callosum and simple visuomotor integration. Neuropsychologia 33(8):923–936. https://doi.org/10.1016/0028-3932(95)00031-W
17. Burdet E, Milner TE (1998) Quantization of human motions and learning of accurate movements. Biol Cybern 78(4):307–318. https://doi.org/10.1007/s004220050435
18. Calhoun V, Miller R, Pearlson G, Adalı T (2014) The chronnectome: time-varying connectivity networks as the next frontier in fmri data discovery. Neuron 84(2):262–274. https://doi.org/10.1016/j.neuron.2014.10.015
19. Cavaco S, Anderson SW, Allen JS, Castro-Caldas A, Damasio H (2004) The scope of preserved procedural memory in amnesia. Brain: J Neurol 127(Pt 8):1853–1867. https://doi.org/10.1093/brain/awh208
20. Chein JM, Schneider W (2005) Neuroimaging studies of practice-related change: fMRI and meta-analytic evidence of a domain-general control network for learning. Cogn Brain Res 25(3):607–623. https://doi.org/10.1016/j.cogbrainres.2005.08.013
21. Chen Z, Kosovicheva A, Wolfe B, Cavanagh P, Gorea A, Whitney D (2018) Unifying visual space across the left and right hemifields. Psychol Sci 29(3):356–369. https://doi.org/10.1177/0956797617735534
22. Cohen JD, Perlstein WM, Braver TS, Nystrom LE, Noll DC, Jonides J, Smith EE (1997) Temporal dynamics of brain activation during a working memory task. Nature 386(6625):604–608. https://doi.org/10.1038/386604a0
23. Crespo-Eguílaz N, Magallón S, Narbona J (2014) Procedural skills and neurobehavioral freedom. Front Hum Neurosci. Jun 20; 8:449. https://doi.org/10.3389/fnhum.2014.00449. PMID 24999324; PMCID PMC4064662.
24. Deco G, Vidaurre D, Kringelbach ML (2021) Revisiting the global workspace orchestrating the hierarchical organization of the human brain. Nat Hum Behav 5:497–511. https://doi.org/10.1038/s41562-020-01003-6
25. Doyon J, Penhune V, Ungerleider LG (2003) Distinct contribution of the cortico-striatal and cortico-cerebellar systems to motor skill learning. Neuropsychologia 41(3):252–262. https://doi.org/10.1016/S0028-3932(02)00158-6
26. Dehaene S, Kerszberg M, Changeux JP (1998) A neuronal model of a global workspace in effortful cognitive tasks. Proc Natl Acad Sci USA 95(24):14529–14534
27. Deardorff A (2020) Assessing the impact of introductory programming workshops on the computational reproducibility of biomedical workflows. https://doi.org/10.1101/2020.03.09.983429
28. Eickhoff SB, Constable RT, Yeo BTT (2018) Topographic organization of the cerebral cortex and brain cartography. Neuroimage 170:332–347
29. Eagleman D (2008) Human time perception and its illusions. Curr Opin Neurobiol 18(2):131–136. https://doi.org/10.1016/j.conb.2008.06.002
30. Escobar WA (2013) Quantized visual awareness. Front Psychol 4. https://doi.org/10.3389/fpsyg.2013.00869
31. Felleman DJ, Van Essen DC (1991) Distributed hierarchical processing in the primate cerebral cortex. Cereb Cortex 1:1–47
32. Friston KJ, Stephan KE (2007) Free-energy and the brain. Synthese 159(3):417–458. https://doi.org/10.1007/s11229-007-9237-y
33. Friston K, Kilner J, Harrison L (2006) A free energy principle for the brain. J Physiol Paris 100(1–3):70–87. https://doi.org/10.1016/j.jphysparis.2006.10.001. Epub 2006 Nov 13 PMID: 17097864

34. Glasser MF et al (2016) A multi-modal parcellation of human cerebral cortex. Nature 536:171–178
35. Ghosh S, Fujita D, Bandyopadhyay A (2015) An organic jelly made fractal logic gate with an infinite truth table. Sci Rep 5(1):11265
36. Ghosh S, Sahu S, Agrawal L, Shiga T, Bandyopadhyay A (2016) Inventing a co-axial atomic resolution patch clamp to study a single resonating protein complex and ultra-low power communication deep inside a living neuron cell. J Integr Neurosci 15(04):403–433
37. Hagmann P et al (2008) Mapping the structural core of human cerebral cortex. PLoS Biol 6:e159
38. Hubert V, Beaunieux H, Chételat G, Platel H, Landeau B, Danion J-M, Viader F, Desgranges B, Francis E (2007) The dynamic network subserving the three phases of cognitive procedural learning. Human Brain Mapping 28(12):1415–1429. https://doi.org/10.1002/hbm.20354
39. Hillebrand A et al (2016) Direction of information flow in large-scale resting-state networks is frequency-dependent. Proc Natl Acad Sci USA 113:3867–3872
40. Huntenburg JM, Bazin PL, Margulies DS (2018) Large-scale gradients in human cortical organization. Trends Cogn Sci 22:21–31
41. Hugdahl K (2000) Lateralization of cognitive processes in the brain. Acta Physiol (Oxf) 105(2–3):211–235
42. Henrich J, Heine S, Norenzayan A (2010) The weirdest people in the world? Behav Brain Sci 33(2–3):61–83. https://doi.org/10.1017/s0140525x0999152x
43. Haggard P, Clark S, Kalogeras J (2002) Voluntary action and conscious awareness. Nat Neurosci 5(4):382–385. https://doi.org/10.1038/nn827
44. Hall P, Fong G (2015) Temporal self-regulation theory: a neurobiologically informed model for physical activity behavior. Front Human Neurosci 9. https://doi.org/10.3389/fnhum.2015.00117
45. Hutchison R, Womelsdorf T, Allen E, Bandettini P, Calhoun V, Corbetta M, Chang C et al (2013) Dynamic functional connectivity: promise, issues, and interpretations. Neuroimage 80:360–378. https://doi.org/10.1016/j.neuroimage.2013.05.079
46. Julius MS, Adi-Japha E (2016) A developmental perspective in learning the mirror-drawing task. Front Human Neurosci 10. https://doi.org/10.3389/fnhum.2016.00083
47. Karolis VR, Corbetta M, Thiebaut de Schotten M (2019) The architecture of functional lateralisation and its relationship to callosal connectivity in the human brain. Nat Commun Mar 29; 10(1):1417. https://doi.org/10.1038/s41467-019-09344-1. PMID 30926845; PMCID PMC6441088.
48. Kiefer AW, Gualberto Cremades J, Myer GD (2014) Train the brain: novel electroencephalography data indicate links between motor learning and brain adaptations. J Novel Physiotherapies 4(2). https://doi.org/10.4172/2165-7025.1000198
49. Kidgell DJ, Frazer AK, Pearce AJ (2017) The effect of task complexity influencing bilateral transfer. Int J Exerc Sci 10(8):1174–1183
50. Klatzky RL (1994) On the relation between motor imagery and visual imagery. Behav Brain Sci 17(2):212–213. https://doi.org/10.1017/S0140525X00034178
51. Knightly AE (1934) The process of learning as shown by the mirror- drawing experiment. University of Massachusetts, Amherst
52. Knowlton BJ, Siegel ALM, Moody TD (2017) Procedural learning in humans. In Eichenbaum H, Byrne JH (eds) Learning and memory: a comprehensive reference, 2nd ed. Elsevier, pp 295–312. https://doi.org/10.1016/B978-0-12-809324-5.21085-7
53. Kowalik ZJ, Wrobel A, Rydz A (1996) Why does the human brain need to be a nonlinear system? Behav Brain Sci 19(2):302–303. https://doi.org/10.1017/S0140525X0004276X
54. Krigolson OE, Cheng D, Binsted G (2015) The role of visual processing in motor learning and control: insights from electroencephalography. Vis Res 110:277–285. https://doi.org/10.1016/j.visres.2014.12.024
55. Kunisato Y (2019) Introduction to reproducible psychological research. https://doi.org/10.31234/osf.io/x8js5

56. Lange RK, Braun C, Godde B (2006) Coordinate processing during the left-to-right hand transfer investigated by EEG. Exp Brain Res 168:547–556. https://doi.org/10.1007/s00221-005-0117-8
57. Lancaster JL, Laird AR, Eickhoff SB, Martinez MJ, Fox PM, Fox PT (2012) Automated regional behavioral analysis for human brain images. Front Neuroinform 6:23. https://doi.org/10.3389/fninf.2012.00023
58. Landau A, Elwan D, Holtz S, Prinzmetal W (2012) Voluntary and involuntary attention vary as a function of impulsivity. Psychon Bull Rev 19(3):405–411. https://doi.org/10.3758/s13423-012-0240-z
59. Lin CL, Shaw FZ, Young KY, Lin CT, Jung TP (2012) Neuroimage 60(4):2258–2273. https://doi.org/10.1016/j.neuroimage.2012.02.008. Epub 2012 Feb 13
60. Logan GD (1988) Toward an instance theory of automatization. Psychol Rev 95(4):492–527. https://doi.org/10.1037/0033-295X.95.4.492
61. Lobier M, Siebenhuhner F, Palva S, Palva JM (2014) Phase transfer entropy: a novel phase-based measure for directed connectivity in networks coupled by oscillatory interactions. Neuroimage 85:853–872
62. Magallón S, Narbona J, Crespo-Eguílaz N (2016) Acquisition of motor and cognitive skills through repetition in typically developing children. PLOS ONE 11(7):e0158684. https://doi.org/10.1371/journal.pone.0158684
63. Marzi CA (2010) Asymmetry of interhemispheric communication. Wiley Interdiscip Rev Cogn Sci 1(3):433–438. https://doi.org/10.1002/wcs.53
64. Markov NT et al (2014) Anatomy of hierarchy: feedforward and feedback pathways in macaque visual cortex. J Comp Neurol 522:225–259
65. Mak JN, Chan RH, Wong SW (2013) Spectral modulation of frontal EEG activities during motor skill acquisition: task familiarity monitoring using single-channel EEG. Annu Int Conf IEEE Eng Med Biol Soc 2013:5638–5641. https://doi.org/10.1109/EMBC.2013.6610829. PMID: 24111016
66. Miall RC, Cole J (2007) Evidence for stronger visuo-motor than visuo-proprioceptive conflict during mirror drawing performed by a deafferented subject and control subjects. Exp Brain Res 176(3):432–439. https://doi.org/10.1007/s00221-006-0626-0
67. Nowicka A, Grabowska A, Fersten E (1996) Interhemispheric transmission of information and functional asymmetry of the human brain. Neuropsychologia 34(2):147–151. https://doi.org/10.1016/0028-3932(95)00064-X
68. Neurosynth (n.d.) Interhemispheric transfer. Retrieved from https://neurosynth.org/analyses/terms/interhemispheric/
69. Newell B, Shanks D (2014) Unconscious influences on decision making: a critical review. Behav Brain Sci 37(1):1–19. https://doi.org/10.1017/s0140525x12003214
70. Papaya (n.d.) Papaya viewer. Retrieved from https://rii-mango.github.io/Papaya/
71. Pradhan B, Bandyopadhyay A, Pal AJ (2004) Molecular level control of donor/acceptor heterostructures in organic photovoltaic devices. Appl Phys Lett 85(4):663–665
72. Poldrack R (2019) The costs of reproducibility. Neuron 101(1):11–14. https://doi.org/10.1016/j.neuron.2018.11.030
73. Pattanayak A, Dutta T, Pranjal P, Singh P, Sahoo P, Sarkar S, Bandyopadhyay A (2022) Meta-analysis of fMRI for emotional and cognitive states shows hierarchical invariant optimization in brain. Proc Trends Electron Health Inform 376:255–265
74. Prete G, Fabri M, Foschi N, Tommasi L (2017) Asymmetry for symmetry: right-hemispheric superiority in bi-dimensional symmetry perception. Symmetry 9(5):76. https://doi.org/10.3390/sym9050076
75. Pouratian N, Sheth SA, Martin NA, Toga AW (2003) Shedding light on brain mapping: advances in human optical imaging. Trends Neurosci 26(5):277–282
76. Reddy S, Sonker D, Singh P, Saxena K, Singh S, Chhajed R et al (2018) A brain-like computer made of time crystal: could a metric of prime alone replace a user and alleviate programming forever? Soft Comput Appl:1–43

77. Roebers CM, Kauer M (2009) Motor and cognitive control in a normative sample of 7-year-olds. Dev Sci 12(1):175–181. https://doi.org/10.1111/j.1467-7687.2008.00755.x
78. Russell P (2013) The brain book, 1st ed. Routledge, London. https://doi.org/10.4324/9780203713600
79. Rorden C, Brett M (2000) Stereotaxic display of brain lesions. Behav Neurol 12(4):191–200
80. Schmidt MF (2008) Using both sides of your brain: the case for rapid interhemispheric switching. PLoS Biology 6(10):e269. https://doi.org/10.1371/journal.pbio.0060269
81. Simmons JP, Nelson LD, Simonsohn U (2011) False-positive psychology. Psychol Sci 22(11):1359–1366. https://doi.org/10.1177/0956797611417632
82. Smyth S, Casey D, Cooney A, Higgins A, McGuinness D, Bainbridge E, Murphy K et al (2016) Qualitative exploration of stakeholders' perspectives of involuntary admission under the mental health act 2001 in Ireland. Int J Mental Health Nurs 26(6):554–569. https://doi.org/10.1111/inm.12270
83. Sun Y, Collinson S, Suckling J, Sim K (2018) Dynamic reorganization of functional connectivity reveals abnormal temporal efficiency in schizophrenia. Schizophr Bull 45(3):659–669. https://doi.org/10.1093/schbul/sby077
84. Sochat V, Eisenberg I, Enkavi A, Li J, Bissett P, Poldrack R (2016) The experiment factory: standardizing behavioral experiments. Front Psychol 7. https://doi.org/10.3389/fpsyg.2016.00610
85. Saxena K, Singh P, Sahoo P, Sahu S, Ghosh S, Ray K, Fujita D et al (2020) Fractal, scale free electromagnetic resonance of a single brain extracted microtubule nanowire, a single tubulin protein and a single neuron. Fractal Fract 4(2):11
86. Singh P, Saxena K, Singhania A, Sahoo P, Ghosh S, Chhajed R, Fujita D et al (2020) A self-operating time crystal model of the human brain: can we replace entire brain hardware with a 3D fractal architecture of clocks alone? Information 11(5):238
87. Saxena K, Singh P, Sahoo P, Ghosh S, Krishnanda D, Ray K, Fujita D, Bandyopadhyay A (2022) All basics that are wrong with the current concept of time crystal: learning from the polyatomic time crystals of protein, microtubule, and neuron. Proc Trends Electron Health Inform 376:243–254
88. Singh P, Ray K, Bandyopadhyay A (2022) Biological antenna to the humanoid bot: electromagnetic resonances in biomaterials. Studies in rhythm engineering (SRE). Springer Singapore. https://doi.org/10.1007/978-981-16-9677-0
89. Singh P et al (2021a) Electrophysiology using coaxial atom probe array: live imaging reveals hidden circuits of a hippocampal neural network. J Neurophysiol 125(6):2107–2116
90. Singh P et al (2021b) Cytoskeletal filaments deep inside a neuron are not silent: they regulate the precise timing of nerve spikes using a pair of vortices. Symmetry 13(5):821. https://doi.org/10.3390/sym13050821
91. Singh P et al (2021c) Quaternion, octonion to dodecanion manifold: stereographic projections from infinity lead to a self-operating mathematical universe. In: Singh P, Gupta RK, Ray K, Bandyopadhyay A (eds) Proceedings of international conference on trends in computational and cognitive engineering. Advances in intelligent systems and computing, vol 1169. Springer, Singapore, pp 55–77
92. Singh P et al (2021d) A space-time-topology-prime, stTS metric for a self-operating mathematical universe uses dodecanion geometric algebra of 2–20 D complex vectors. In: Ray K, Roy KC, Toshniwal SK, Sharma H, Bandyopadhyay A (eds) Proceedings of international conference on data science and applications. Lecture notes in networks and systems, vol 148. Springer, Singapore, pp 1–31
93. Soderstrom NC, Bjork RA (2015) Learning versus performance. Persp Psychol Sci 10(2):176–199. https://doi.org/10.1177/1745691615569000
94. SOCR (n.d.) SOCR brain viewer. https://socr.umich.edu/HTML5/BrainViewer/
95. Swinnen SP (2002) Intermanual coordination: from behavioural principles to neural-network interactions. Nat Rev Neurosci 3(5):348–359. https://doi.org/10.1038/nrn807. PMID: 11988774

96. Thut G (2014) Modulating brain oscillations to drive brain function. PLoS Biology 12(12):e1002032. https://doi.org/10.1371/journal.pbio.1002032
97. Vicari S (2005) Do children with developmental dyslexia have an implicit learning deficit? J Neurol Neurosurgery Psychiatry 76(10):1392–1397. https://doi.org/10.1136/jnnp.2004.061093
98. van den Heuvel MP, Sporns O (2011) Rich-club organization of the human connectome. J Neurosci 31:15775–15786
99. Waters RH, Sheppard R (1952) The mirror drawing experiment: a brief historical note. J Gen Psychol 46(1):63–72. https://doi.org/10.1080/00221309.1952.9710638
100. Willingham DB (1998) A neuropsychological theory of motor skill learning. Psychol Rev 105(3):558–584. https://doi.org/10.1037/0033-295X.105.3.558
101. Willingham DB, Koroshetz WJ (1993) Evidence for dissociable motor skills in Huntington's disease patients. Psychobiology 21(3):173–182. https://doi.org/10.3758/BF03327132
102. Wong SWH, Chan RHM, Mak JN (2014) Spectral modulation of frontal EEG during motor skill acquisition: a mobile EEG study. Int J Psychophysiol 91(1):16–21. https://doi.org/10.1016/j.ijpsycho.2013.09.004
103. Wyatt T (2020) Reproducible research into human chemical communication by cues and pheromones: learning from psychology's renaissance. Philos Trans R Soc B Biol Sci 375(1800):20190262. https://doi.org/10.1098/rstb.2019.0262
104. Yarkoni T, Poldrack RA, Nichols TE, Van Essen DC, Wager TD (2011) Large-scale automated synthesis of human functional neuroimaging data. Nature Meth 8(8):665–670. https://doi.org/10.1038/nmeth.1635
105. Zhu H, Sun Y, Wang F (2013) Electroencephalogram evidence for the activation of human mirror neuron system during the observation of intransitive shadow and line drawing actions. Neural Regener Res 8(3):251–257. https://doi.org/10.3969/j.issn.1673-5374.2013.03.007
106. Zhang W, Su Y, Tian B, Fei D (2022) Statistical assumptions and reproducibility in psychology: data mining based on open science. Front Psychol 13.https://doi.org/10.3389/fpsyg.2022.905977

Chapter 3
Universal Thermal Markers of Human Emotions: Geometric-Biological Invariants for Emotional Transitions

3.1 Introduction

3.1.1 The Problem of Self-Reporting and Transition to Thermal Imaging to Capture Emotion

Understanding human emotions is like decoding a complex puzzle with many pieces. In the world of psychology, experts have tried various methods to unravel the mystery of how people express their feelings. They've used tools like asking individuals about their emotions, observing their behavior, and measuring their body's responses [30]. These methods, however, have some limitations. Sometimes, people may not tell the whole truth because they want to seem socially acceptable, and observers might see things differently based on their own perspectives. Additionally, some of these methods can be a bit intrusive.

So, researchers have been on a quest to find better ways to understand our emotional states. They've turned to our body's signals for help. You see, our bodies give off important clues about how we're feeling, and these signals can help us learn more about our emotions. But, traditional techniques like EEG (which looks at brain activity) and ECG (which examines heart activity) involve physically connecting sensors to our bodies. This can be a bit uncomfortable and might not capture the full story of our emotions.

Now, here's where things get interesting. Scientists have started using a method called thermal imaging to peek into our emotional world ([47, 53]; Principal component analysis, PCA; [32]; Functional Infrared Thermal Imaging, FITI). Instead of touching us, this method simply looks at our body heat from a distance. It turns out that the heat patterns on our faces can tell a lot about how we're feeling.

But here's the twist—up until now, scientists have treated our entire face as one big emotion messenger. For example, when we're upset, more blood flows to our face, making it warmer. On the flip side, when we're happy or relaxed, less blood

T. Dutta and A. Bandyopadhyay, *Emotion, Cognition and Silent Communication: Unsolved Mysteries*, Studies in Rhythm Engineering,
https://doi.org/10.1007/978-981-99-9334-5_3

goes to our face, making it cooler. But what if different parts of our face had their own emotional stories to tell? Like, what if fear made our nose cooler? That's what some people have wondered when they study facial expressions.

So, as we dive deeper into understanding emotions, we're not just decoding feelings, but also exploring how our own bodies can give us the answers, one warm or cool clue at a time.

3.1.1.1 The Role of Autonomic Nervous System in Defining an Emotion

Skin temperature can serve as a valuable indicator of various physiological processes within the body. These temperature changes can reflect underlying shifts in the individual's physiological state. Through the analysis of captured thermal images, a wealth of physiological signals can be extracted, encompassing parameters such as heart rate, respiration rate, and skin conductance. Notably, thermal imaging holds a distinct advantage as it offers a non-invasive and contactless approach to gather this vital data.

The autonomic nervous system (ANS) holds a central role in governing biological thermoregulation and heat-related expressions. It is responsible for the management of unconscious bodily functions, including heart rate, respiratory patterns, tissue metabolism, perspiration, respiration, and cutaneous blood perfusion. This intricate network enables the body to convey emotional cues through thermal variations. In scientific circles, thermal infrared imaging, also referred to as functional infrared imaging (fITI), has emerged as an indispensable tool for investigating the diverse subdivisions of the ANS [32].

In the realm of computing thermal infrared signals derived from bioheat, research endeavors have predominantly centered on the analysis of facial regions. This preference stems from the face's distinct advantages—it is readily observable, aiding in the facilitation of social communication and interpersonal interactions.

3.1.1.2 Subcutaneous Vasoconstriction and Emotional Sweating

Our bodies are like secret agents when it comes to showing emotions, and they have two clever ways to do it using temperature changes. Let's meet these undercover agents:

Agent 1: Subcutaneous Vasoconstriction

Imagine you're in a scary situation—your body has a superhero response to protect you. It sends out a special agent called epinephrine. This agent tells your blood vessels to tighten up under your skin. It's like putting a lock on them to prevent too much blood from flowing, which could be dangerous [38, 55]. This protective move mainly happens in the parts of your body that are most exposed, like your face and hands. Once the threat is gone, your blood vessels relax, and your body slowly warms up again. This happens because your parasympathetic system, which is like your body's "chill" mode, takes over [52]. Here's the cool part—these temperature changes don't happen instantly, they take at least 5 s to kick in [38].

Agent 2: Emotional Sweating

Now, picture this: When you're feeling all sorts of emotions, your sympathetic nervous system springs into action. It sends signals to your sweat glands, telling them to produce more sweat. This extra sweat isn't meant to make you uncomfortable; it's actually a smart move by your body. It helps your skin become more flexible and reduces any friction, especially in areas that have a lot of contact with the environment [37, 57, 70].

But here's where it gets even cooler. Scientists have used thermal imaging to find out that different parts of your face heat up or cool down when you shift between different emotional states. It's like your face has its own secret language of warmth, and it's a reliable way to figure out what's going on inside your emotional world. So, next time you wonder how someone's feeling, just take a look at their face—it might be revealing more than they think [37, 57, 70].

3.1.2 How Detecting Emotional Transition Has Become an Intractable Problem

Despite the promising potential of thermal imaging in the realm of studying emotional states, there exists a noticeable research gap concerning the unique and distinct transitions between emotional states experienced by each individual. This study endeavors to address this void by harnessing thermal imaging technology to discern localized shifts in emotional states as reflected in the human face. Existing methodology to measure emotions is often subjective [10]. However, there is a need for objective measures of emotion [10]. One study suggests using the Choquet integral method and l-fuzzy measure, a subjective decision-making technique, to infer emotions on a personalized emotional space [36].

Gaining a deeper comprehension of emotional responses is of paramount importance, particularly in the context of crafting effective interventions for individuals grappling with emotional dysregulation or psychiatric disorders. Traditional self-report measures, while frequently employed to gauge emotional responses, are inherently subjective and susceptible to various biases. Furthermore, inducing and quantifying emotional responses through visual stimuli, as aptly expressed by the idiom "one man's meat is another man's poison" [39], remains a complex endeavor.

In pursuit of more objective assessments, researchers have turned to biophysical measures, encompassing galvanic responses, oximeter responses, and pulse responses, to quantify emotional reactions. However, the intricate interplay between these biophysical metrics and emotional responses has yet to be fully elucidated. The present study seeks to bridge this knowledge gap by exploring the intricate relationship between emotional responses and biophysical measures. This investigation is conducted through a comprehensive thermal study, supplemented by the incorporation of three additional parameters that contribute to the creation of invariants.

It is worth noting that our previous work has already underscored the efficacy of geometric invariants within various brain signals as a potent avenue for estimating human cognition [54]. This prior research serves as an essential foundation for the current study's approach to understanding emotional states and their objective measurements.

The Importance of Dynamic Thermal Imaging: Emotions are a fundamental part of human experience. They play a vital role in our social interactions, decision-making, and overall well-being. However, understanding and measuring emotions can be challenging, especially when it comes to individual differences. Traditional Approaches to Emotion Detection: Traditional approaches to emotion detection have relied on self-reporting and static thermal imaging of the face. Self-reporting is subjective and prone to error, while static thermal imaging cannot capture the dynamic nature of emotional transitions. Dynamic Thermal Imaging: Dynamic thermal imaging of the face is a promising new approach for detecting and measuring emotions. It can capture the subtle temperature changes that occur on the face as emotions shift. Individual-Specific Emotional Transitions: Researchers at the University of California, San Diego, have used dynamic thermal imaging of the face to investigate individual-specific emotional transitions. They found that distinct and unique geometric changes occur as emotions shift within a group of people.

Invariant Geometric Shapes: The researchers identified 23 facial regions (Fig. 3.2) that exhibit temperature changes during these emotional transitions. These regions form an invariant geometric shape for the group.

Region-Specific Invariants: Each individual possesses a unique combination of these regions, which serve as markers for their emotional expressions. These region-specific invariants remain consistent across various emotional states for each individual. Implications: The findings of this study have important implications for our understanding of individual differences in emotional expression. They also suggest that dynamic thermal imaging of the face could be a valuable tool for developing more accurate and personalized emotion detection systems.

Future research should focus on validating these findings using larger and more diverse samples. Additionally, researchers should investigate the potential of dynamic thermal imaging of the face to detect and measure emotions in real-world settings.

Potential Applications: Dynamic thermal imaging of the face could have a wide range of potential applications, including:

- Mental health diagnosis and treatment: Dynamic thermal imaging could be used to develop more accurate and personalized diagnostic tools for mental health conditions such as anxiety, depression, and autism. It could also be used to monitor the effectiveness of mental health treatments.

- Neuromarketing: Dynamic thermal imaging could be used to measure consumers' emotional responses to products, advertisements, and other stimuli. This information could be used to develop more effective marketing campaigns.
- Human–computer interaction: Dynamic thermal imaging could be used to develop more intuitive and user-friendly human–computer interfaces. For example, it could be used to develop computers that can adapt to the user's emotional state.

Overall, dynamic thermal imaging of the face is a promising new technology with the potential to revolutionize the way we detect and measure emotions.

In this chapter, we aim to investigate individual-specific emotional transitions by utilizing dynamic thermal imaging of the face. This approach represents a significant departure from the static thermal image-based methods typically used for detecting emotions. Unlike previous studies, we challenge the conventional practice of relying on self-reporting and instead propose that distinct and unique geometric changes occur as emotions shift within a group of people. We have identified 23 facial regions that exhibit temperature changes during these emotional transitions, forming an invariant geometric shape for the group. Additionally, we have discovered that each individual possesses a unique combination of these regions, which serve as markers for their emotional expressions. These region-specific invariants remain consistent across various emotional states for each individual.

This study serves as an extension of our prior research endeavors [54, 71], wherein we employed geometric analysis techniques to scrutinize the logical responses of our test subjects. The objective was to delve into how individuals categorized as introverts and extroverts employ distinct ocular behaviors during transitions between emotional states. Our previous findings illuminated the discernible disparities in how these two primary personality classifications utilize their eyes as a means of conveying their emotions in the context of thermal transitions.

Furthermore, our earlier investigations unveiled intriguing insights into how individuals exhibiting varying psychological dispositions—including those characterized as aggressive, subdued, or exhibiting composure—employ specific facial regions to articulate their emotional states. Building upon this foundational research and without deviating from the scope of thermal imaging, we had previously postulated the existence of geometric invariants governing the realm of emotions and perceptions. Within this context, we embark on an exploration of time management mechanisms for the expression of emotions, aiming to construct a 3D arrangement of temporal entities or a temporal architecture, which may offer a tangible representation of the human cognitive state.

Given the inherent absence of quantifiable mathematical or physical attributes for measuring, identifying, or distinguishing psychological states of the mind, our pursuit is driven by the quest to identify scientific parameters that can facilitate the quantification of various facets of human psychology, including emotions, perceptions, cognition, and even the complex realm of deception [56].

The application of thermal imaging in this context represents a straightforward yet potent investigative tool. As we endeavor to map the 3D geometric constructs corresponding to distinct cognitive states, the potential practical implications of our work extend into diverse fields, encompassing psychology, psychiatry, and neurology. This pursuit holds the promise of enhancing our understanding of psychological phenomena while contributing valuable tools for assessment and diagnosis within these critical domains.

3.1.3 *Facial Expressions of Infants*

Facial expressions are an important aspect of human communication and can convey a wide range of emotions. One interesting aspect of facial expressions is the effect they have on facial skin temperature. In a study conducted by Nakanishi et al. [51], it was found that facial skin temperature decreases in infants with joyful expressions. The study focused on measuring the skin temperature of the forehead in infants while they displayed different facial expressions. The researchers used a non-contact infrared thermometer to measure the temperature changes. They found that when infants displayed joyful expressions, their facial skin temperature decreased compared to when they displayed neutral expressions.

This finding is significant because it suggests that facial expressions are not only a visual representation of emotions but also have physiological effects on the body. The decrease in facial skin temperature during joyful expressions could be attributed to changes in blood flow and vascular activity in the face. It is possible that the activation of certain facial muscles during joyful expressions leads to increased blood flow to the face, resulting in a decrease in skin temperature [51]. The study also highlights the importance of studying facial expressions in infants. Infants are not able to verbally communicate their emotions, so facial expressions serve as an important means of understanding their emotional state. By studying the physiological changes associated with different facial expressions, researchers can gain insights into the emotional development of infants.

Understanding the relationship between facial expressions and physiological changes in infants can have practical implications in various fields. For example, in the field of pediatrics, this knowledge can help healthcare professionals assess the emotional well-being of infants who may not be able to express their emotions verbally. It can also be useful in the development of interventions and therapies for infants with emotional or developmental disorders [51].

3.2 Experimental Protocols Adopted for Classifying Emotions

3.2.1 General Experimental Methodology

Participants:

The study involved 287 participants aged between 18 and 35 years. Participants were recruited from management institutes and provided informed consent before participating in the study.

Instruments:

The study used a thermal imaging camera that captured the heat signature of the face in the dark room. Participants were asked to sit in a chair with their face in a comfortable and relaxed position. Participants were instructed to remain calm and relaxed during the thermal imaging process. Galvanic response, oximeter reading, and pulse rates were measured before and after each thermal measurement.

Procedure:

Experimental Procedure

A comprehensive investigation was conducted utilizing thermal imaging technology to assess the thermal signatures of participants' faces across various emotional states. The study involved 287 participants, drawn from diverse management schools, who provided informed consent to participate. Prior to the experimental sessions, participants were instructed to recollect and select significant memory events associated with emotions of happiness and anger/grief that were known to evoke these specific emotional states.

Emotional State Induction

Participants were asked to experience six distinct emotional states, namely happiness, surprise, fear, disgust, anger, and sadness. To facilitate the elicitation of these emotions, participants were guided to recall relevant memory events associated with each emotion. Each emotional state was experienced for a two-minute duration.

Thermal Imaging Data Collection

Thermal imaging cameras were employed to capture the thermal patterns on the faces of the participants. This process was carried out over a two-minute interval for each emotional state. The thermal camera recorded temperature data from twelve predefined facial regions, including the nose, eye corners, and eyes.

Biophysical Measures and Data Collection

In addition to thermal imaging, several biophysical measures were recorded simultaneously. These measures encompassed galvanic responses, oximeter responses, and pulse responses. Each participant's biophysical data was represented through a pair of plots reflecting maximum and minimum changes. The temporal aspect was depicted on the horizontal axis, measured in seconds from the baseline mental state where the participant experienced neither happiness nor sadness. The vertical axis represented the change in measured values.

Data Analysis

Data analysis commenced by calculating the area under the segmented b-spline curve for each participant, denoting the response latency of biophysical measures. Subsequently, response differences between the emotional states of anger and happiness were plotted in ascending order, with the area under the curve serving as the basis for comparative analysis. Additionally, the number of peaks in the transition pattern was counted.

To ensure the comparability of participants, steps were taken to verify that individuals exhibited similar biological characteristics concerning thermal, galvanic, and stress parameters.

Thermal Emission Analysis

To further explore psychophysical self-similarity, alterations in the area of thermal emission on the face were examined, alongside variations in intensity. The analysis revealed distinctive features across the facial regions of the 287 participants. In most instances, moderately bright regions exhibited the highest likelihood of change corresponding to shifts in emotional states. However, some individuals displayed exceptionally bright eyes, for which minimal changes were observed in response to emotional transitions. Conversely, regions characterized by extreme brightness or darkness demonstrated limited fluctuations in response to changes in emotional states within the brain.

3.2.2 Typical Cases Explaining Challenges in Human Subject Study

Analysis of Human Subjects

Human Subject 1 (Male): The examination of Human Subject 1 revealed an entire facial region characterized by excessive brightness, making it challenging to generate an activity map. This observation is indicative of elevated tension levels, potentially associated with high blood pressure and reduced oxygen levels. Moreover, heightened blood flow to the brain was inferred from this brightness. Notably, researchers identified a distinctive three-corner shape, akin to a triangle, formed by the convergence of the center of the forehead and the two eye corners. This unique configuration was deemed a combinatorial marker specific to this subject. Consequently, researchers focused their analysis on the brightest facial region during emotional transitions, rather than treating all components equally. The investigation of videos featuring other subjects underscored the importance of both dynamic and static facial aspects in the research.

Human Subject 2 (Female): The analysis of Human Subject 2 was distinguished by the significance of the neck region, although its intermittent visibility posed certain challenges. Researchers concentrated on moderately bright and dark facial areas, taking into consideration alterations that occurred during emotional transitions. Notably, a darkened nose region was interpreted as an indicator of reduced blood flow or metabolism. Interestingly, interviews with individuals who exhibited a notably cold nose, possibly due to heightened brain activity induced by intense meditation, suggested a correlation. The collective response of the neck and nose, similar to the triangular pattern observed in Human Subject 1's forehead, emerged as a noteworthy finding.

Human Subject 3 (Male): Human Subject 3 displayed a distinctive pattern of isolated thermal dominance within facial regions, signifying a lack of interconnectedness among these areas. Specifically, heightened brightness was observed at the inner corners of the eyes, forming isolated areas of thermal activity separate from the tranquil forehead region marked by corner brightness. This localized thermal agitation was primarily concentrated in the rear regions of the eye corners, traditionally associated with emotional responses. Consequently, the localized blood flow pattern, combined with insights from an interview with the subject, suggested an absence of typical emotional responses in this individual. Furthermore, the isolated thermal activity on the face, corroborated by the interview, raised the possibility of deficiencies in the cranial sensory systems responsible for emotional expression. These findings collectively indicate a complex and potentially concerning emotional expression pattern in this particular subject.

Cortex Area 38 (Brodmann's Regions): It is noteworthy to highlight that Cortex Area 38, encompassing Brodmann's regions, emerges as the pivotal area of influence on human facial expressions. This specific cortical junction integrates functions related to hearing, vision, memory, emotional awareness, and reactions, underscoring its significance in shaping emotional responses reflected in facial expressions.

3.3 Reevaluating Universal Thermal Markers of Human Emotion

In this study, we present six distinct emotional states as depicted in Fig. 3.1a. Conventionally, scientific literature has held the notion that a single standardized thermal image of the face corresponds to each emotional state, and this concept is assumed to be universally applicable to all individuals (please refer to Table 1 in the article by Ioannou [34]). However, our research took a more focused approach. We delved into a specific age group within a particular ethnic population and uncovered a fascinating

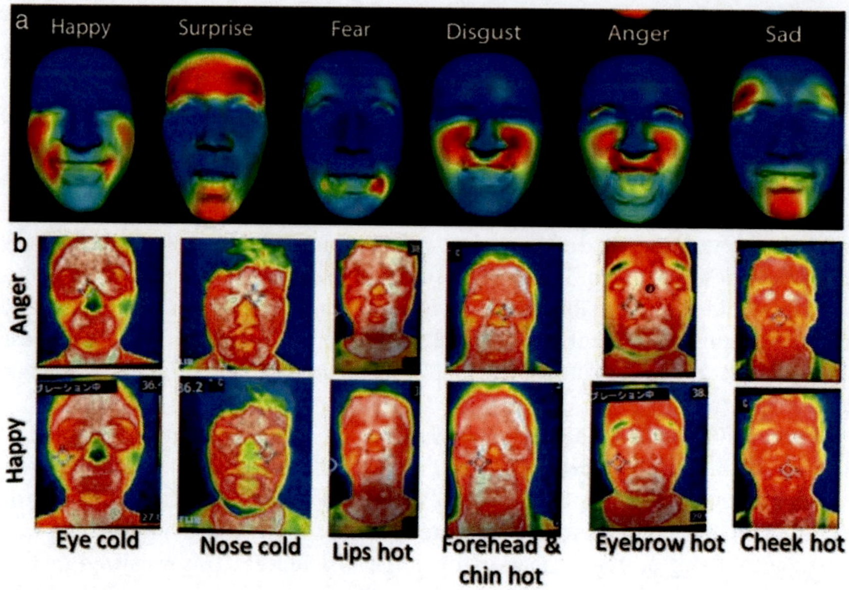

Fig. 3.1 Comparison between existing idea and key finding. **a** Six infrared images illustrating various moods of a person. **b** Two rows depicting different emotional states. The top row demonstrates the changes in different facial parts when a person transitions from anger (top row) to happiness (bottom row). Each column represents infrared images of the same person captured during two distinct mental states

revelation. Even among this select group, individuals displayed diverse thermal emissions across various facial regions while expressing identical emotional states, as vividly illustrated in Fig. 3.1b. Our investigative efforts were primarily centered on two prominent emotional expressions, namely happiness and anger. This focus allowed us to meticulously scrutinize the intricate changes in facial thermal patterns that occur when human subjects recollect their most joyous moments and their most infuriating experiences.

3.3.1 *Classification of Human Emotional Expressions*

To delve deeper into our exploration, we conducted an experiment involving human participants. Our aim was to prompt these individuals to recollect moments from their past experiences and memories, with a specific focus on events that had evoked intense happiness. Importantly, we reassured them that we wouldn't probe into the specific details of these memories, granting them the freedom to select deeply personal and meaningful instances.

Each participant was allotted a two-minute timeframe for this task. During this period, we closely monitored a key parameter: the latency time, which denotes the

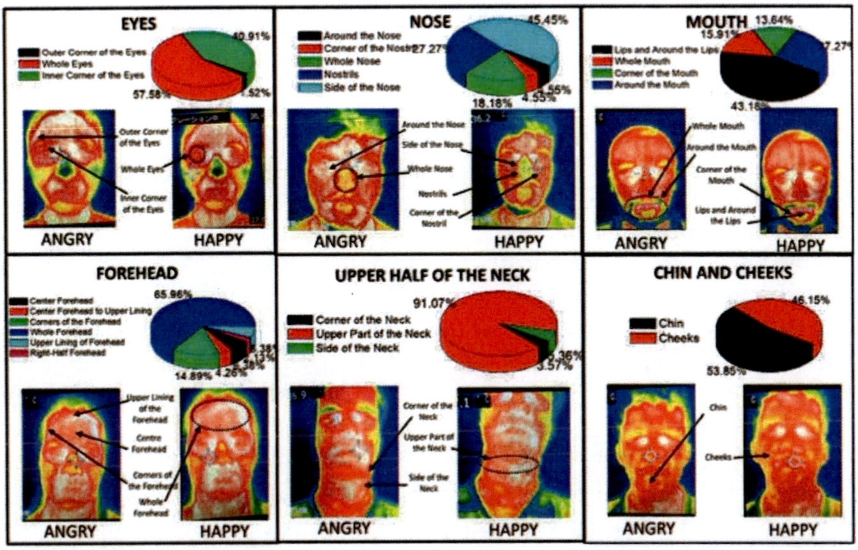

Fig. 3.2 It presents a concise summary of the statistical analysis conducted on facial components involved in expressing emotions. The figure consists of six panels, each representing the eyes, nose, mouth, forehead, upper half of the neck, and chin/cheeks. These panels depict the observed changes in these facial parts during the transition from anger to happiness, as captured in infrared images. Each panel is divided into two sections. The upper section features a pie chart that visually represents the proportion of changes within subcategories of the respective facial part. The pie chart utilizes distinct colors to represent specific local facial regions. In the lower section of each panel, two infrared images of a subject are displayed, one for the emotional state of anger and the other for happiness. The subcategories within the facial part are indicated with arrows. In the infrared images, the color scheme corresponds to temperature readings: White regions represent 27 °C, yellow regions represent 24 °C, red regions represent 25 °C, and green–blue regions represent 22 °C

duration it took for the participant to commence contemplating the selected memory. As soon as we discerned discernible changes through our thermal imaging camera, we meticulously recorded the precise moment. Additionally, we diligently measured the duration of happiness and anger experienced by the participants during these recollections. These time intervals held significant importance in our efforts to gauge the intensity of their emotional responses.

The results of our study have unveiled a fascinating revelation: Emotional states do not imprint a permanent signature on the human face. Instead, the expression of emotions varies significantly depending on the individual. To comprehensively categorize these variations, we have classified humans into multiple categories based on the thermal signatures of their emotional states, ultimately identifying six primary classes and several subclasses associated with these expressions.

The Six Primary Emotional States (Refer to Fig. 3.1a):

1. **Happiness**
2. **Anger**

3. **Surprise**
4. **Fear**
5. **Disgust**
6. **Sadness**

Traditionally, it has been widely accepted in the scientific community that each emotional state possesses a single, standardized thermal image applicable to the entire human population (refer to Table 1 in the article by Ioannou et al. [34]). However, our meticulous research, which focused on a specific age group within a particular ethnic demographic, has debunked this notion. Even within this specific population, we have observed diverse thermal emissions from various facial regions when individuals express the same emotional state (as illustrated in Fig. 3.1b).

Our investigation primarily delved into two emotional expressions: happiness and anger. This focus allowed us to closely scrutinize the dynamic changes in facial thermal patterns that occur as human subjects vividly recall memories of their most euphoric and infuriating moments.

Thermal Signatures of the Nose and Their Implications (Refer to Fig. 3.2):

Notably, we observed that the temperature of the human nose varies significantly from person to person, with differences of approximately 7–8 °C. Some individuals possess a notably cold nose, while others experience a warm sensation. However, it is essential to refrain from prematurely attributing the use of the nose in emotional expression solely based on its temperature.

Instead, the key lies in observing thermal radiation that remains moderate or close to room temperature. It is under these conditions that individuals are more likely to employ their nose as an emotional indicator. Conversely, consistently bright or consistently dark regions across facial components make it unlikely for them to exhibit changes corresponding to emotions.

This principle holds true for all six primary emotional classes.

3.3.2 Subcategorization of Emotions Based on Facial Components

Here are six primary categories:

1. **Eyes (Refer to Fig. 3.2)**: Within this category, we observed three distinct modes of change: Inner corners changed for 40.91% of subjects, whole eyes changed for 57.58%, and heating up or cooling down of the external end of the eye was observed in a mere 1.5% of cases.
2. **Nostrils (Refer to Fig. 3.2)**: Expressing emotions using both sides of the nose was the most prominent method, accounting for 45.45% of cases. However, asymmetry between nostrils, with one nostril being wetter than the other, posed a unique challenge.

3. **Lips (Refer to Fig. 3.2)**: The lips exhibited intricate patterns of thermal emissions, with five distinct regions undergoing changes in intensity when transitioning between emotional states.
4. **Forehead (Refer to Fig. 3.2)**: The forehead displayed six unique patterns or subclasses of thermal activity. These included vertical and horizontal parallel lines, both symmetric and asymmetric variations, encompassing 65.95% of subjects.
5. **Upper Half of the Neck (Refer to Fig. 3.2)**: Often overlooked, this region exhibited three modes of thermal radiation variations during emotional transitions.
6. **Cheek and Chin with Upper Neck (Refer to Fig. 3.2)**: Here, we identified a duality of contradictory changes in intensity as individuals transitioned between negative and positive emotional states, akin to our observations in other facial regions.

Let's break down our findings into more digestible portions:

Eye Changes: Our study identified three distinct ways in which the eyes express emotions. For 40.91% of subjects, the inner corners of the eyes changed, while for 57.58%, the entire eyes underwent alterations. Notably, very few individuals, only 1.5%, exhibited heating up or cooling down at the outer corners of the eyes during experiences of anger or happiness. Despite our relatively modest dataset of 287 participants, our research challenges the notion that there's a one-size-fits-all emotional signature for the entire human population. Instead, we identified at least six classes, comprising 23 subclasses (Fig. 3.2), of thermal signatures associated with expressing emotions.

The Intriguing Nose: When it comes to the nose, we made some intriguing observations. The thermal radiation emitted by the nose varies significantly among individuals, with temperature differences of 7–8 °C not uncommon. However, it's hasty to conclude that a person's nose can solely convey their emotions based on its temperature. Simply observing a very cold or very hot nose does not necessarily indicate emotional expression.

In reality, when nasal thermal radiation remains moderate and close to average room temperature, without significant fluctuations, it's more likely that the individual is using their nose to express emotions. Conversely, if the eyes consistently appear bright and white across their entire region, it becomes highly unlikely for them to exhibit emotion-related changes. This principle applies across all six identified emotional classes. Interestingly, none of the subclasses were observed in the brightest or coolest conditions among the selected facial features in Fig. 3.2. It's the moderate intensity of radiation that allows us to detect these variations. Therefore, it's plausible that facial parts emitting moderate levels of radiation are also suitable for expressing emotions.

We identified five subcategories related to using the nose to express emotions, with the most prominent method being the utilization of both sides of the nose, accounting for 45.45% of cases. However, using the nose as an indicator of emotions comes with challenges, primarily due to the fact that one nostril tends to be wetter than the other. Human pairs of nostrils, ears, and eyes operate slightly out of sync with each other to enhance sensitivity. This synchronization is beneficial in most cases but poses a challenge with the nose. When one nostril becomes wetter than the other, it significantly affects the thermal signature.

To address this issue, we carefully waited until the asymmetry between the two nostrils subsided. Afterward, we determined that the final contribution of the nostrils as an indicator of emotions was 27.27%. Additionally, we found that other areas surrounding the nostril corners, the entire nose, and its immediate surroundings also played a role in expressing emotions for some individuals.

The Role of Lips: Moving on to the lips, the third facial component involved in expressing emotions, we observed significant changes, particularly when individuals had a mustache. Emotional transitions, such as those from happiness to anger, resulted in complex patterns of thermal emissions. In our examination of various subjects, we identified five regions around the lips that underwent changes in intensity during these transitions. In some cases, the entire lip region became highly active, affecting both intensity and the area of active radiations. Conversely, the lip regions sometimes remained relatively still, with the elliptic region encompassing the lips regulating emotional expression. These two complementary selections were observed in the majority of subjects expressing emotions through their mouths.

Forehead Patterns: Let's explore the fourth component, the forehead, which exhibited six distinct patterns or subclasses. The forehead displayed intricate patterns and specific geometric shapes in regions emitting high thermal energy. Notably, the majority of human subjects (65.95%) exhibited three vertical parallel lines spanning the entire forehead when it played a significant role. Additionally, 14.89% of subjects experienced activation or deactivation of thermal emission in the corners of the forehead during emotional transitions. Some individuals (6.38%) displayed three parallel horizontal lines on the forehead. Importantly, both horizontal and vertical parallel lines were observed, and they could be perpendicular to each other. The remaining three options represented asymmetric parts of the three primary geometric shapes where thermal activation was observed.

Upper Half of the Neck: The fifth component involved in facial expression of emotions is the upper half of the neck, often overlooked during human interactions as a primary mode of expression. However, this region exhibited prominent variations in thermal signatures as revealed by thermal imaging. Depending on the region of activation, we identified three modes of thermal radiations.

Cheek, Chin, and Upper Neck: Our sixth and final component for emotional expression included the cheek and chin, along with the upper neck. In this region, we observed a duality of contradictory changes in intensity during transitions from negative to positive emotions and vice versa, similar to what we found in other facial regions. Through a careful analysis that involved matching two types of images—emotion-driven thermal changes in the face and epinephrine (adrenaline)-driven thermal changes—we confirmed and estimated the degree of emotional intensity. It was found that the neck and cheek complemented each other in this context, aligning with previous findings.

This approach provides a physiological understanding of emotions, shedding light on how they manifest in the body, specifically in the face. It's crucial to recognize that emotions are complex and multifaceted, and the 23 facial expressions we've captured represent just one way of expressing them. Countless combinations of different patterns in thermal images are possible. The potential exists to develop an algorithm for automated facial recognition of emotions using thermal imaging. However, gaining deeper insights into the changes in facial components requires a more profound understanding of the physiological processes involved in emotional expression. These insights hold the promise of enhancing our understanding of how emotions impact our bodies, a topic we'll delve into further in Fig. 3.3.

Relationship Between Emotional Response and Biophysical Measures (Refer to Fig. 3.5):

In our pursuit to comprehend the connection between emotional responses and biophysical measures, we analyzed galvanic responses, oximeter responses, and

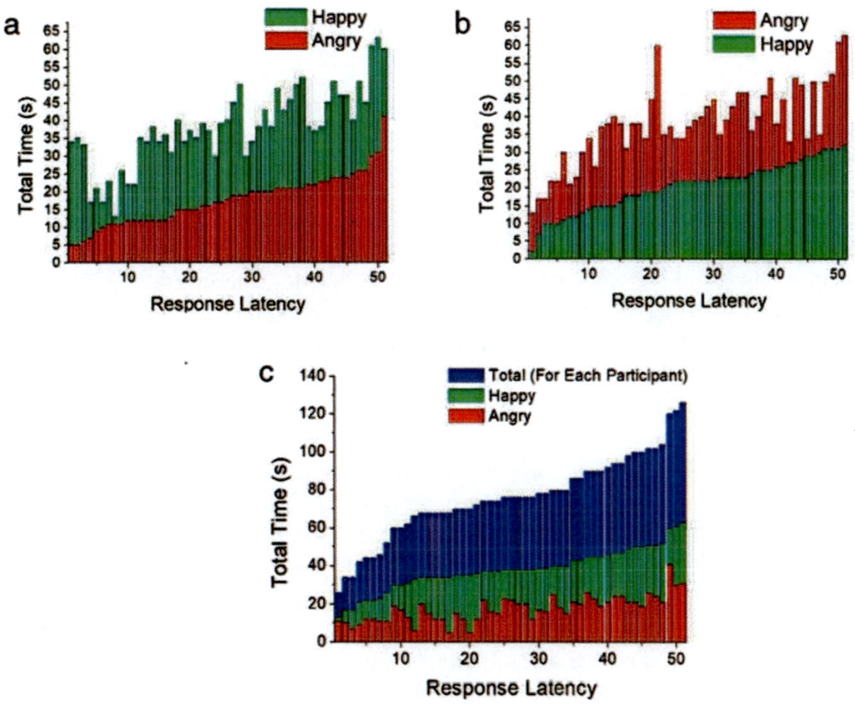

Fig. 3.3 Response delay in emotional state activation: This figure examines the time it takes for individuals to respond to emotional states, specifically anger (represented by red) and happiness (represented by green). **a** The stacked column graph displays the ranking of response delays in ascending order for the transition from anger to happiness. **b** The stacked column graph shows the ranking of response delays in ascending order for the transition from happiness to anger. **c** The stacked column graph presents the ranking of response delays in ascending order, considering the total time taken to transition between both anger and happiness (represented by blue)

pulse responses. These measures allowed us to determine individual response latencies and identify six distinct classes of people who respond similarly when expressing emotions.

Final Thoughts

Our approach provides valuable physiological insights into emotions, particularly in how they manifest within the human face. While our study illuminates 23 distinct facial expressions, it is essential to acknowledge that emotions are complex and multifaceted, and there may exist countless combinations of different thermal image patterns. This research paves the way for the development of automated algorithms for facial recognition of emotions using thermal imaging, but a deeper understanding of the physiological processes involved in emotional expression remains crucial.

To gain further insights into these processes, we have initiated experiments that instruct subjects to recall emotionally charged memories, facilitating a deeper exploration of the underlying mechanisms governing emotional expression. Our findings not only contribute to the field of psychology but also lay the foundation for potential advancements in various domains, including psychology, psychiatry, and neurology.

3.3.3 Detection of Geometric Invariants of Emotion

Additionally, we made intriguing observations regarding variations in temperature on different facial regions. Some individuals exhibited a notably chilly nose, while others had a considerably warm one. The debate surrounding these "cold" and "hot" noses is a subject of significant discussion and investigation [50, 55].

Understanding the connection between static temperature and emotion remains an elusive challenge. Surprisingly, for certain individuals, the nose emerged as a primary indicator of emotional outbursts, although this pattern didn't hold true universally. In some cases, the entire eye region displayed heightened warmth and brightness, while for others, it was confined to the inner or outer corners of the eyes. Similarly, some individuals exhibited increased brightness in their lips, while others displayed these patterns in their eyebrows or various sections of the forehead. It's noteworthy that forebrain areas play a pivotal role in receiving information for initiating threat engagement strategies [40].

Our exploration unveiled a fascinating aspect not evident in static images, as depicted in Fig. 3.1b. We found that high-intensity points from nearby facial components combined to form a geometric shape that remained consistent, even as thermal images changed throughout the day and in different environmental conditions. In a similar vein, Nakayama et al.'s study [50] involved meticulous pixel-by-pixel analysis of temperature profiles. While conventional methods often rely on nonparametric approaches to compare mean temperatures across phases and conditions, our approach incorporated two innovative techniques. Firstly, we connected high-intensity regions to construct a geometric shape. Secondly, we mapped boundaries of iso-intensity regions within the thermal map. These strategies provided a direct and computationally efficient means to derive invariants, effectively sidestepping the need for resource-intensive image processing procedures.

Our study centered on the use of infrared imaging to investigate the dynamic changes in facial components involved in expressing emotions. We specifically focused our attention on the eyes, nose, mouth, forehead, upper portion of the neck, and chin/cheeks. The transitions across 23 different expressions of emotions as they shifted from anger to happiness are thoughtfully summarized in Fig. 3.2. Each of the six panels within Fig. 3.2 is divided into two sections: The upper part features a pie chart depicting the distribution of changes in subcategories of facial parts, color-coded to represent local facial regions. The lower section of each panel presents a pair of infrared images for a subject experiencing two emotional states, anger and happiness, with subcategories thoughtfully annotated with arrows.

3.4 Latency-Emotion Duration Relationship in Thermal Imaging: Insights from Coalition Game Theory

This section delves into the intriguing relationship between latency in emotional expression and the duration of emotions, particularly the transition from happiness to anger and vice versa. We present Fig. 3.3a, b, illustrating this relationship, along with Fig. 3.3c, depicting the correlation between latency time and emotion duration. This particular study is based on a dataset of 283 subjects, and it uncovers valuable insights into the temporal aspects of emotional expression.

Understanding the temporal dynamics of emotional expression is a crucial aspect of emotion research. Emotions are complex phenomena that encompass not only the intensity and type of emotional experience but also the time it takes for an emotion to manifest. In this article, we explore the relationship between latency in emotional expression and the subsequent duration of that emotion.

In our study, we employed thermal imaging technology to capture the thermal patterns associated with emotional transitions. We tracked 23 facial parts as geometric combinations of regions of interest (ROIs) to ensure accurate measurement. To minimize errors introduced by participant movement or human error, we adopted a technique known as Coalition Game Theory, as introduced by Dowdall et al. [17].

Coalition Game Theory: Coalition Game Theory is a powerful technique used for precise motion tracking of a target surface, even when it's partially obscured or exhibits irregularities. The method relies on effective communication between different ROIs or trackers, allowing for the continuous extraction of infrared data. This approach ensures the consistency and reliability of measurements across frames, making it particularly well-suited for capturing thermal data in the context of emotional expression analysis.

Our analysis of the dataset revealed a compelling correlation between latency time and the duration of emotional expression. Figure 3.3a, b visually represent this relationship, showcasing how the transitional sequence of emotions, specifically from happiness to anger and vice versa, aligns with latency times. Interestingly, we observed that a longer latency time generally corresponded to a more extended expression of the emotion, as visually depicted in Fig. 3.3c. However, it's worth noting that, in our study, the maximum observed latency time did not exceed 50 s.

The findings of our study have significant implications for the field of emotion research and the practical applications of thermal imaging technology in emotion recognition. The correlation between latency and emotion duration suggests that individuals who take more time to express an emotion tend to experience that emotion for a more extended period. This observation has the potential to inform various domains, including psychology, neuroscience, and human–computer interaction.

Furthermore, the use of Coalition Game Theory in conjunction with thermal imaging technology demonstrates its effectiveness in minimizing errors related to participant movement or human error. This method ensures the consistent and reliable extraction of thermal data from facial ROIs, enhancing the precision of emotional expression analysis.

In conclusion, our research sheds light on the temporal dynamics of emotional expression, particularly in the context of transitions from happiness to anger and vice versa. The correlation between latency and emotion duration uncovered in our study contributes valuable insights to the field of emotion research. Additionally, the successful application of Coalition Game Theory for thermal data extraction underscores its potential for advancing the accuracy of emotion recognition systems. Our findings open up new avenues for the study of emotions and their temporal aspects, with implications for both scientific understanding and practical applications.

3.5 Mapping Emotional Patterns: Insights into the Universality of Emotional Expression

3.5.1 Regulations of Emotional Expressions

This part delves into the intriguing realm of emotional expression, investigating the temporal dynamics and universality of emotions across various facial components. Our research challenges the conventional belief in specific thermal signatures for distinct emotional states and offers novel insights into the regulation of emotional expression in humans. Chotard et al. [12] demonstrated that infrared thermal imaging can detect changes in skin temperatures of monkey and ape faces, indicating positive and negative emotions. This research highlights the sensitivity of facial thermal imaging in detecting affective states induced by visual stimuli, making it comparable to other traditional methods such as facial EMG and skin conductance. Note that they do not talk about differences among humans. Utilizing thermal imaging technology, we explore how the brain's processing time influences the manifestation of emotions. Our findings indicate that, regardless of the facial region involved, humans exhibit consistent emotional expression durations, with the eyes emerging as the dominant organ for emotional expression. This study has significant implications for the field of psychology, shedding light on the temporal aspects of emotional experience.

The study of emotions has long fascinated scholars across various disciplines, and one intriguing aspect of this complex phenomenon is how emotions are expressed through facial cues. Commonly held beliefs have suggested the existence of specific thermal signatures associated with particular emotional states. However, our research challenges this notion and seeks to unravel the temporal dynamics of emotional expression, offering fresh perspectives on the universality of emotions.

To investigate the regulation of emotional expression, we conducted a comprehensive study utilizing thermal imaging technology. Our primary objective was to determine whether specific facial components, such as the cheeks, chin, upper half of the neck, forehead, mouth, nose, or eyes, exhibit unique temporal patterns in emotional expression. We analyzed a limited sample of human subjects to extract meaningful insights into the manifestation of emotions. Our study yielded results that challenge conventional wisdom in the field of emotional expression. Regardless of the facial component under consideration, we observed remarkable consistency in the time taken by the brain to express emotional changes. This finding suggests a certain universality in the temporal dynamics of emotional expression, a phenomenon that holds significant importance in psychology (see Fig. 3.4b).

Analyzing the total time taken by all human subjects, we made interesting discoveries. The eyes emerged as the most dominant organ in expressing emotions, depicted by the prominent yellow stack in Fig. 3.4a. Following the eyes, the neck exhibited a significant influence on emotional expression (pink stack). In contrast, the cheeks (dark blue) and the nose (red) were found to be less favored by humans for expressing emotions. Notably, Fig. 3.4a also highlighted a striking difference between happiness and sadness. Happiness was expressed for a considerably longer duration compared to sadness. The height of the stacks representing the sum of emotional response times was notably greater for happiness than for sadness. In essence, human subjects tended to sustain an expression of happiness for a more extended period than sadness.

Figure 3.4a, c display similar variations, representing the cumulative expression types of all 287 subjects. We used this graph to explore if there are any overarching patterns in how emotions are expressed across the entire human population. It's worth noting that while emotional expressions manifest in various facial features, all of this processing occurs within the brain, where the amygdala plays a crucial role in directing muscle movements in the cerebellum, where the intricate patterns of brain rhythms emerge. In Fig. 3.4c, it appears that there's an equal preference for expressions involving the mouth and forehead, as well as an equal preference for the cheeks and chin. However, it's important to recognize that drawing definitive conclusions about the entire human population's generic emotional expressions based on a sample of just 287 subjects would be premature.

Our findings challenge the conventional belief in specific thermal signatures for distinct emotional states, instead pointing to a universal temporal pattern in emotional expression across various facial components. The dominance of the eyes in emotional expression is a noteworthy discovery, emphasizing the significance of ocular cues

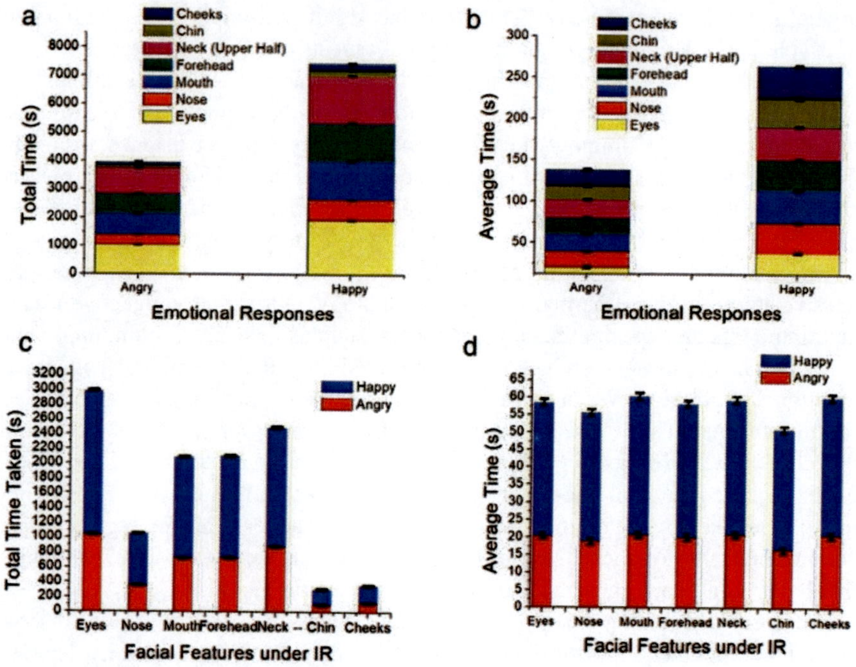

Fig. 3.4 Comparison of facial features and response latency for anger and happiness: This figure examines different facial features and their time delays in expressing anger and happiness. Panel **a**: Represents the total time taken (by adding values from 120 subjects) for emotional expression. Panel **b**: Represents the average time taken (based on 287 value sets) for the duration of emotional expression. In these two panels, the time taken to express anger and happiness is compared, and the variations in different facial features for each case are plotted in a stacked manner. The cumulative time taken for each facial feature to express anger and happiness is shown in distinct stacked modes. Panel **c**: Compares the cumulative facial features in terms of total time. Panel **d**: Compares the cumulative facial features in terms of average time. In these panels, the stacked columns represent the different facial features, and the length of each column indicates how emotions of anger and happiness vary across these features

in understanding emotions. The observed difference in emotional expression duration between happiness and sadness raises intriguing questions about the underlying mechanisms of these emotions. It suggests that happiness may be more enduring, while sadness tends to be relatively fleeting.

Our research uncovers new dimensions in the study of emotional expression. The consistent temporal patterns we observed across various facial components challenge conventional beliefs and highlight the universality of emotional expression. The dominance of the eyes in emotional expression and the difference in duration between happiness and sadness offer valuable insights into the complexities of human emotions. This study significantly contributes to the field of psychology, offering a deeper understanding of the temporal aspects of emotional experience and paving the way for further exploration into the fascinating world of emotions.

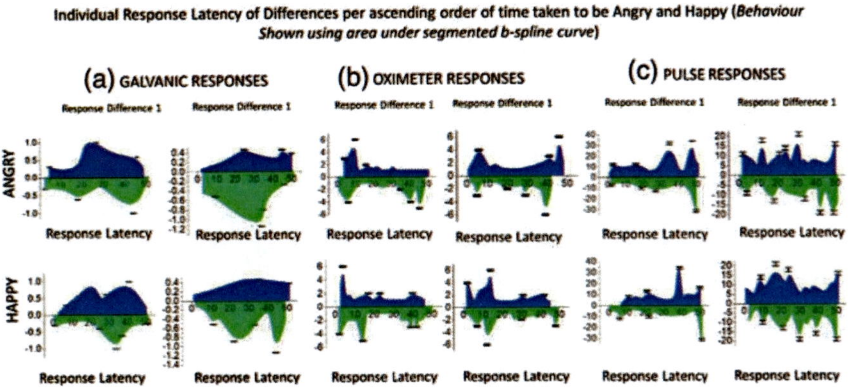

Fig. 3.5 Individual ranking of biophysical measure response latency: This figure examines the response latency of biophysical measures, specifically galvanic potential (measured in Ohms), oxygen variation measured with an oximeter (measured as a percentage), and pulse count (number). The response latencies are depicted as areas under segmented b-spline curves. For each type of response in each panel, separate plots are provided to show the maximum and minimum changes. These changes represent the difference in measured parameter values when a subject is in a normal, non-emotional state compared to when a specific emotion is expressed. **a** Displays the ascending order of galvanic response differences for anger and happiness. **b** Shows the ascending order of oximeter response differences for anger and happiness. **c** Depicts the ascending order of pulse response differences for anger and happiness. The response latency is determined by plotting the response differences in ascending order for anger and happiness among the 287 participants and fitting a segmented b-spline curve. Comparative analysis is conducted based on the area under the curves, with positive shifts (blue) and negative shifts (green) shaded differently

3.5.2 The Influence of Emotional Transitions in Negotiations

Negotiations are ubiquitous in human interactions, spanning diverse domains from business transactions to personal relationships [23]. While the focus in negotiation literature has traditionally emphasized rational decision-making and strategic planning, the emotional dimension of negotiations is gaining prominence. Emotional transitions, defined as shifts in emotional states during negotiations, wield substantial influence over the negotiation process and its outcomes. **Positive emotional transitions** denote shifts from negative emotions, such as anger or frustration, to positive emotions like happiness or contentment within the negotiation context. Extensive research underscores the constructive impact of positive emotional transitions on negotiation outcomes. Negotiators experiencing these shifts tend to exhibit cooperative behavior, engage in innovative problem-solving, and establish rapport with their counterparts. These elements synergize to foster more productive negotiations, culminating in mutually advantageous agreements. In contrast, **negative emotional transitions**, wherein negotiators transition from positive emotions to negative ones, can be detrimental to the negotiation process. Transitions from happiness to anger, for instance, often give rise to heightened competitiveness, hostility, and adversarial strategies. These shifts frequently impede effective communication and compromise, leading to less satisfactory outcomes for all parties involved.

Emotional Intelligence and Proficient Emotional Management: Emotional intelligence stands as a pivotal skill for negotiators, encompassing the capacity to recognize and regulate one's emotions, as well as comprehend the emotions of others. Negotiators equipped with high emotional intelligence are better poised to navigate emotional transitions and uphold a constructive atmosphere during negotiations. They possess the ability to defuse potential conflicts, cultivate trust, and steer the negotiation process toward cooperation.

3.5.3 Physiological and Affective Computing Through Thermal Imaging: A Comprehensive Survey

Physiological and affective computing have gained significant attention in recent years due to their potential applications in various fields, including healthcare, robotics, and human–computer interaction. Thermal imaging, also known as thermal infrared imaging (fITI), has emerged as a promising technique for non-invasive monitoring of physiological and affective states. This article provides a comprehensive survey of the current state of research in physiological and affective computing through thermal imaging. Ioannou et al. [34] conducted a review of 23 experimental procedures that employed fITI for investigations of affective nature. The study highlighted the adopted experimental protocols and the thermal changes that occurred in selected regions of interest in both human and non-human subjects. This research demonstrates the potential of fITI in capturing physiological responses associated with emotions and arousal.

Applications in Computational Psychophysiology: Cardone [14] discussed the principal achievements of thermal infrared imaging in computational psychophysiology. The review emphasized the capability of fITI for providing ubiquitous and unwired monitoring of psychophysiological activity and affective states. This capability makes fITI a valuable tool for studying the relationship between physiological responses and psychological processes.

Advancements in Affective Computing: Filippini et al. [20] highlighted the market entry of smaller and low-cost thermal cameras, which has paved the way for thermal infrared imaging applications outside laboratory environments. These advancements have expanded the potential of fITI in affective computing, enabling cutting-edge applications in the field. The integration of deep learning approaches with bio-signal analysis has further enhanced the automated analysis of affective states.

Correlation with Psychometric Indexes: Cardone and Pinti [15] explored the correlation between individual thermal signatures and psychometric indexes. The study aimed to assess whether specific personality traits lead to interindividual differences in the facial thermal signature of autonomic activity or affective state.

This research opens up new possibilities for using fITI as a tool for psychometric assessment.

Integration with Human–Robot Interaction: Filippini et al. [21] investigated the use of thermal infrared imaging-based affective computing in educational robots. The study aimed to enrich the interaction between robots and humans by incorporating real-time assessment of the interlocutor's psychophysiological state. This integration has the potential to enhance the effectiveness of human–robot interaction and improve the adaptability of robots to human emotions.

Future Directions and Challenges: The use of thermal infrared imaging in human stress research has shown promising results [18]. High-resolution thermal imaging provides indices of sympathetic activity through contact-free recording of facial tissues. However, challenges such as the standardization of experimental protocols and the development of robust analysis techniques need to be addressed for wider adoption of fITI in affective computing.

Thermal infrared imaging has emerged as a valuable tool for physiological and affective computing. The ability to capture physiological responses associated with emotions and arousal, along with advancements in computational psychophysiology and affective computing, has expanded the potential applications of fITI. Future research should focus on addressing challenges and standardizing protocols to further enhance the effectiveness and reliability of fITI in affective computing.

3.6 Emotional Responses in Human Subjects, Namely Galvanic Responses, Oximeter Responses, and Pulse Responses

3.6.1 Non-thermal Responses of the Human Subjects

In our study, we used various biophysical measures to analyze emotional responses in human subjects. These measures included galvanic responses, oximeter responses, and pulse responses. Figure 3.5 presents these measures with two plots each, showing the maximum and minimum changes. The horizontal axis represents time in seconds for the changes to occur, while the vertical axis indicates the change in the measured value from the normal mental state where subjects are neither happy nor sad. We also examined the differences between positive and negative shifts from the normal emotional state.

Our results indicated that the latency-based shift of biological variables could serve as a marker for identifying groups of people who respond similarly when expressing emotions. Notably, the area under the curve was significantly higher for maximum changes compared to minimum changes in all three biophysical measures. We observed a range of one to six peaks in the shift pattern, with six being the maximum. This suggests the presence of approximately six distinct classes of human subjects who respond to emotional transitions in specific ways. Furthermore, our findings revealed an asymmetry between the patterns of positive and negative shifts from the normal emotional state, with negative shifts being less pronounced than positive ones.

This study sheds light on the relationship between emotional responses and biophysical measures. Analyzing the area under the curve and counting peaks in the shift pattern aids in comparative analysis and the identification of distinct response classes among individuals. The observed asymmetry in positive and negative shifts suggests that individuals may experience positive or negative mental states to varying degrees. While there were individual differences, we consistently observed maximum changes in galvanic responses, oximeter responses, and pulse responses at specific latency intervals. This observation underscores the psychological significance of delayed emotional responses, where biological parameters exhibit their maximum and minimum changes depending on the subjects and their response classes.

To rank individual response latency in biophysical measures, we used the area under the segmented b-spline curve. We plotted response differences for anger and happiness in ascending order for our 287 participants and fitted a segmented b-spline curve to obtain response latency. This approach allowed us to conduct comparative analysis. Panels a, b, and c depict galvanic response differences, oximeter response differences, and pulse response differences, respectively, all presented in ascending order of anger and happiness.

We also noted an interesting similarity in the geometric features of peaks between positive and negative shifts, despite differences in delay along the horizontal axis. This consistent geometric variation occurred across all the biological measures, including galvanic response, blood oxygen levels, and pulse rate. It further supports the notion that the latency-based shift of biological variables can be used as a marker for groups of people who respond similarly when expressing emotions.

Lastly, we counted the number of peaks in the shift pattern and found that six was the maximum number. This suggests the existence of approximately six response classes among human subjects when it comes to emotional transitions. The findings underscore the importance of biophysical measures like galvanic response, oximeter response, and pulse response in understanding emotional responses. Additionally, they indicate the potential for using these measures in developing personalized emotion recognition and management systems. Figure 3.5 summarizes the key insights from our study.

Figure 3.5 underscores the significance of biophysical measures like galvanic response, oximeter response, and pulse response in comprehending emotional responses. Our findings further propose that the latency-based shift of biological variables could serve as an indicator for identifying groups of individuals who exhibit similar emotional expression patterns. Moreover, we observed that the maximum number of peaks in the shift pattern was six, indicating the potential existence of around six distinct classes of human subjects who exhibit unique responses during emotional transitions. These study results highlight the promise of biophysical measures in unraveling the complexities of emotional responses and may hold implications for the development of personalized emotion recognition and management systems. Figure 3.5 demonstrates the use of physiological measurements and design concepts to evaluate the emotional impact objectively [33].

3.6.2 Emotion Perception from Dynamic and Static Body Expressions: A Comprehensive Review

Emotion perception plays a crucial role in human communication and social interaction. While facial expressions are often considered the primary channel for conveying emotions, recent research has highlighted the importance of body expressions in emotion recognition [1]. This article aims to provide a comprehensive review of the study conducted by on emotion perception from dynamic and static body expressions in point-light and full-light displays [3]. Additionally, relevant studies that support and expand upon the findings of will be discussed.

Automated classification of humans faces an interesting twist. Atkinson conducted a study to develop a set of dynamic and static whole-body expressions of basic emotions and assess their impact on emotion classification and intensity ratings [3]. The researchers hypothesized that more exaggerated whole-body movements would lead to more accurate emotion classification and higher ratings of emotional intensity. The study utilized forced-choice emotion classification tasks and found that both dynamic and static body expressions were successfully recognized by viewers [3]. These findings support previous research that demonstrated the effectiveness of static body expressions and dynamic point-light displays in conveying emotions [1].

While we have concentrated in this chapter on the face, a similar classification and rigorous emotion study could be performed on the whole body. In Chaps. 5 and 8, we would see that holding hand changes the brain signals significantly and which hand is held, left or right would change the whole thermal energy exchange between the two bodies. Aviezer conducted a study that challenged the standard models of emotion by demonstrating that peak face expressions of highly intense situations cannot be discriminated solely based on positivity and negativity [4]. This finding emphasizes the importance of considering body cues, rather than facial expressions, in discriminating between intense positive and negative emotions. Meeren investigated the rapid perceptual integration of facial expression and emotional body language [45]. Their findings indicated that human bodies capture awareness in normal observers and that fearful bodies reduce extinction in neglect patients. This suggests that body expressions play a significant role in emotional communication and perception.

How big an object is and how much resolution does it have are two image recognition parameters that have direct impact in providing the threshold information for the advent of an emotional state. Bachmann reviewed neuroimaging research on the perception of emotional states derived from biological human motion [8]. They found that while emotional expressions can be identified from minimal kinematic information provided by point-light displays (PLDs), the use of full-light displays leads to slightly better recognition accuracy rates for specific emotional content. This supports the notion that both point-light and full-light displays contribute to emotion perception from body expressions.

In conclusion, the study conducted by provides valuable insights into emotion perception from dynamic and static body expressions in point-light and full-light displays [3]. The findings highlight the importance of body cues in conveying emotions and challenge the notion that facial expressions alone are sufficient for accurate emotion recognition. The reviewed studies further support and expand upon the findings of body expressions emphasizing the role of body expressions in emotional communication and perception [1, 4, 8, 45]. Understanding the mechanisms underlying emotion perception from body expressions can have implications for various fields, including psychology, neuroscience, and social communication.

3.6.3 Theory of a Mind

Theory of mind refers to the ability to understand and attribute mental states to oneself and others, including beliefs, desires, intentions, and emotions [35]. It involves the cognitive processes that allow individuals to infer and predict the thoughts, feelings, and intentions of others in order to understand and explain their behavior [19]. The development of theory of mind is a complex process that relies on various brain regions with distinct computational roles [35].

Research has shown that theory of mind is closely related to social cognition and has significant implications for social functioning [19]. It has been found that social cognition, particularly theory of mind, is strongly associated with functional outcomes in various populations, such as individuals with schizophrenia [19]. The ability to understand and interpret the mental states of others is crucial for successful social interactions and relationships [42].

Neuroimaging studies have provided insights into the neural basis of theory of mind. Different brain regions have been implicated in theory of mind processes, including the prefrontal cortex, temporoparietal junction, and amygdala [28, 44]. These regions are involved in various aspects of theory of mind, such as mentalizing, perspective-taking, and emotional processing [22]. The amygdala, for example, may play a role in the development of theory of mind, but its involvement in mentalizing per se is still under investigation [28].

The development of theory of mind occurs throughout childhood and adolescence [60]. As children grow older, they become increasingly able to integrate belief information in moral judgment, indicating the maturation of theory of mind abilities [72]. The neural development of theory of mind networks has been found to be related to behavioral changes in reasoning about others' minds [60].

In conclusion, theory of mind is a cognitive process that allows individuals to understand and attribute mental states to oneself and others. It plays a crucial role in social cognition and has implications for social functioning. The development of theory of mind involves various brain regions and is associated with changes in behavior. Further research is needed to fully understand the neural mechanisms underlying theory of mind and its impact on social interactions.

3.6.4 Automated Emotion Detection Systems

Our study proposes a smart-thermal system for detecting human emotions using thermographic images. This work is supported by the research of Akazue, who investigated the effect of thermal stimuli on the emotional perception of images [2]. They highlight the importance of understanding the emotional responses elicited by different thermal properties and presentation techniques. It is essential to discuss the potential and limitations of thermal infrared imaging in psychophysiology, emphasizing the role of subcutaneous blood flow and perspiration patterns in observing affective nature [34].

Thermal cameras are expanding their application base, describing the nature of thermal radiation and the technology behind thermal cameras [29]. reviews the use of thermography as a non-invasive detection modality for breast tumors, highlighting its potential in biomedical research [48]. explore the use of thermal imaging and smartphone photoplethysmography for the detection of perceived mental stress [16]. They discuss the limitations of head movement during data collection.

For example, the thermographic assessment of tumor growth in mouse xenografts, demonstrated the potential of thermographic imaging in monitoring tumor response to anticancer drugs [69]. focused on the classification of driver's emotion using multimodal camera sensors, including thermal imaging [41]. They compare their approach with previous studies that used accelerometers, gyro-sensors, EEG, or ECG sensors. Finally, they present work on fine alignment of thermographic images for robotic inspection, enhancing the deployability of thermographic inspections in industrial environments [46].

These works provide a limited glimpse of the use of thermal imaging in various domains [11], including emotion detection, psychophysiology, biomedical research, stress detection, tumor monitoring, driver emotion classification, and industrial inspections. However, one could notice that all research carried out thus far finds a generic conclusion, and the human subject variations are almost never taken into account. For example Marqués-Sánchez et al. [43] explored the potential of thermal infrared imaging as a tool to evaluate emotional competences in nursing students. The participants were exposed to different emotional stimuli, including anger-inducing scenarios. The authors observed an increase in facial temperature in response to anger-inducing stimuli, however, variations among nursing students were never considered. However, as we have shown above, there are fundamental hardware-level changes, i.e., the blood vessel configurations are very different for different humans. Thus, two humans could never experience similar emotions; however, if we classify humans into 23 categories, then we could intuitively find homogeneous human subjects of our choice.

3.7 Conclusion

3.7.1 Rethinking Emotional Expression: Insights from Dynamic Thermal Imaging

Our understanding of human emotional expression has long been shaped by static visual observations. Additionally, self-report measures can be used to evaluate the subjective emotional experience of an individual [24]. It is important to note that the measurement of emotions has often relied on scales assessing broad dimensions of affect rather than discrete emotions [31]. Furthermore, the development of verbal knowledge facilitates the expansion of emotion concept representations from a dichotomy to a multidimensional organization [49]. Overall, by developing objective measures of emotion, we can enhance our understanding of emotions and their impact on individuals. However, recent advancements in technology and scientific research are challenging this traditional perspective. In a groundbreaking study, we delve into the world of dynamic thermal imaging, revealing that the complexities of emotional states are far beyond what the naked eye can perceive. This research not only expands our understanding of human emotions but also opens up new avenues for the fields of psychology, neuroscience, and medicine.

The Limitations of Static Observations

Traditionally, the study of human emotional expressions has primarily relied on static visual observations of facial features. While these observations have provided valuable insights into basic emotions such as happiness, anger, and sadness, they often fall short when it comes to capturing the full spectrum of emotional experiences. Emotions are dynamic and multifaceted, and our understanding of them should reflect this complexity.

Dynamic Thermal Imaging: A New Perspective

Our research takes a bold step by utilizing dynamic thermal imaging to explore the intricate world of human emotions. Unlike traditional static observations, dynamic thermal imaging allows us to track and analyze changes in thermal energy patterns on the human face during emotional experiences. This approach offers a more nuanced and comprehensive view of how emotions manifest physiologically. In one interesting work, the role of the skin of a body could be seen as a "mental and subjective skin" that reflects internal emotional experiences [64], and individuals with higher levels of empathy showed greater changes in facial temperature in response to emotional stimuli.

Discovering Six Classes of Human Responders

One of the most striking findings of our study is the identification of approximately six distinct classes of human subjects, each exhibiting unique responses during emotional transitions. Within these classes, we further identified 23 subcategories (Fig. 3.2), highlighting the rich diversity of emotional experiences among individuals. This discovery challenges the conventional notion of universal emotional expressions and suggests that our responses to emotions are far more individualized than previously thought.

Implications for Psychology, Neuroscience, and Medicine

The implications of our research are profound. By recognizing the complexity and individuality of emotional responses, we can reshape the way we approach psychology, neuroscience, and medicine. For psychologists, this means moving beyond simplistic categorizations and considering a more personalized approach to emotional well-being. Neuroscientists can explore how these distinct response classes are reflected in neural activity, shedding light on the neural underpinnings of emotions. In the realm of medicine, understanding individualized emotional responses can lead to more effective treatments for conditions such as depression and anxiety.

The Future of Emotional Research: AI Detecting Emotions

While our study marks a significant step forward, it also opens up new questions and avenues for future research. We have only scratched the surface of understanding the intricacies of emotional responses. Further studies can delve deeper into the specific characteristics of each response class, exploring how they relate to personality traits, life experiences, and cultural influences. There was a significant decrease in facial skin temperature in response to joyful expressions [51]. This suggests that facial skin temperature can be a physiological marker of positive emotions in infants. In the meantime, several image analysis-based products are coming to market related to automated emotional state detection. Furthermore, Cruz-Albarran et al. present a smart-thermal system of thermographic images for human emotion detection. The key findings of the study indicate that the smart-thermal system was able to accurately detect and classify human emotions based on the thermal patterns observed in the images [13]. The system achieved a high level of accuracy in identifying emotions such as happiness, sadness, anger, and surprise. This suggests that thermal imaging technology has the potential to be a reliable and non-invasive method for emotion detection.

Beyond Thermal Imaging: Unveiling the Mysteries of Consciousness

In the scholarly work ([9], Chap. 7), the challenges of creating a brain model with functional cognitive responses represented as a 3D architecture of clocks are discussed. This ambitious endeavor aims to unravel the mysteries of consciousness by replicating the brain's intricate workings. Bandyopadhyay highlights the need for innovative materials and manufacturing techniques to create time crystals capable of mimicking the biological brain's functions [61, 65]. Additionally, the development of advanced algorithms and software is crucial in replicating specific brain components.

3.7.2 *Polyatomic Time Crystal Model of the Brain: A Fractal Logic in the Hardware*

The polyatomic time crystal model of the brain, as proposed by Reddy et al. [61] and Saxena et al. 65] and Singh et al. [59], offers an intriguing perspective on brain function. In this model, recurring events are converted into 3D clock assembly, from that clock assembly, we convert it into 3D geometric shapes; and several geometric shapes spread over 12 dimensions represent a cognitive behavior [67, 68]. Cognitive, perceptive, and emotional responses and feelings could be quantified by considering repeated instances of events that lead to such experiences. Every recurring event is a clock embedded on the surface of a sphere. The origin of these clocks could be anywhere in the fractal hardware of brain. Based on this perspective, we proposed Fractal Information Theory, FIT [5], which expanded our previous proposal that brain is a resonance chain that holds a triplet of triplet symmetry in the distribution of resonance frequencies [25].

The neuron is not empty inside. The Pradhan et al. [58] study emphasizes the crucial role of membrane microtubule junctions in neurons, highlighting the significance of heterojunctions for signal transfer within a fractal structure. This fractal, scale-free electromagnetic resonance observed in a single brain-extracted microtubule nanowire, a single tubulin protein, and a single neuron suggests the coexistence of a triplet of triplet resonance bands or 9 distinct frequency or time domains that function in unison [66]. The electromagnetic resonance is not part of the mainstream neuroscience as of now, though electromagnetic resonance of proteins was measured in the 1930s. It is not that we always see prime number of resonance peaks whenever we do the measurements of biomaterials, we find that prime number of peaks exchange energy together, as we see continuous changes in the intensity and relative phase of these resonance peaks [63, 65]. The triplet of triplet has a profound scale-free impact that could transfer quantum information one part to another [63]. The concept of employing quantum tunneling for reading and writing information within such a network has been proposed [7]. It means even at the molecular level, the symmetry of primes works. Furthermore, it has been postulated that a fractal network could effectively process information in a logical manner [26]. Thus, at the molecular scale where direct control of free will is not observed, when one makes a journey, they don't find the location of mind anywhere. This is the fractal nature of mind.

The temporal management of mind is also fractal in nature. Research conducted by Agrawal et al. [6] involving the simultaneous measurement of neuron firing at the neuron membrane and the microtubular bundle inside a neuron reveals that microtubules are capable of processing firing signals at a rate approximately 10,000 times faster than the membrane. This finding challenges the notion that microtubules remain silent in the brain's information processing [62]. Notably, even non-connected neurons lacking a direct connection through the synaptic junctions are capable of exchanging energy, as demonstrated by imaging the neural circuits using a scanning dielectric microscope (SDM) [27]. Here we have mapped the face and potential difference of a face with emotions, look similar to SDM or thermal camera.

3.7.3 Quantification of Human Emotions Amplifies by Classification of Human Subjects

Our research challenges the traditional understanding of human emotional expression, emphasizing the need for a more dynamic and individualized perspective. The use of dynamic thermal imaging has revealed the existence of distinct response classes among individuals, suggesting that our emotional experiences are highly personalized. This breakthrough has far-reaching implications for psychology, neuroscience, and medicine, paving the way for more effective approaches to emotional well-being and mental health. As we venture further into the mysteries of consciousness, the possibilities for future research and discovery are boundless, promising to reshape our understanding of what it means to be human.

Furthermore, our study unveils the intriguing possibility that each individual possesses a unique thermal signature associated with their emotional states. This revelation carries profound significance for tailoring personalized interventions aimed at assisting individuals who struggle with emotional regulation. In the field of education, Marqués-Sánchez et al. [43] explore the use of thermal infrared imaging to evaluate emotional competencies in nursing students. The authors aim to apply a quantifiable methodology that can easily monitor the emotional behavior of students and potentially improve the learning-teaching processes. This study demonstrates the potential of thermal imaging as a non-invasive and objective tool for assessing emotional states in educational settings. As we look to the future, there is immense potential for further research to build upon our discoveries. Investigating the neural underpinnings of these geometric thermal signatures and delving into the intricate relationship between thermal and visual facial expressions of emotions could offer even deeper insights. Moreover, these findings hold promise in the realm of emotional disorder diagnosis and treatment, offering a more holistic understanding of emotional states.

In summation, our study not only pioneers the utilization of thermal imaging in the exploration of emotional states but also underscores the intricate nature of emotional expression on the human face.

References

1. Abramson L, Marom I, Petranker R, Aviezer H (2017) Is fear in your head? a comparison of instructed and real-life expressions of emotion in the face and body. Emotion 17(3):557–565. https://doi.org/10.1037/emo0000252

2. Akazue M, Halvey M, Baillie L, Brewster S (2016) The effect of thermal stimuli on the emotional perception of images. https://doi.org/10.1145/2858036.2858307
3. Atkinson A, Dittrich W, Gemmell A, Young A (2004) Emotion perception from dynamic and static body expressions in point-light and full-light displays. Perception 33(6):717–746. https://doi.org/10.1068/p5096
4. Aviezer H, Trope Y, Todorov A (2012) Body cues, not facial expressions, discriminate between intense positive and negative emotions. Science 338(6111):1225–1229. https://doi.org/10.1126/science.1224313
5. Agrawal L, Chhajed R, Ghosh S, Ghosh B, Ray K, Sahu S, Fujita D (2018) Fractal information theory (FIT)-derived geometric musical language (GML) for brain-inspired hypercomputing. Soft Comput Theories Appl 548:343–372
6. Agrawal L, Sahu S, Ghosh S, Shiga T, Fujita D, Bandyopadhyay A (2016) Inventing atomic resolution scanning dielectric microscopy to see a single protein complex operation live at resonance in a neuron without touching or adulterating the cell. J Integr Neurosci 15(04):435–462
7. Bandyopadhyay A, Miki K, Wakayama Y (2006) Writing and erasing information in multilevel logic systems of a single molecule using scanning tunneling microscope. Appl Phys Lett 89(24)
8. Bachmann J, Munzert J, Krüger B (2018) Neural underpinnings of the perception of emotional states derived from biological human motion: a review of neuroimaging research. Front Psychol 9. https://doi.org/10.3389/fpsyg.2018.01763
9. Bandyopadhyay A (2020) Nanobrain: the making of an artificial brain from a time crystal. Boca Raton CRC Press, Taylor and Francis. ISBN: 9781439875490. https://doi.org/10.1201/9780429107771
10. Bonanno G, Papa A, Lalande K, Westphal M, Coifman K (2004) The importance of being flexible. Psychol Sci 15(7):482–487. https://doi.org/10.1111/j.0956-7976.2004.00705.x
11. Clay-Warner J, Robinson DT (2014) Infrared thermography as a measure of emotion response. Emotion Rev 7(2) (April 2015):157–162. https://doi.org/10.1177/1754073914554783
12. Chotard H, Ioannou S, Davila-Ross M (2018) Infrared thermal imaging: positive and negative emotions modify the skin temperatures of monkey and ape faces. Am J Primatol 80(5). https://doi.org/10.1002/ajp.22863
13. Cruz-Albarran IA, Benitez-Rangel JP, Osornio-Rios RA, Moralez-Hernandez LA (2017) Human emotions detection based on a smart thermal system of thermographic images. Infrared Phys Technol 81:250–261. https://doi.org/10.1016/j.infrared.2017.01.002
14. Cardone D (2017) New frontiers for applications of thermal infrared imaging devices: computational psychopshysiology in the neurosciences. Sensors 17(5):1042. https://doi.org/10.3390/s17051042
15. Cardone D, Pinti P (2015) Thermal infrared imaging-based computational psychophysiology for psychometrics. Comput Math Methods Med 2015:1–8. https://doi.org/10.1155/2015/984353
16. Cho Y, Julier S, Bianchi-Berthouze N (2019) Instant stress: detection of perceived mental stress through smartphone photoplethysmography and thermal imaging. Jmir Mental Health 6(4):e10140. https://doi.org/10.2196/10140
17. Dowdall J, Pavlidis TI, Tsiamyrtzis P (2006) Coalition tracking in facial infrared imaging and beyond. IEEE Conf Comp Vis Pattern Recogn:134–144. https://doi.org/10.1109/CVPRW.2006.55
18. Engert V, Grant J, Cardone D, Tusche A, Singer T (2014) Exploring the use of thermal infrared imaging in human stress research. PLoS ONE 9(3):e90782. https://doi.org/10.1371/journal.pone.0090782
19. Fett AK, Maat A (2011) GROUP Investigators. Social cognitive impairments and psychotic symptoms: what is the nature of their association? Schizophr Bull. 2013 Jan;39(1):77–85. https://doi.org/10.1093/schbul/sbr058. Epub 2011 Jun 22. PMID 21697150; PMCID PMC3523914
20. Filippini C, Crosta A, Palumbo R, Perpetuini D, Cardone D, Ceccato I, Domenico A et al (2022) Automated affective computing based on bio-signals analysis and deep learning approach. Sensors 22(5):1789. https://doi.org/10.3390/s22051789

21. Filippini C, Spadolini E, Cardone D (2019) Thermal imaging based affective computing for educational robot. https://doi.org/10.3390/proceedings2019027027
22. Frith C, Frith U (2006) The neural basis of mentalizing. Neuron 50(4):531–534. https://doi.org/10.1016/j.neuron.2006.05.001
23. Filipowicz A, Barsade S, Melwani S (2011) Understanding emotional transitions: the interpersonal consequences of changing emotions in negotiations. J Pers Soc Psychol 101(3):541–556. https://doi.org/10.1037/a0023545. PMID: 21639649
24. Ferreira H, Saraiva M (2019) Subjective and objective measures, pp 143–159. https://doi.org/10.1007/978-3-319-96722-6_9
25. Ghosh S, Sahu S, Fujita D, Bandyopadhyay A (2014) Design and operation of a brain like computer: a new class of frequency-fractal computing using wireless communication in a supramolecular organic, inorganic systems. Information, 5, 28–99
26. Ghosh S, Fujita D, Bandyopadhyay A (2015) An organic jelly made fractal logic gate with an infinite truth table. Sci Rep 5(1):11265
27. Ghosh S, Sahu S, Agrawal L, Shiga T, Bandyopadhyay A (2016) Inventing a co-axial atomic resolution patch clamp to study a single resonating protein complex and ultra-low power communication deep inside a living neuron cell. J Integr Neurosci 15(04):403–433
28. Gallagher H, Frith C (2003) Functional imaging of 'theory of mind.' Trends Cogn Sci 7(2):77–83. https://doi.org/10.1016/s1364-6613(02)00025-6
29. Gade R, Moeslund T (2013) Thermal cameras and applications: a survey. Mach Vis Appl 25(1):245–262. https://doi.org/10.1007/s00138-013-0570-5
30. Hessler C, Mohamed A (2018) Using thermal images and physiological features to model human behavior: a survey. In: 2018 IEEE conference on multimedia information processing and retrieval (MIPR) (2018), pp 278–281
31. Harmon-Jones C, Bastian B, Harmon-Jones E (2016) The discrete emotions questionnaire: a new tool for measuring state self-reported emotions. PLoS ONE 11(8):e0159915. https://doi.org/10.1371/journal.pone.0159915
32. Ioannou S (2020) "Functional Infrared Thermal Imaging: A Contemporary Tool in Soft Tissue Screening." Scientific Reports 10: https://doi.org/10.1038/s41598-020-66397-9
33. Iacono E, Becchimanzi C, Brischetto A (2022) Emotional design: affective evaluation methods to assess the emotional response of 6–11 years children. https://doi.org/10.54941/ahfe1001785
34. Ioannou S, Gallese V, Merla A (2014) Thermal infrared imaging in psychophysiology: potentialities and limits. Psychophysiology 51(10):951–963. https://doi.org/10.1111/psyp.12243
35. Koster-Hale J, Saxe R (2013) Theory of mind: a neural prediction problem. Neuron 79(5):836–848. https://doi.org/10.1016/j.neuron.2013.08.020
36. Kwon I, Lee S (2014) Fuzzy integral-based emotion inference on personalized v-a emotional space. J Syst Inf Technol 16(2):138–148. https://doi.org/10.1108/jsit-11-2013-0057
37. Kamei T, Tsuda T, Kitagawa S, Naitoh K, Nakashima K, Ohhashi T (1998) Physical stimuli and emotional stress-induced sweat secretions in the human palm and forehead. Anal Chim Acta 365:319–326
38. Kistler A, Mariazouls C, Berlepsch VK (1998) Fingertip temperature as an indicator for sympathetic responses. Int J Psychophysiol 29:35–41
39. Kosonogov V et al (2017) Facial thermal variations: a new marker of emotional arousal. PLoS ONE 12(2017)
40. Lacroix L, Spinelli S, Heidbreder CA, Feldon J (2000) Differential role of the medial and lateral prefrontal cortices in fear and anxiety. Behav Neurosci 114:1119–1130
41. Lee K, Yoon H, Song J, Park K (2018) Convolutional neural network-based classification of driver's emotion during aggressive and smooth driving using multi-modal camera sensors. Sensors 18(4):957. https://doi.org/10.3390/s18040957
42. Lincoln S, Mukerji C, Dodell-Feder D, Riccio A, Hooker C (2021) The neural basis of social cognition in typically developing children and its relationship to social functioning. Front Psychol 12. https://doi.org/10.3389/fpsyg.2021.714176

43. Marqués-Sánchez P, Liébana-Presa C, Benítez-Andrades J, Gundín-Gallego R, Álvarez-Barrio L, Rodríguez-Gonzálvez P (2020) Thermal infrared imaging to evaluate emotional competences in nursing students: a first approach through a case study. Sensors 20(9):2502. https://doi.org/10.3390/s20092502
44. McCleery J, Surtees A, Graham K, Richards J, Apperly I (2011) The neural and cognitive time course of theory of mind. J Neurosci 31(36):12849–12854. https://doi.org/10.1523/jneurosci.1392-11.2011
45. Meeren H, Heijnsbergen C, Gelder B (2005) Rapid perceptual integration of facial expression and emotional body language. Proc Natl Acad Sci 102(45):16518–16523. https://doi.org/10.1073/pnas.0507650102
46. Mineo C, Montinaro N, Fustaino M, Pantano A, Cerniglia D (2022) Fine alignment of thermographic images for robotic inspection of parts with complex geometries. Sensors 22(16):6267. https://doi.org/10.3390/s22166267
47. Nguyen H, Kotani K, Chen F, Le B (2013) Estimation of human emotions using thermal facial information. In: Proceedings SPIE 9069, Fifth international conference on graphic and image processing (ICGIP 2013), 90690O (10 January 2014); https://doi.org/10.1117/12.2050206
48. Ng E (2009) A review of thermography as promising non-invasive detection modality for breast tumor. Int J Therm Sci 48(5):849–859. https://doi.org/10.1016/j.ijthermalsci.2008.06.015
49. Nook E, Sasse S, Lambert H, McLaughlin K, Somerville L (2017) Increasing verbal knowledge mediates development of multidimensional emotion representations. Nat Hum Behav 1(12):881–889. https://doi.org/10.1038/s41562-017-0238-7
50. Nakayama K, Goto S, Kuraoka K, Nakamura K (2005) Decrease in nasal temperature of rhesus monkeys (Macaca mulatta) in negative emotional state. J Physiol Behav 84:783–790. https://doi.org/10.1016/j.physbeh.2005.03.009
51. Nakanishi R, Imai-Matsumura K (2008) Facial skin temperature decreases in infants with joyful expression. Infant Behav Dev 31(1):137–144. https://doi.org/10.1016/j.infbeh.2007.09.001
52. Nhan BR, Chau T (2010) Classifying affective states using thermal infrared imaging of the human face. IEEE Trans Biomed Eng 57:979–987
53. Nguyen H, Kotani K, Chen F, Le B (2014) A thermal facial emotion database and its analysis. In: Klette R, Rivera M, Satoh S (eds) Image and video technology. PSIVT 2013. Lecture notes in computer science, vol 8333. Springer, Berlin, Heidelberg. https://doi.org/10.1007/978-3-642-53842-1_34
54. Pattanayak A, Dutta T, Pranjal P, Singh P, Sahoo P et al (2022) Meta-analysis of fMRI for emotional and cognitive states shows hierarchical invariant optimization in brain. In: Kaiser MS, Bandyopadhyay A, Ray K, Singh R, Nagar V (eds) Proceedings of trends in electronics and health informatics. Lecture notes in networks and systems, vol 376. Springer, Singapore, pp 255–265
55. Pavlidis I, Levine J, Baukol P (2001) Thermal imaging for anxiety detection. IEEE 2:315–318. https://doi.org/10.1109/ICIP.2001.958491
56. Pavlidis I, Eberhardt NL, Levine J (2002) Human behavior: seeing through the face of deception. Nature 415:35
57. Porges SW (1992) Vagal tone: a physiologic marker of stress vulnerability. Pediatrics 90:498–504
58. Pradhan B, Bandyopadhyay A, Pal AJ (2004) Molecular level control of donor/acceptor heterostructures in organic photovoltaic devices. Appl Phys Lett 85(4):663–665
59. Singh P, Saxena K, Sahoo P, Ghosh S, Chhajed R, Ray K, Daisuke Fujita and Anirban Bandyopadhyay, A self-operating time crystal model of the human brain: Can we replace entire brain hardware with a 3D fractal architecture of clocks alone? Information 11, 238 (2020). https://doi.org/10.3390/info11050238
60. Richardson H, Lisandrelli G, Riobueno-Naylor A, Saxe R (2018) Development of the social brain from age three to twelve years. Nat Commun 9(1). https://doi.org/10.1038/s41467-018-03399-2
61. Reddy S, Sonker D, Singh P, Saxena K, Singh S, Chhajed R et al (2018) A brain-like computer made of time crystal: could a metric of prime alone replace a user and alleviate programming forever? Soft Comput Appl:1–43

62. Singh P, Sahoo P, Saxena K, Manna JS, Ray K, Ghosh S, Fujita D et al (2021) Cytoskeletal filaments deep inside a neuron are not silent: they regulate the precise timing of nerve spikes using a pair of vortices. Symmetry 13(5):821
63. Singh P, Ray K, Bandyopadhyay A (2022) Biological antenna to the humanoid bot: electromagnetic resonances in biomaterials. Studies in rhythm engineering (SRE). Springer Singapore. https://doi.org/10.1007/978-981-16-9677-0
64. Salazar-López E, Domínguez E, Juárez Ramos V, de la Fuente J, Meins A, Iborra O, Gálvez G, Rodríguez-Artacho MA, Gómez-Milán E (2015) The mental and subjective skin: emotion, empathy, feelings and thermography. Conscious Cogn 34(2015):149–162. https://doi.org/10.1016/j.concog.2015.04.003
65. Saxena K, Singh P, Sahoo P, Ghosh S, Krishnananda D, Ray K, Fujita D, Bandyopadhyay A (2022) All basics that are wrong with the current concept of time crystal: learning from the polyatomic time crystals of protein, microtubule, and neuron. Proc Trends Electron Health Inform 376:243–254
66. Saxena K, Singh P, Sahoo P, Sahu S, Ghosh S, Ray K, Fujita D et al (2020) Fractal, scale-free electromagnetic resonance of a single brain-extracted microtubule nanowire, a single tubulin protein, and a single neuron. Fractal Fract 4(2):11
67. Singh P, Sahoo P, Saxena K, Ghosh S, Sahu S, Ray K, Fujita D, Bandyopadhyay A (2021) Quaternion, octonion to dodecanion manifold: stereographic projections from infinity lead to a self-operating mathematical universe. Proc Int Conf Trends Comput Cogn Eng 1169:55–77
68. Singh P, Sahoo P, Saxena K, Ghosh S, Sahu S, Ray K, Fujita D, Bandyopadhyay A (2021) A space-time-topology-prime, stTS metric for a self-operating mathematical universe uses Dodecanion geometric algebra of 2–20 D complex vectors. Proc Int Conf Data Sci Appl 148:1–31
69. Song C, Appleyard V, Murray K, Frank T, Sibbett W, Cuschieri A, Thompson A et al (2007) Thermographic assessment of tumor growth in mouse xenografts. Int J Cancer 121(5):1055–1058. https://doi.org/10.1002/ijc.22808
70. Vetrugno R, Liguori R, Cortelli P, Montagna P (2003) Sympathetic skin response: basic mechanisms and clinical applications. Clin Auton Res 13:256–270
71. Viechtbauer W, Domínguez M, Penn D, Os J, Krabbendam L (2011) The relationship between neurocognition and social cognition with functional outcomes in schizophrenia: a meta-analysis. Neurosci Biobehav Rev 35(3):573–588. https://doi.org/10.1016/j.neubiorev.2010.07.001
72. Young L, Cushman F, Hauser M, Saxe R (2007) The neural basis of the interaction between theory of mind and moral judgment. Proc Natl Acad Sci 104(20):8235–8240. https://doi.org/10.1073/pnas.0701408104

Chapter 4
Are Errors Indicative of Learning Strategies: Use It or Minimize It?

4.1 Introduction

4.1.1 Embracing Errors: Utilize It Instead of Eliminating Them

4.1.1.1 The Classical Learning Process: The Acquisition, Retention, and Retrieval of Information

The learning process is a complex phenomenon that involves the acquisition, retention, and retrieval of information [16]. It is influenced by various factors, including the teaching methods employed and the interaction between teachers and learners [43]. Active learning, which involves engaging students in activities that promote critical thinking and problem-solving, has been shown to be effective in enhancing learning outcomes [50]. In addition, the use of robots as teaching tools has gained attention in recent years [69]. Teachable robots have been developed to understand human teaching behavior and improve their learning capabilities [69]. Human–robot interaction plays a crucial role in the teaching–learning process, and several issues need to be addressed to ensure effective learning and comfort in this interaction [40, 70].

The COVID-19 pandemic has necessitated a shift to online learning, which has its own challenges [49]. Blended learning, which combines online and face-to-face instruction, has been found to be effective in vocational education [41]. Learning analytics, the use of data to inform educational design, can support teachers in designing effective learning experiences [46]. It provides insights into student performance and engagement, allowing teachers to make data-informed decisions [46]. Furthermore, learning design can be improved by analyzing content and patterns of interaction in networked learning environments [38]. This analysis helps identify indicators that support learners in staying focused on the learning process.

T. Dutta and A. Bandyopadhyay, *Emotion, Cognition and Silent Communication: Unsolved Mysteries*, Studies in Rhythm Engineering,
https://doi.org/10.1007/978-981-99-9334-5_4

The learning process is not limited to traditional education settings. It also extends to individuals with special needs, such as students with mental retardation [1]. Platforms for online learning have been developed specifically for these students, taking into account their unique conditions [1]. Additionally, the concept of sustainability learning emphasizes the importance of learning processes in addressing sustainability challenges at various levels [32]. Educational psychology plays a crucial role in understanding memory, conceptual processes, and individual differences in learning. It provides insights into the development of new strategies for learning processes in humans [21]. In conclusion, the learning process is a multifaceted phenomenon that involves various factors and contexts. Active learning, human–robot interaction, online learning, blended learning, learning analytics, and educational psychology are all important areas of research that contribute to our understanding of the learning process.

4.1.1.2 Errors in Learning: The Science of Mistakes Are Neglected by Brain Researchers

In the quest for excellence and improved performance, every individual aspires to minimize errors [11]. The difference in performance levels between amateurs and experts can often be attributed to their error rates. Amateurs tend to make more mistakes, while experts exhibit greater proficiency with fewer errors. Consequently, the concept of using errors as a measure of learning holds widespread appeal. Traditionally, scholars have approached errors from two main perspectives. The first perspective, rooted in the beliefs of behaviorists and cognitive psychologists, asserts that errors should be avoided due to their potential negative consequences [35]. According to this view, errors can disrupt the learning process and should therefore be minimized or controlled. On the other hand, the second perspective regards errors as valuable opportunities for learning, often referred to as 'error friendliness' [26]. This viewpoint suggests that errors can serve as stepping stones to acquiring better knowledge and understanding. However, the concept of actively learning from errors has received relatively less attention in research [34].

Embracing errors and adapting to them can lead to the acquisition of superior knowledge. The theories of negative knowledge and impasse-driven learning are examples where error minimization becomes an integral aspect of the learning process. In this context, we propose that, for the sake of rapid and efficient learning, learners may intentionally choose to make errors as part of a well-considered strategy. To explore this concept further, our study employed procedural learning to investigate motor learning tasks. The results revealed that groups of subjects sometimes deliberately overlooked specific types of errors when confronted with multiple potential errors in their learning process. This approach underscores the complex interplay between errors and learning strategies, shedding light on the nuanced dynamics of error tolerance and adaptation in the pursuit of knowledge acquisition. By examining errors through these different lenses, we gain a deeper understanding of their role in the learning process and the strategies that individuals employ to enhance their performance and expertise.

Strategies adopted by human in learning are complex and go beyond the simple stimulus-performance-reinforcement paradigm. Research suggests that humans engage in higher-order cognitive functioning while acquiring new skills or in learning a new task [18, 45]. Minimizing errors is a key sign of effective learning. Experimental research also points out that errors can reveal the strategies learners use. The types of errors made can give valuable insights into a person's learning style. Being adaptable and responsive to errors can lead to acquiring better knowledge. The theories of negative knowledge and impasse-driven learning show how error minimization can be a part of the learning process. In fact, we suggest that for swift and efficient learning, a learner might deliberately choose to make errors as part of a well-thought-out strategy.

4.1.1.3 Management of Errors: Making Errors Intentionally to Learn Faster and Better?

In a study using procedural learning for motor tasks, we discovered that groups of participants decided to overlook specific types of errors when they had multiple errors to contend with. This goes against the common belief that errors should always be avoided. It turns out that intentionally allowing a mix of errors can actually help master motor tasks more quickly and efficiently. Considering errors as a positive attribute and giving greater importance to one class of errors over others can significantly enhance learning ability. Since existing learning theories do not account for deliberately making meaningful errors to accelerate learning, some researchers have already proposed a new theory: Intentional Error Committing for Learning (IECL theory). Intentional Error Committing for Learning (IECL) theory is a cognitive theory of learning that proposes that people learn the best when they make mistakes. The theory was developed by psychologist John Anderson and his colleagues in the 1980s [2].

IECL theory is based on the idea that people learn by developing and refining mental models of the world. These mental models are used to make predictions about how the world works and to guide behavior. When a person makes a mistake, it means that their mental model is incorrect in some way. This error provides the person with an opportunity to update and improve their mental model. IECL theory has been supported by a large body of research. For example, studies have shown that students who are encouraged to make mistakes and learn from them perform better than students who are discouraged from making mistakes [17]. IECL theory has also been used to develop effective teaching methods, such as error-based instruction and problem-based learning [3].

Here are some of the key principles of IECL theory:

1. Learning is most effective when it is challenging. When people are challenged, they are more likely to make mistakes. However, these mistakes provide opportunities for learning.
2. Making mistakes is a natural part of the learning process. Everyone makes mistakes, and they are an essential part of how we learn.
3. Learning is most effective when people are actively engaged in the learning process. This means that people should be encouraged to ask questions, explore ideas, and make predictions [20].
4. Feedback is essential for learning. Feedback helps people to identify their mistakes and to learn from them [36].
5. IECL theory has important implications for education and training. It suggests that educators should create learning environments that are challenging and supportive. Educators should also provide feedback to students on their performance so that they can learn from their mistakes [54].

IECL theory is a powerful theory of learning that has the potential to improve education and training. By understanding and applying the principles of IECL theory, educators can create learning environments that help students to learn more effectively.

4.1.2 ASMR, A Meaningless Noise that Tweaks Better

4.1.2.1 What is ASMR?

Autonomous sensory meridian response (ASMR) is a recently-coined term to describe a pleasurable tingling sensation that occurs on the scalp and neck in response to specific auditory and visual stimuli. ASMR is a unique sensory experience characterized by tingling sensations that typically begin in the scalp and move down the spine in response to specific auditory or visual stimuli. Common ASMR triggers include soft-spoken speech, tapping, whispering, and various visual cues. These stimuli are often described as being "whispery," "soft," or "personal" and may include sounds such as tapping, scratching, or crinkling. The experience of ASMR is often accompanied by feelings of relaxation, well-being, and focus. Some people also report experiencing a sense of "peace" or "calm" during ASMR. The prevalence of ASMR is unknown, but it is estimated that millions of people around the world experience it. ASMR has become increasingly popular in recent years, with the rise of online ASMR videos and communities [37].

While ASMR's primary association has been with relaxation and stress reduction, its potential impact on cognitive functions such as mathematical problem-solving and pattern recognition remains a relatively unexplored area. Individuals who experience ASMR may find that it enhances their focus, reduces stress, and aids in recognizing patterns and solving mathematical problems. However, the extent to which

ASMR can reliably and consistently produce these effects remains to be investigated. Future research should employ controlled experiments to assess the impact of ASMR on cognitive functions and elucidate the mechanisms underlying any observed enhancements.

4.1.2.2 ASMR and Cognitive Enhancement

Several studies have suggested that ASMR may have cognitive benefits beyond relaxation. Some individuals report improved focus, concentration, and creativity while experiencing ASMR triggers. It is plausible that ASMR could enhance mathematical problem-solving skills and aid in pattern recognition, as these cognitive processes require heightened attention and mental clarity. Neurologically, ASMR has been found to have an impact on brain connectivity. Individuals with chronic conditions, such as pain syndromes, may use ASMR to manage brain states, indicating a potential role in regulating cognitive processes [14]. Additionally, functional magnetic resonance imaging (fMRI) studies have shown that ASMR videos have a distinct effect on the neural activity of individuals with ASMR, suggesting that it may influence cognitive processing [56].

Enhanced Concentration: ASMR experiences often promote a state of heightened concentration and attention. This heightened focus may assist individuals in maintaining their concentration while working on complex mathematical problems or recognizing intricate patterns. Attention and focus: One study found that ASMR can improve attention and focus in people with ADHD. The study participants were randomly assigned to listen to either an ASMR video or a control video. The ASMR group showed significant improvements in attention and focus, as measured by a standardized test.

Stress Reduction: Reduced stress levels induced by ASMR experiences can potentially improve overall cognitive function, making it easier for individuals to approach mathematical tasks with a clearer mind. Lower stress levels have been linked to improved problem-solving abilities and cognitive performance. Another study found that ASMR can improve focus in people with generalized anxiety disorder (GAD). The study participants were randomly assigned to listen to either an ASMR video or a control video. The ASMR group showed significant improvements in focus, as measured by a standardized test. Another study found that ASMR can reduce stress in people with chronic pain. The study participants were randomly assigned to listen to either an ASMR video or a control video.

Sleep Quality: A few studies have shown that ASMR can improve sleep quality in people with insomnia. One study found that ASMR can help people fall asleep faster. The study participants were randomly assigned to listen to either an ASMR video or a control video. The ASMR group fell asleep significantly faster than the control group.

Sensory Stimulation: Brain Plasticity and Sign Language: ASMR triggers involve various sensory stimuli, such as gentle tapping, soft-spoken explanations, or visual patterns. These sensory experiences may help activate different parts of the brain, potentially facilitating creative thinking and novel problem-solving approaches. Another interesting area of research is the relationship between ASMR and brain plasticity. Brain plasticity is the ability of the brain to change and adapt in response to experience. Sign language is a good example of brain plasticity, as it has been shown to lead to changes in the brain's auditory and visual cortex [25, 48]. It is possible that ASMR could also lead to changes in the brain's plasticity. For example, one study found that ASMR can lead to increased activity in the brain's reward system. This suggests that ASMR could be used to promote neuroplasticity and improve cognitive function.

Pattern Recognition: ASMR experiences often involve repetitive sounds and movements, which could help train the brain in recognizing and deciphering patterns. This skill is essential in mathematics, as it enables individuals to identify relationships and solve complex problems more efficiently.

Motor Learning and Enhancing Interhemispheric Transfer [67]: Impact of interference task in activities performed using the same hemisphere and across hemispheres: One interesting area of research on ASMR is the impact of interference tasks on activities performed using the same hemisphere and across hemispheres. For example, one study found that ASMR can interfere with tasks that are typically performed by the left hemisphere, such as language processing. However, ASMR can also facilitate tasks that are typically performed by the right hemisphere, such as spatial processing. This suggests that ASMR may have the potential to be used to improve or impair cognitive function, depending on the specific task being performed. Further research is needed to explore this possibility. The nature of interference task has an impact on the nature of task: Another interesting area of research is the impact of the nature of the interference task on the nature of the ASMR task. For example, one study found that ASMR was more effective at reducing anxiety when the interference task was also anxiety-inducing. This suggests that ASMR may be more effective at reducing negative emotions when it is paired with a similar emotion.

4.2 Attention, Concentration, and Meditative Mental State

4.2.1 Different Theories of Attention and Neurological Disorders

4.2.1.1 One-Factor Theory of Attention

Attention, often described as the cognitive spotlight that selects and filters information from our environment, is a multifaceted phenomenon. Over the years, various theories have been proposed to elucidate the mechanisms governing attention, each offering a unique perspective on this intricate cognitive process. Among these theories, the one-factor theory of attention has emerged as a compelling framework that simplifies the understanding of attention while maintaining its explanatory power. The one-factor theory of attention posits that attention is a unified construct governed by a single underlying factor [15]. In essence, it suggests that attention is a unitary concept, and its various forms, such as selective attention, sustained attention, and divided attention, can be explained by a common mechanism. This theory challenges the notion that attention is a complex, multifaceted phenomenon with distinct processes for different types of attention. Instead, it simplifies the concept by suggesting that a single factor can account for the variations in attentional processes.

Insights from Neurological Patients: To gain insights into the validity of the one-factor theory of attention, researchers have turned to neurological patients with attention disorders. These patients provide a unique opportunity to explore the relationship between attentional deficits and their behavioral manifestations. One fascinating finding from studies involving neurological patients is the correlation between the degree of attention loss and the frequencies of their reactions. It appears that the measurement of reaction frequencies alone can serve as a reliable indicator of the extent of attentional impairment in these individuals.

Lack of Patterning due to Etiology: Another noteworthy observation is the absence of distinct patterns of reactions based on the etiology of the attention disorder. In other words, regardless of the underlying cause of attentional deficits, the behavioral manifestations, as reflected in reaction frequencies, exhibit a consistent pattern. This finding challenges the notion that different types of attentional disorders result in distinct behavioral patterns.

Support for the One-Factor Theory: The results gleaned from studies involving neurological patients lend support to the one-factor theory of attention. By demonstrating a clear correlation between attention loss and reaction frequencies and by highlighting the lack of distinct patterning based on etiology, these findings bolster the notion that attention is a unitary construct governed by a common mechanism.

The one-factor theory of attention offers a simplified yet powerful framework for understanding the complex processes that govern our attention. Insights from studies involving neurological patients, where the correlation between attention loss and reaction frequencies is evident, provide compelling support for this theory. By embracing the one-factor theory of attention, we move closer to demystifying the intricate workings of our cognitive spotlight, paving the way for a deeper understanding of attentional processes and their implications for human cognition and behavior.

4.3 Perception of Errors in Learning: The Intriguing Crossroads of Brain Hemispheres

The human brain is a remarkable organ, divided into two hemispheres that often work in tandem to process information and make decisions. However, when tasked with performing activities that require one hemisphere to override its counterpart, errors can emerge. In this article, we delve into the intriguing world of errors in learning, focusing on the perceptual intricacies that arise when a single human subject is asked to use their left-hand fingers to reply to decisions made in the right hemisphere and vice versa. This phenomenon not only sheds light on the brain's fascinating capabilities but also provides valuable insights for psychology researchers and neuroscientists.

Understanding Hemispheric Specialization

The brain's lateralization is a well-documented phenomenon, with each hemisphere specializing in certain functions. For most right-handed individuals, the left hemisphere is responsible for analytical thinking, language processing, and logical reasoning, while the right hemisphere tends to excel in creativity, spatial awareness, and emotional processing.

Challenging the Brain's Natural Order

When individuals are asked to perform tasks that challenge the natural order of hemispheric specialization, errors can occur. One intriguing experiment involves instructing a subject to use their left-hand fingers to reply to decisions that were processed in the right hemisphere and, conversely, to use their right-hand fingers to reply to decisions that originated in the left hemisphere. This seemingly simple task can elicit a cascade of errors and provide valuable insights into perception and cognition.

Perception of Errors: A Complex Dance

Perception of errors in learning involves a complex interplay between cognitive processes and sensory feedback. When a subject is asked to use their non-dominant hand to reply to a decision made in the opposite hemisphere, several intriguing phenomena can unfold:

1. **Confusion and Slower Response Times**: The brain's natural tendency is to process information in a manner that aligns with its specialized hemisphere. When this pattern is disrupted, individuals may experience confusion and slower response times as they attempt to override their brain's default mode.
2. **Perceptual Errors**: Subjects may perceive their responses as erroneous, even when they are correct. This heightened self-awareness can lead to a heightened perception of errors, contributing to the challenge of the task.
3. **Learning and Adaptation**: Over time, individuals can learn to adapt to the task's demands. With practice, response accuracy tends to improve as the brain's plasticity allows for the development of new neural pathways.

Implications for Psychology and Neuroscience

The study of errors in learning, particularly when it involves challenging the brain's natural hemispheric specialization, offers several valuable implications for psychology researchers and neuroscientists:

1. **Insights into Brain Plasticity**: These experiments provide insights into the brain's remarkable plasticity and its ability to adapt to novel tasks and challenges.

2. **Perceptual Awareness**: Understanding how individuals perceive errors in such tasks sheds light on the brain's self-awareness and its role in decision-making processes.
3. **Neural Adaptation**: Researchers can investigate the neural mechanisms that underlie the learning and adaptation process when subjects are tasked with challenging cognitive activities.
4. **Clinical Applications**: These findings may have implications for clinical populations, such as individuals recovering from brain injuries or undergoing cognitive rehabilitation.

The perception of errors in learning, especially when the brain is asked to challenge its natural hemispheric specialization, unveils the intricate dance of cognition and perception. While errors may initially abound, the brain's remarkable plasticity and ability to adapt allow individuals to improve their performance over time. For psychology researchers and neuroscientists, this phenomenon serves as a captivating window into the brain's inner workings and its capacity for learning and change. As we continue to explore the boundaries of cognitive research, errors in learning stand as a testament to the brain's enduring capacity for growth and adaptation.

4.4 Unraveling the Brain's Symphony: A Comparative Study of Learning Modes

4.4.1 Brain as a Musical Device, Redefining Time

A musical note could be written on a phase sphere as a circle. So, many such spheres of various diameters nested together represent a music. These nested clocks together are called a polyatomic time crystal. The polyatomic time crystal model of the biomaterials including the whole human brain, proposed by Reddy et al. [52], Saxena et al. [57] and Singh et al. [60], offers a fascinating perspective on brain function. It suggests that cognitive, perceptual, and emotional responses and feelings can be quantified by examining recurring events that lead to these experiences. Each recurring event serves as a clock on the surface of a sphere, with their origins scattered throughout the brain's fractal hardware. Thus, brain's symphony that is linked to the brain hardware also stems the cognitive, perceptive, and emotional responses. Based on these observations, we proposed a new theory namely Fractal Information Theory, FIT [5], which suggests that information does not grow linearly side by side in the brain, rather, fractally, within and above.

The fractal nature of mind originates from the hardware. Intriguingly, the study by Pradhan et al. [51] underscores the vital role of membrane microtubule junctions in neurons, emphasizing the importance of heterojunctions in signal transmission within the brain's fractal structure. This structure exhibits scale-free electromagnetic resonance within a single microtubule nanowire extracted from the brain, a single tubulin protein, and a single neuron, revealing the existence of a triplet of triplet resonance bands or 9 distinct frequency/time domains operating in harmony [59]. Extending resonance chain that connects every single component in the brain [27], we have mapped how electromagnetic interaction could link distant components and identify right neurons without using chemicals [61]. This triplet of triplet resonance has profound implications, even at the molecular level, demonstrating the symmetry of primes and suggesting the potential for quantum tunneling in information processing [10]. Additionally, the idea that a fractal network could efficiently process information logically has been proposed [28], highlighting the fractal nature of the mind.

The temporal organization of the mind also exhibits a fractal nature. Research by Agrawal et al. [4] involving simultaneous measurements of neuron firing at the neuron membrane and within microtubular bundles reveals that microtubules can process firing signals at speeds approximately 10,000 times faster than the membrane. This challenges the notion that microtubules play a passive role in the brain's information processing [64], providing a key support to neuron's clock system. Interestingly, even non-connected neurons, without direct synaptic junctions, can exchange energy, as demonstrated by imaging neural circuits using a scanning dielectric microscope (SDM) [29]. This imaging resembles mapping facial expressions and emotional states, akin to SDM or a thermal camera.

4.4.2 *The Human Brain is a Complex and Fascinating Organ, Continuously Evolving to Adapt to Various Learning Modes and Challenges*

In the realm of psychology and neuroscience, researchers have long been intrigued by the mechanisms underlying learning and how different conditions can impact the learning process. In this section, we embark on an unprecedented journey to compare three distinct modes of learning: under autonomous sensory meridian response (ASMR), under the steady beating of clocks, and under normal conditions. This comparative study sheds light on the intricate interplay of brain functions and lateralization, offering a fresh perspective for psychologists and neuroscientists.

A Novel Approach to Learning

In our study, we introduced a unique learning test protocol that challenges participants with tasks requiring mathematical ability and pattern or image recognition. What sets this study apart is the introduction of three perturbations for each task in different brain hemispheres, specifically the left and right brain. Additionally, participants were asked to respond using the opposite side of their brain, creating a multifaceted learning environment with 12 distinct variables ($3 \times 2 \times 2$).

Key Insights from the Comparative Study

Our research has revealed several noteworthy insights into the brain's learning processes:

1. **Clarity in EEG Data**: Unlike mirror drawing tasks that involve a delicate balance between finger movements and visual cognition, our learning test protocol presents a more straightforward scenario. This clarity in task design results in EEG data that is less prone to ambiguity and confusion.
2. **Deeper Lateralization**: Traditional question–answer-based psychological tests often fail to expose the deep lateralization of brain functions. In contrast, our protocol, which requires participants to switch between hemispheres or engage in a perfect flip-flop, highlights the brain's ability to shift its focus seamlessly.
3. **Impact of External Stimuli**: Comparing learning modes under ASMR, clock beats, and normal conditions unveils the influence of external stimuli on the learning process. ASMR, known for its calming effects, may lead to enhanced concentration and improved cognitive performance.
4. **Hemispheric Specialization**: The study's design also allows us to explore the unique capabilities of each hemisphere. We can discern how mathematical tasks, for instance, may activate different neural networks when processed by the left versus the right hemisphere.
5. **Potential for Future Applications**: These findings offer exciting prospects for applying this knowledge to educational settings and cognitive rehabilitation. Tailoring learning experiences to the brain's innate lateralization may lead to more effective teaching strategies and therapeutic interventions.

The comparative study of learning modes under ASMR, clock beats, and normal conditions represents a groundbreaking exploration into the brain's capacity for adaptation and lateralization. The introduction of perturbations, distinct tasks, and cross-hemispheric responses has unveiled a multifaceted landscape of cognitive processes. Unlike traditional psychological tests, this study provides a deeper understanding of how the brain navigates learning challenges and the influence of external stimuli. As psychologists and neuroscientists continue to unravel the brain's symphony, this research opens new avenues for enhancing learning experiences and advancing our knowledge of brain lateralization and cognition.

4.5 The Process Model of Errors: Unveiling the Foundations of Learning

Errors are an intrinsic part of the human experience, integral to the process of learning and growth. In the realm of psychology, understanding the process of errors and their classification is crucial for unraveling the intricate pathways of cognition. In this article, we explore the process model of errors, focusing on the classification of errors into categories such as cognitive errors and perceptual errors. We delve into how the process model of errors varies for different types of errors and how these errors serve as the bedrock of learning. Additionally, we differentiate between error-based learning and the innate universal learning mechanisms we are born with.

4.5.1 Understanding the Process Model of Errors

1. **Classification of Errors**: Errors in human behavior and cognition can be classified into various categories, including cognitive errors, perceptual errors, motor errors, and memory errors. Each type of error involves distinct cognitive processes and mechanisms.
2. **Cognitive Errors**: Cognitive errors are related to how we process information, make judgments, and solve problems. These errors can stem from biases, faulty reasoning, or incorrect assumptions. Understanding the cognitive processes underlying these errors is essential for improving decision-making and problem-solving skills.
3. **Perceptual Errors**: Perceptual errors occur when our sensory perception misinterprets information from the external world. These errors can include misjudging distances, misperceiving shapes, or misinterpreting sensory input. Studying perceptual errors sheds light on the intricacies of sensory processing and perception.

4.5.2 Variations in the Process Model of Errors

The process model of errors varies depending on the type of error being examined:

1. **Cognitive Errors**: Understanding cognitive errors involves investigating the thought processes, biases, and heuristics that lead to erroneous conclusions. Researchers in this field often explore decision-making, critical thinking, and reasoning abilities.
2. **Perceptual Errors**: Investigating perceptual errors requires a focus on sensory perception, neural processing, and the factors that influence our perception of the world. This research often involves studying visual illusions, auditory misperceptions, and the impact of environmental factors on perception.

Building the Foundation of Learning

Errors play a pivotal role in the learning process:

1. **Identifying Gaps in Knowledge**: Errors highlight areas where our understanding or skills are lacking. When we make an error, it signifies a gap in our knowledge or abilities, prompting us to seek solutions and acquire new information.
2. **Trial-and-Error Learning**: Many learning processes involve trial-and-error, where errors serve as valuable feedback. Through repeated attempts and

corrections, individuals refine their skills and knowledge, ultimately achieving mastery.

Error-Based Learning Versus Universal Learning

1. **Error-Based Learning**: This type of learning occurs through the recognition and correction of errors. Individuals learn from their mistakes, adapting their behavior and cognition to avoid similar errors in the future. Error-based learning is a conscious and deliberate process.
2. **Universal Learning**: At birth, humans possess innate mechanisms for learning, such as language acquisition, social bonding, and basic motor skills. These universal learning processes are hardwired into the brain and do not rely on the recognition of errors. Instead, they facilitate the acquisition of fundamental skills and knowledge necessary for survival and interaction.

The process model of errors offers a fascinating lens through which psychologists can explore the intricacies of human cognition and perception. By classifying errors into categories like cognitive and perceptual, researchers gain insights into the specific cognitive processes and mechanisms involved in each type of error. Errors are not merely setbacks; they are stepping stones in the journey of learning and growth. Error-based learning, which involves recognizing and rectifying errors, complements the universal learning mechanisms we are born with, shaping our understanding of the world and our ability to adapt to it. Ultimately, the study of errors in psychology is a testament to the resilience and adaptability of the human mind.

4.6 Beyond Facts and Programs: Unraveling the Brain's Unique Learning Journey

The human brain, with its intricate neural networks and astonishing adaptability, has long been a subject of fascination and exploration for psychologists and neuroscientists. Traditionally, the understanding of brain learning has centered around the acquisition of facts and the execution of programs, akin to a Turing machine. However, in recent years, a paradigm-shifting perspective has emerged, challenging this conventional wisdom. In this article, we delve into the remarkable idea that the brain's learning process extends beyond facts and linear programming. Instead, it involves the acquisition of invariants, the exploration of geometric shapes, and the utilization of a temple-like structure known as the metric of primes. This unique learning journey opens new frontiers in our comprehension of the brain's cognitive abilities.

Breaking Free from Linear Learning

In the conventional model of learning, the brain is likened to a Turing machine, a linear processor of bits and bytes, where facts are stored and algorithms are executed.

However, mounting evidence suggests that the brain's learning process is far more intricate and nuanced.

1. **Fractal Tape of Learning**: The brain is proposed to follow a fractal tape, with each cell of this tape containing another tape inside. These tapes are circular, defying the linear nature of conventional learning. This fractal structure hints at a profound departure from the storage of mere facts.
2. **Learning Invariants**: Instead of memorizing isolated facts, the brain appears to specialize in learning invariants—patterns and geometric shapes that remain constant regardless of spatio-temporal conditions. These invariants serve as the foundation for a deeper understanding of the world.

Beyond Spatio-Temporal Events

One of the most intriguing aspects of this new perspective on brain learning is the transcendence of spatio-temporal events.

1. **Exploring Geometric Shapes**: The brain engages in the exploration of geometric shapes formed by neural signals. These shapes are not confined to the limitations of a 6D spatio-temporal environment but exist in a realm beyond, transcending the constraints of time and space.
2. **Temporal Invariance**: The brain's focus on learning invariants is particularly noteworthy. These invariants, which remain constant regardless of changing conditions, are not limited to spatio-temporal events. They represent a deeper layer of understanding that defies conventional linear learning.

The Metric of Primes: A Temple of Symmetry

At the heart of this unique learning journey lies the metric of primes—a temple-like structure that houses all possible symmetry-breaking rules.

1. **Projecting Invariants**: When the brain needs to reconstruct 6D spatio-temporal events, it calls upon the metric of primes to project the learned invariants onto this temple of symmetry. This process generates a dynamic understanding of the world, rooted in the unchanging principles of symmetry.
2. **An Unprecedented Learning Paradigm**: This approach to learning marks a profound departure from traditional models. Instead of rote memorization or linear processing, the brain navigates a symphonic landscape of invariants and geometric shapes, leveraging the metric of primes to construct a rich and ever-evolving understanding of the universe.

The brain's learning journey is a mesmerizing odyssey that transcends conventional notions of linear fact storage and program execution. Instead, it involves the acquisition of invariants, the exploration of geometric shapes, and the utilization of a temple-like structure known as the metric of primes, phase prime metric or PPM or Purusha. This unique perspective revolutionizes our understanding of cognition, offering a glimpse into the brain's capacity to navigate a multidimensional and ever-evolving universe of knowledge. As we continue to unravel the mysteries of the brain, this groundbreaking paradigm shift invites us to explore the brain's true potential for learning and adaptation.

4.7 Unveiling the Role of Noise in Brain Signal Transmission and Cognitive Function Through Fractal Information Theory

In the quest to comprehend the intricacies of the human brain, one enduring challenge has been to understand the role of noise in brain signal transmission and cognitive function, even microtubule harvests thermal noise to make it a meaningful signal [58]. Traditionally, our understanding of information theory was shaped by the pioneering work of Claude Shannon in the 1950s, leading to the development of Shannon Information Theory. However, recent advancements in the field have introduced a paradigm-shifting perspective—Fractal Information Theory (FIT)—which has transformed our perception of how noise influences brain function [5]. In this article, we explore the evolving role of noise in brain signal transmission and cognition, guided by the principles of FIT.

Shannon Information Theory: A Historical Framework

Claude Shannon's Information Theory provided the foundation for understanding the transmission of information in communication systems. It focused on concepts such as data compression, signal-to-noise ratio, and the efficient encoding of messages. While Shannon's theory has been immensely valuable in fields like telecommunications and information technology, its application to neuroscience left questions unanswered. Shannon Information Theory (SIT), which was first proposed in the 1950s, has been the dominant framework for understanding information transmission in communication systems and biological brains. SIT views information as a measure of the uncertainty associated with a random variable. However, SIT has certain limitations, such as its inability to capture the long-range correlations and fractal structure that are ubiquitous in real-world data. In 2016, a new theory of information, called Fractal Information Theory (FIT), was proposed. FIT builds on the foundations of SIT, but it also incorporates the concept of fractals. Fractals are geometric objects that are self-similar at different scales. They are found in nature, from snowflakes to coastlines, and they are also present in many biological signals, including brain signals. FIT has the potential to provide a more accurate and comprehensive understanding of information transmission in the brain. In this article, we will discuss the role of noise in brain signal transmission and cognitive function in the context of FIT.

Challenging the Status Quo with FIT

Fractal Information Theory, introduced in 2016, challenges the traditional notions of information theory by incorporating fractal geometry into the framework. FIT is based on the recognition that complex systems, including the human brain, exhibit fractal-like patterns that defy the assumptions of linear information theory.

1. **Fractals in Brain Dynamics**: The brain's activity is not a smooth, linear process but rather a complex interplay of fractal-like patterns. FIT acknowledges the self-similar structures present in neural activity, allowing for a more accurate representation of brain dynamics.
2. **Role of Noise**: In FIT, noise is not considered an interference but an integral part of information processing. Noise is viewed as a carrier of crucial information that influences cognitive function and decision-making. **Noise in Brain Signal Transmission**: Noise is a ubiquitous feature of brain signal transmission. It can arise from a variety of sources, such as thermal fluctuations, metabolic activity, and electromagnetic interference. Noise can interfere with the transmission of information in the brain, leading to cognitive impairments such as attention deficits and memory loss. SIT views noise as a random process that is independent of the signal. However, FIT takes a different approach. FIT views noise as a fractal process that is correlated with the signal. This means that the noise in a brain signal is not random, but rather it exhibits a fractal structure. The fractal structure of noise in brain signals has been demonstrated by a number of studies. For example, one study found that the noise in EEG signals exhibits a fractal power spectrum. This means that the noise is distributed across a wide range of frequencies, with a power-law relationship between frequency and power.

The Role of Noise in Brain Signal Transmission:

1. **Stochastic Resonance**: FIT proposes that the brain's sensitivity to weak signals can be enhanced by the presence of noise. This phenomenon, known as stochastic resonance, suggests that the addition of noise can actually improve the detection and transmission of signals within the brain.
2. **Enhanced Information Processing**: Noise is not merely an unwanted disturbance; it contributes to the richness of information processing within the brain. By embracing noise, FIT reveals that the brain can efficiently extract valuable information from chaotic input.

Cognitive Function and Noise:

1. **Optimal Cognitive States**: FIT suggests that the brain operates optimally within a certain range of noise levels. Too little noise can lead to rigid thinking and reduced adaptability, while excessive noise can overwhelm cognitive processes. The brain seeks a balance where noise facilitates flexible cognition. **Fractal Information Theory and Cognitive Function**: FIT has the potential to provide a new perspective on the relationship between noise and cognitive function. In the context of SIT, noise is seen as a nuisance that interferes with the transmission of information. However, in the context of FIT, noise can also be seen as a source of information. Fractal noise has been shown to play a role in a variety of cognitive processes, including attention, memory, and decision-making. For

example, one study found that fractal noise in EEG signals is associated with increased attention performance. Another study found that fractal noise in EEG signals is associated with improved memory performance.
2. **Adaptation and Learning**: Noise-driven adaptability plays a crucial role in learning and cognitive flexibility. In this context, noise serves as a catalyst for neural plasticity, allowing the brain to adapt to changing environments and challenges.

The transition from Shannon Information Theory to Fractal Information Theory has revolutionized our understanding of the role of noise in brain signal transmission and cognitive function. FIT offers a fresh perspective on noise as an essential component of information processing within the brain, challenging the notion that noise is merely disruptive. Instead, noise is a carrier of vital information, shaping the brain's adaptability, learning, and cognitive flexibility. As we continue to explore the brain's intricate dynamics, FIT invites us to embrace noise as a source of wisdom and a fundamental aspect of the cognitive processes that define our humanity. FIT is a new theory of information that has the potential to provide a more accurate and comprehensive understanding of information transmission in the brain. FIT views noise as a fractal process that is correlated with the signal. This has implications for our understanding of the relationship between noise and cognitive function. In the context of FIT, noise can no longer be seen as a nuisance that interferes with the transmission of information. Rather, it can be seen as a source of information that plays a role in a variety of cognitive processes. Future research should investigate the role of fractal noise in brain signal transmission and cognitive function in more detail. This research could lead to new insights into the development of treatments for cognitive impairments such as attention deficits and memory loss.

4.8 Polyatomic Time Crystals and Their Potential for Finding Clusters of Recurring Events and Integrating Them into an Information Structure

4.8.1 Unit of Information in Nature, Integration of Information, and Human Cognition

In the realm of quantum physics and information theory, a groundbreaking concept has emerged that promises to revolutionize our understanding of time, information, and the very fabric of reality itself. This concept is the polyatomic time crystal (PTC), a multidimensional assembly of clocks that functions as a continuous engine, generating intricate outputs from a deceptively simple structure [52]. Polyatomic time crystals (PTCs) are a new type of time crystal that is made up of multiple atoms where each atom is a clock or discrete time piece. PTCs have the potential to be used in a variety of applications, including the development of new computing technologies and the study of consciousness. One potential application of PTCs is in the area of information processing. PTCs could be used to find clusters of recurring events and integrate them into an information structure. This could be useful for a variety of tasks, such as identifying patterns in data, developing new algorithms, and creating new forms of artificial intelligence. This article explores the fascinating world of PTCs, their representation using a 3D clock assembly, and their pivotal role in a novel framework called geometric musical language (GML). Furthermore, we delve into the transformative potential of PTCs in deciphering the recurrent events that shape the universe and their integration into an information structure through the symmetry of primes.

The Essence of a Polyatomic Time Crystal: A time crystal, in its simplest form, can be visualized as a clock—a device that measures the passage of time. However, the polyatomic time crystal elevates this concept to a multidimensional level. Imagine a 3D assembly of clocks, each synchronized in a manner that they function as a continuous engine, producing intricate and ever-changing outputs from a seemingly straightforward structure. The complexity emerging from this assembly defies conventional notions of timekeeping.

Geometric Musical Language (GML): Bridging the Gap: The translation of natural events into the realm of PTCs represents a profound shift in our understanding of information and time. GML serves as the protocol through which all natural events can be reimagined as PTCs. In this process, clocks are transformed into dots at the center of spheres that represent the clocks, resulting in a representation of geometric shapes. As these geometric shapes evolve, they reveal the rules governing geometric symmetry breaking—an intricate dance of patterns and forms.

The Metric of Primes: Unveiling Symmetry: The journey into the realm of PTCs culminates in the emergence of a metric of primes, known as phase prime metric (PPM) or simply Purusha. This metric encapsulates the symmetries inherent in the PTCs and the evolving geometric shapes. It serves as a key to understanding the underlying structure of reality, as it governs the transformations and transitions within the PTCs.

Mapping the Learning Process: Perhaps, one of the most compelling aspects of PTCs and the associated PPM is their ability to map the learning process of conscious or near-conscious entities. By analyzing the evolution of geometric shapes and their symmetries, researchers can gain insights into how learning occurs, how information is acquired, and how patterns and structures emerge within the fabric of existence. Geometric Musical Language (GML) and the Phase Prime Metric (PPM): Geometric musical language (GML) is a protocol for presenting all information in nature using a 3D assembly of clocks. Once natural events are rewritten as PTCs, we replace clocks with dots at the center of the spheres that represent the clocks and we get geometric shapes. As geometric shapes change, we get the rules of geometric symmetry breaking, which could be represented using the symmetry of the primes. The phase prime metric (PPM) is a metric of primes that is derived from the symmetry of the primes. The PPM can be used to map the learning process of any conscious or nearly conscious entity.

The brain, a marvelously intricate organ, constantly exchanges signals crucial for cognitive processes like memory, attention, and decision-making. Recent interest has surged in understanding the impact of noise on brain signal transmission. Noise, essentially unwanted or random signal fluctuations, isn't always detrimental; it can have positive implications for brain function. A study discovered that both autonomous sensory meridian response (ASMR), a tingling sensation triggered by specific sounds or stimuli and random noise can enhance cognitive task efficiency. These phenomena boost connectivity between different brain regions, contributing to improved cognitive performance. Conversely, another study found that periodic or repetitive sounds can hinder task efficiency, disrupting the brain's information processing. These findings illuminate the nuanced role of noise in brain signal transmission, which varies based on the type of noise and the specific cognitive task at hand.

Potential Applications of PTCs, GML, and PPM

How PTCs could be used to find clusters of recurring events and integrate them into an information structure: One way to use PTCs to find clusters of recurring events is to use them to create a network of clocks. Each clock in the network would represent a different event. The clocks would be connected to each other in such a way that they would synchronize with each other when the corresponding events occurred. Once the clocks are synchronized, it would be possible to identify clusters of recurring events by looking for groups of clocks that are synchronized with each other. This could be done by using a variety of techniques, such as pattern recognition and machine learning. Once clusters of recurring events have been identified, they could be integrated into an information structure. This could be done by creating a database of the events and their relationships to each other. The database could then be used to develop new algorithms and artificial intelligence systems.

PTCs, GML, and PPM have the potential to be used in a variety of applications, including:

- Information processing: PTCs could be used to develop new computing technologies that are more efficient and powerful than current technologies.
- Artificial intelligence: PTCs could be used to develop new forms of artificial intelligence that are more intelligent and capable than current AI systems.
- Consciousness: PTCs, GML, and PPM could be used to study the nature of consciousness and to develop new theories of consciousness.

Integrated Information Structure

The exploration of polyatomic time crystals and their integration into geometric musical language represents a profound leap forward in our comprehension of the universe. PTCs, with their multidimensional clock assemblies, provide a unique window into the complexity hidden within seemingly simple structures. Through the symmetry of primes and the emergence of the phase prime metric (PPM), we gain access to the very essence of reality itself. As we continue to unravel the mysteries of PTCs and their role in shaping our understanding of information and time, the possibilities are limitless. From mapping the learning processes of conscious beings to deciphering the recurrent events that define our existence, PTCs and GML offer a tantalizing glimpse into the future of science, consciousness, and the fundamental nature of our universe.

4.8.2 Experimental Verification by Cognitive Tests and ASMR Effect on Learning

Experimental Method

Figure 4.1 shows the schematic presentation of the experimental protocol. In this schematic, we depict the left brain performing calculations and transmitting the results to the right brain, which, in turn, initiates a response carried out by the left brain. Additionally, image recognition occurs within the right brain, and the information is subsequently relayed to the left brain, enabling a response from the right hand. Our experimental protocol has been thoughtfully designed to ensure that human subjects refrain from speaking, as any vocalization would significantly alter the electroencephalogram (EEG) readings. By maintaining silence, the speech control region remains inactive, allowing us to detect changes exclusively in both the left and right brain regions.

We conducted experiments involving straightforward cognitive tasks like mathematical calculations and pattern recognition, anticipating clear evidence of interhemispheric signal transfer. Yet, the brain's signal transmission pathways proved to be intricate, defying simple routes from the left brain to the right and vice versa. This

Fig. 4.1 Schematic that shows left brain doing calculation and send it to the right brain so that response is sent by left brain. Also, image recognition happens in the right brain which is sent to the left brain so that right hand could respond

complexity highlights that there are no universal rules governing signal transmission in the brains of all individuals. Nevertheless, we made a significant discovery applicable to all subjects we studied: ASMR and random noise enhance the efficiency of various cognitive tasks, while periodic or repetitive sounds diminish task efficiency universally.

Mathematics primarily occurs in the left brain, where we process math problems and request answers to be displayed by the right brain, typically executed using the left hand. This brain transfer of mathematical results is observable in real-time. Conversely, art-related processes take place in the right brain, while language processing is predominantly a left brain function. In an experiment, we present images alongside scrambled sentences, requiring participants to match explanations to various pictures. When a person successfully connects the explanation, we witness the real-time transfer from the right brain to the left brain. We also introduce periodic noise and autonomous sensory meridian response (ASMR) to examine their effects. Notably, both positive and negative effects observed are temporary due to the brain's plasticity. We identify two key factors contributing to human-to-human variation: the way the brain shifts patterns between different parts, such as from left to right and vice versa, and the individual's capacity for plasticity revival, where they can nullify both positive and negative activations.

We used a variety of cognitive tasks to assess the effects of noise on brain signal transmission. These tasks included: mathematical calculations, pattern recognition, working memory, attention, and decision-making. We also used EEG to measure the electrical activity of the brain during the cognitive tasks.

Results: ASMR Performs Better Than Periodic or Ordered Music

Figure 4.2 elegantly illustrates ASMR alongside periodic clock-like beating signals. Our utilization of these distinctive, unpredictably random patterns proves to be immensely beneficial for enhancing cognitive activities in the brain. We found that ASMR and random noise enhanced the efficiency of the cognitive tasks, while periodic or repetitive sounds diminished task efficiency. For example, in the mathematical calculations task, participants who listened to ASMR or random noise were able to solve more problems correctly than participants who listened to periodic or repetitive sounds. Similarly, in the pattern recognition task, participants who listened to ASMR or random noise were able to identify more patterns correctly than participants who listened to periodic or repetitive sounds. We also found that ASMR and random noise increased the connectivity between different brain regions. This increased connectivity is thought to be responsible for the improvement in cognitive function.

Discussion

Our enlightening EEG signal findings, as evidenced by Fig. 4.3, reveal a striking clarity. They vividly demonstrate how, during mathematical tests, signals elegantly traverse from one part of the brain to another. This suggests that, with careful experiment design, we possess the potential to minimize the distraction of simultaneous

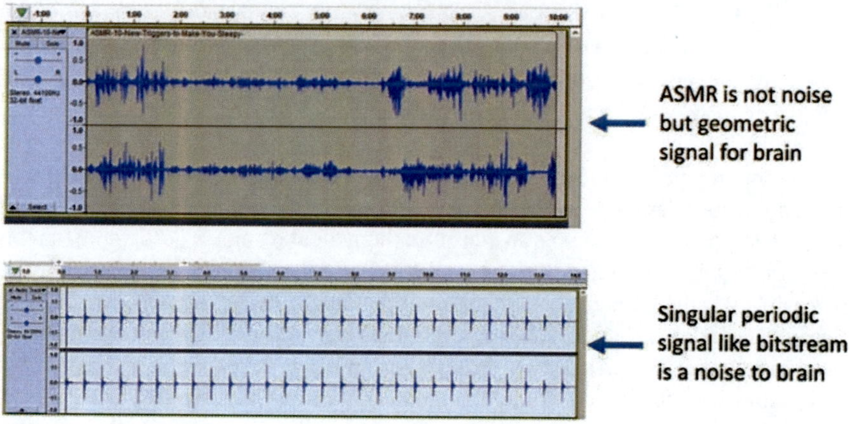

Fig. 4.2 Audacity sound profiles for ASMR (top) and clock (repeating) sounds

activities. Nonetheless, the EEG complexity surges when dealing with pattern recognition and the simultaneous organization of descriptions, as vividly portrayed in Figs. 4.4 and 4.5. It suggests that the role of noise in brain signal transmission is complex and depends on the type of noise and the cognitive task being performed. Figure 4.6 shows ASMR and random noise appear to be beneficial for cognitive function, while periodic or repetitive sounds appear to be detrimental. The mechanisms underlying these effects are not fully understood, but they may involve changes in the brain's connectivity. Further research is needed to investigate the effects of noise on brain signal transmission and cognitive function.

ASMR, or autonomous sensory meridian response, is a sensory phenomenon characterized by a tingling sensation that individuals experience in response to specific audio and visual triggers [13] which was implemented here as a modulating tool.

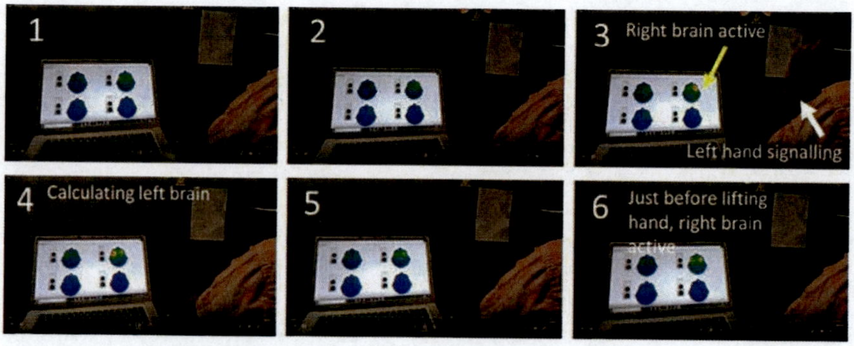

Fig. 4.3 Series of six EEG images showing six steps when a person reads the mathematical problem and solves it in the left brain, we can notice left brain is active, and then using the right brain and left hand, he responds. Image 6 is interesting that just before the hand is lift off, we see that right brain got active

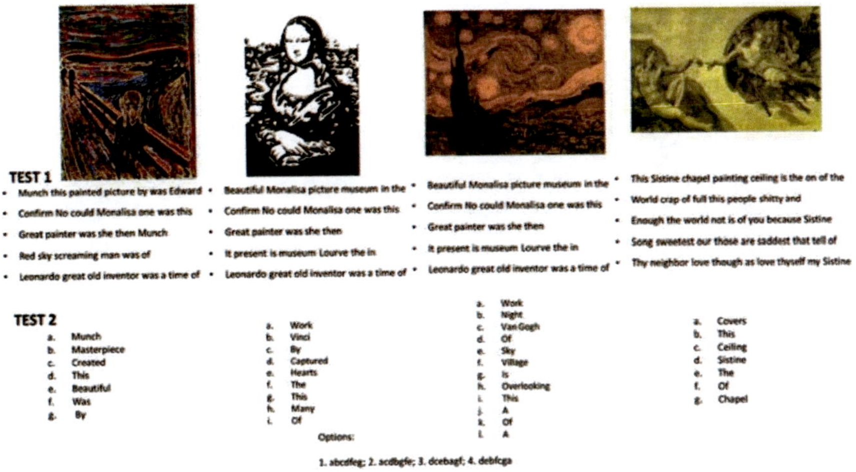

Fig. 4.4 Four images are shown here in four columns. Test 1 is about sentence that represents the picture, and Test 2 is about a single sentence and its words are jumbled up. So, a person has to find which sentence represents the picture most and then arrange the words to make it a right sentence

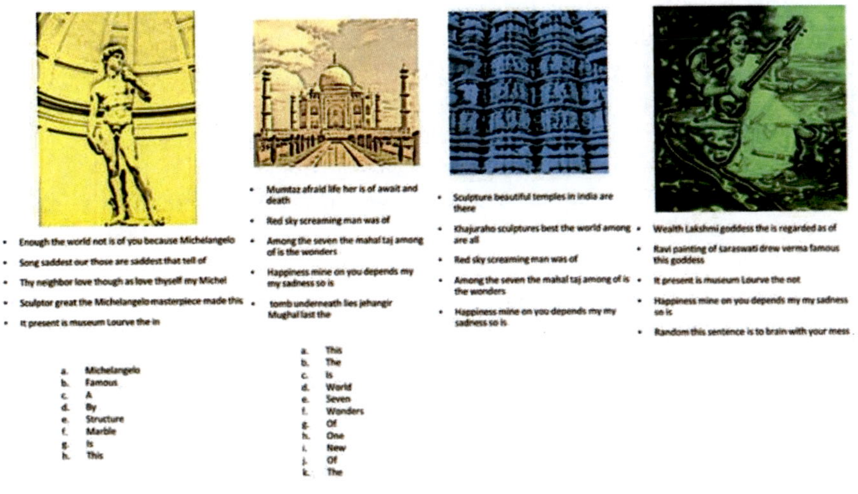

Fig. 4.5 Four images are shown here in four columns. Test 1 is about sentence that represents the picture, and Test 2 is about a single sentence and its words are jumbled up. So, a person has to find which sentence represents the picture most and then arrange the words to make it a right sentence

One study supported that individuals with ASMR tend to be more mindful than those without ASMR [23]. Mindfulness is a state of active and open attention to the present moment, which has been associated with improved cognitive abilities, including learning and attentional performance [22]. Therefore, it is possible that the mindfulness experienced by individuals with ASMR may contribute to enhanced

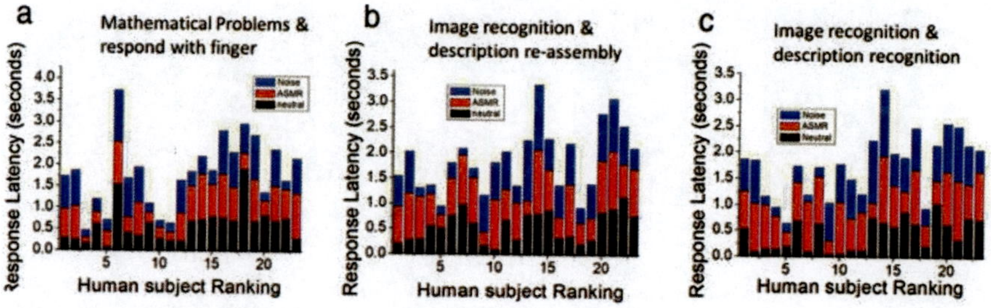

Fig. 4.6 There are three sets of experiments, first mathematics test in Panel **a**, then picture test done in two ways (description of a picture jumbled up in Panel **b**; words of a proper description of an image in Panel **c**). We carry out three experiments in three conditions: when there is no disturbance in the room (black), when there is an ASMR (red), and finally, when we have regular beating that acts as a noise to the brain (blue). Here, we have compared how mathematical calculation in the brain (right) and response using left-hand fingers. Recognizing images using left brain and identifying using right-hand fingers. Horizontal axis depicts 25 human subjects, and vertical axis depicts latency time in seconds. Latency time means time lapse from the seeing the picture and changes in EEG confined to a particular lobe of the brain

learning ability. Questionnaires given to the subjects and their overall attitude showed to have positive effects on mood and cognitive abilities. A study conducted on healthy young adults found that ASMR can improve mood and attention, suggesting a potential benefit for learning [22]. Since we performed the ASMR experiment on the young people who were in the hostel after whole-day work, the boring cognitive tests under ASMR induced relaxation and reduce stress and anxiety, which did create a conducive environment for learning supporting the previous finding [55].

Furthermore, ASMR has been associated with increased absorption and experiential involvement [47]. This heightened ability to become fully engaged with the current experience may facilitate learning by promoting focused attention and deep processing of information. The mindfulness experienced by individuals with ASMR, along with its potential to improve mood, attention, and relaxation, may contribute to enhanced learning ability.

4.8.3 The Structure of Time or Precursor to Time Crystal in ASMR

While ASMR has primarily been studied in the context of sensation and biology, the research on the response time is scarce. Here in this book since we are exploring the possibility of representing all human cognition in terms of a 3D assembly of clocks, the sharpness of a mind to an ASMR signal is important factor. During interhemispheric transfer, there is the highest probability to see the quantization of time [68]. In another study by Simon-Dack et al. [66], the researchers explored the relationship between resting EEG and interhemispheric transfer times. They found that individuals with slower right-to-left interhemispheric transfer times performed better on tasks typically associated with the right hemisphere (RH). This suggests that the dominance of one hemisphere may come at a cost to interhemispheric communication. The study highlights the importance of efficient interhemispheric transfer for optimal cognitive performance. Additionally, Thut et al. [68] conducted an EEG study to investigate interhemispheric motor-level transfer in a simple reaction time task. They found evidence for interhemispheric transfer at the motor level, as indicated by changes in event-related potentials (ERPs) in response to the task. The study provides further support for the existence of interhemispheric communication in motor control.

Interhemispheric synchronization plays a crucial role in the control of bimanual movements, according to the study conducted by Serrien and Brown [65]. The researchers investigated the functional role of interhemispheric synchronization in bimanual timing tasks. They found that when participants performed bimanual tasks, there was a significant increase in interhemispheric synchronization compared to unimanual tasks. This suggests that interhemispheric synchronization is necessary for coordinating the timing of movements between the two hands. The study provides evidence for the importance of interhemispheric communication in motor control.

4.9 Learning from Errors in a Self-Operating Mathematical Universe (SOMU) Using the Phase Prime Metric (PPM)

4.9.1 Structure of Time as We See It in the Brain

The realm of consciousness, learning, and the very fabric of the universe itself have long been subjects of profound intrigue and exploration. In the pursuit of understanding these intricate phenomena, the Self-Operating Mathematical Universe (SOMU) theory of consciousness has emerged as a groundbreaking framework. At its core, SOMU converts all natural events into a mesmerizing 3D clock assembly, which, in turn, transforms into an ever-changing array of geometric shapes spread over 12 dimensions as a manifold [62, 63]. This intricate dance of forms and patterns ultimately converges into the phase prime metric (PPM), resembling a majestic Hindu temple. Within this temple lies the key to harnessing errors in learning and forging a path toward hierarchical thought.

SOMU: A Gateway to Universal Understanding: SOMU theory posits that all natural events, no matter how diverse or complex, can be translated into a unified framework—a 3D clock assembly. This assembly serves as a bridge between the observable world and the hidden structures that underpin our reality. Each clock represents a moment in time, and their collective movement creates a dynamic and ever-evolving tapestry of information.

The Dance of Geometric Shapes: As the 3D clock assembly transforms, it gives rise to a mesmerizing array of geometric shapes. These shapes, intricate and ever-changing, represent the symmetries and patterns that govern the universe. They are the language through which the universe communicates its secrets, and they hold the key to understanding the errors in our learning.

The Phase Prime Metric (PPM): Temple of Knowledge: At the culmination of this transformative journey lies the phase prime metric, a structure reminiscent of a grand Hindu temple. This temple-like metric is not merely an architectural wonder; it is a repository of the symmetries and invariants that shape our reality. Walking a path on the surface of the PPM reveals the intricacies of learning and the role of errors within it.

Harnessing Errors in Learning: The PPM serves as a guide to understanding errors in learning and, crucially, correcting them. If the path on the temple's surface appears broken, convoluted, or lacks the formation of a coherent geometric shape, it signals a learning error. The beauty of this system lies in its ability to identify errors universally, regardless of the context or the nature of the error.

Hierarchical Thought: A Pathway to Enlightenment: The utilization of errors in learning, as guided by the PPM, leads to the development of hierarchical thought. Hierarchical thought is the ability to organize knowledge and concepts in a structured manner, with each piece of information fitting seamlessly into a broader framework. This structured thinking enhances problem-solving abilities, fosters creativity, and ultimately leads to a deeper understanding of the universe.

The PPM can be used to learn from errors in the following way:

1. Identify the path on the surface of the PPM that represents the learning process.
2. Look for breaks or discontinuities in the path. These breaks or discontinuities represent errors in the learning process.
3. Correct the errors in the learning process by modifying the path on the PPM.

This process can be repeated iteratively until the learning process is error-free.

Benefits of Using the PPM to Learn from Errors

There are several benefits to using the PPM to learn from errors:

- The PPM is a universal representation of all possible learning processes. This means that the PPM can be used to learn from errors in any domain, including science, engineering, art, and music.
- The PPM is a hierarchical structure. This means that the PPM can be used to identify and correct errors at different levels of the learning process.
- The PPM is a self-organizing structure. This means that the PPM can automatically correct errors in the learning process without any human intervention.

The phase prime metric (PPM) within the Self-Operating Mathematical Universe (SOMU) theory of consciousness offers a transformative perspective on learning, errors, and the development of hierarchical thought. By converting natural events into a unified 3D clock assembly, transitioning through dynamic geometric shapes, and culminating in the PPM, we unlock the potential to universally learn from errors.

Errors in learning are not stumbling blocks but stepping stones on the path to enlightenment [24]. The PPM, with its temple-like structure, guides us through this journey, helping us identify errors, correct them, and forge a path toward hierarchical thought. In this grand tapestry of knowledge and consciousness, errors become catalysts for growth, and the PPM stands as a beacon of wisdom, lighting the way toward a deeper understanding of the universe and our place within it. The PPM could also be used to develop new methods of teaching and learning. For example, the PPM could be used to create personalized learning plans for students. The PPM could also be used to develop new educational tools and resources. The PPM is a new and exciting technology, and we are only beginning to explore its potential applications. As we learn more about the PPM and how to use it, we can expect to see it revolutionize many different aspects of education and learning.

The phase prime metric (PPM) is a mathematical structure that represents all possible symmetry-breaking events in the universe. It is a 3D clock assembly that is derived from the symmetry of the primes. The PPM can be used to model a wide range of phenomena, including the learning process of any conscious or nearly conscious entity. In a Self-Operating Mathematical Universe (SOMU), all natural events are converted into a 3D clock assembly, which is then transformed into a changing geometric shape, and finally into the final structure of primes, which is the PPM.

4.9.2 How Periodic Signals Disturb the Brain

According to auditory scene analysis theory proposed by Bregman [9], the auditory system segregates sound sources based on their spectral and temporal properties. When a periodic beat is generated repeatedly without changing any parameter, it can be perceived as a single sound source that stands out from the background. This can lead to a sense of disturbance and discomfort for the conscious mind. Periodic signals are sounds or other stimuli that repeat at regular intervals. They can be found in nature, such as the ticking of a clock or the heartbeat, or they can be created by humans, such as the music we listen to or the noise from a factory. While periodic signals can be pleasant and soothing, they can also be very disturbing to the brain. This is because the brain is constantly trying to make sense of the world around it, and periodic signals can disrupt this process. One way that periodic signals can disturb the brain is by creating sensory overload. When the brain is constantly bombarded with the same stimulus, it can become overwhelmed and unable to process other information effectively. This can lead to a variety of problems, including difficulty concentrating, irritability, and anxiety.

Another way that periodic signals can disturb the brain is by interfering with attention. Attention is the process of focusing on a particular stimulus or task. Periodic signals can distract the brain and make it difficult to focus on anything else. This can lead to problems with learning, memory, and performance. Periodic signals can also disturb the brain by triggering the fight-or-flight response. The fight-or-flight response is a physiological reaction to stress or danger. It is characterized by an increase in heart rate, blood pressure, and breathing. When the brain perceives a periodic signal as a threat, it can trigger the fight-or-flight response, even if the signal is not actually dangerous. This can lead to feelings of anxiety, fear, and even panic [44].

There is a growing body of research on the effects of periodic signals on the brain. One study, Adcock et al. [6], found that people who were exposed to a periodic signal for a short period of time showed decreased activity in the prefrontal cortex, a region of the brain that is involved in attention, decision-making, and emotional regulation. Most interestingly, Buchanan et al. [8] found that people who were exposed to a periodic signal for a longer period of time showed increased activity in the amygdala, a region of the brain that is involved in fear and emotion processing. These studies suggest that periodic signals can have a significant impact on the brain, even when people are not consciously aware of them.

4.10 Conclusion: Cognitive Tasks in Brodmann's Regions

4.10.1 Left Brain and Right Brain: Two Tasks Supporting Each Other

Interhemispheric Transfer and the Role of Brodmann's Areas: The human brain is divided into two hemispheres, the left and the right. Each hemisphere is responsible for different functions. The two hemispheres of the brain communicate with each other through a network of nerve fibers called the corpus callosum [30]. This connection allows the two hemispheres to communicate and share information. One of the key functions of the corpus callosum is to support interhemispheric transfer. Interhemispheric transfer is the process of transferring information from one hemisphere of the brain to the other. This process is essential for many cognitive functions, such as language, attention, and memory.

Brodmann's areas are a set of 47 regions in the cerebral cortex that are involved in different cognitive functions. Some Brodmann's areas are located in the left hemisphere, while others are located in the right hemisphere. Recent research has shown that interhemispheric transfer is not as simple as it once was thought to be. It was previously believed that each Brodmann's area was responsible for a specific cognitive function, and that interhemispheric transfer involved the direct transfer of information between two homologous Brodmann's areas in the two hemispheres. This interhemispheric transfer of signals is essential for a variety of functions, including language, vision, and motor control.

One way that interhemispheric transfer affects brain function is by activating different Brodmann's areas. Brodmann's areas are specific regions of the brain that are associated with different functions. For example, Broca's area in the left hemisphere is responsible for speech production, while Wernicke's area in the left hemisphere is responsible for language comprehension. Studies have shown that interhemispheric transfer can activate both Broca's and Wernicke's areas when people are performing language tasks. For example, one study found that people who were asked to listen to a story and then repeat it showed increased activity in Broca's and Wernicke's areas, as well as in the corpus callosum. One could visualize that interhemispheric transfer can activate both Broca's and Wernicke's areas when people are performing language tasks [31, 71]. Another study found that people who were asked to read a sentence and then identify the subject and verb showed increased activity in Broca's and Wernicke's areas, as well as in the corpus callosum. These studies suggest that interhemispheric transfer plays an important role in language processing by activating different Brodmann's areas in both hemispheres of the brain.

Interhemispheric transfer can also activate Brodmann's areas in the visual cortex. For example, one study found that people who were asked to view a picture of a face showed increased activity in the visual cortex of both hemispheres. However, the activity in the left hemisphere was greater than the activity in the right hemisphere. For some, the activity in the right hemisphere was greater than the activity in the left hemisphere. Most importantly, which hemisphere would dominate is difficult to predict since comprehension and problem-solving methods are built in the brain depending on life activities [39].

In addition to language and vision, interhemispheric transfer can also activate Brodmann's areas in the motor cortex [33]. Interhemispheric transfer can also activate Brodmann's areas in the visual cortex [31]. One study found that people who were asked to move their right hand showed increased activity in the motor cortex of both hemispheres. However, the activity in the left hemisphere was greater than the activity in the right hemisphere.

The Implications of the Research on Interhemispheric Transfer: The research on interhemispheric transfer has a number of implications for our understanding of cognitive function and brain injury. For example, the research suggests that interhemispheric transfer is essential for many cognitive functions, such as language, attention, and memory. This means that damage to the corpus callosum or to the Brodmann's areas involved in interhemispheric transfer can lead to significant cognitive impairments. One study found that people who were asked to walk showed increased activity in the motor cortex of both hemispheres. However, the activity in the right hemisphere was greater than the activity in the left hemisphere. These studies suggest that interhemispheric transfer plays an important role in motor control by activating different Brodmann's areas in the motor cortex of both hemispheres.

The research also suggests that interhemispheric transfer is a complex process that involves the activation of multiple Brodmann's areas in both hemispheres. This means that it is difficult to predict the specific cognitive impairments that will result from damage to the corpus callosum or to the Brodmann's areas involved in interhemispheric transfer. The research on interhemispheric transfer also has implications for the development of new treatments for cognitive impairments. For example, researchers are developing new brain-computer interfaces that can be used to bypass damage to the corpus callosum and restore interhemispheric transfer. Overall, the research on interhemispheric transfer is a rapidly developing field with important implications for our understanding of cognitive function and brain injury.

Future Directions for Research: More research is needed to better understand the role of Brodmann's areas in interhemispheric transfer. For example, researchers need to identify the specific Brodmann's areas that are involved in different types of interhemispheric transfer tasks. Additionally, researchers need to understand how the specific Brodmann's areas that are involved in interhemispheric transfer vary from person to person. Researchers are also developing new brain-computer interfaces that can be used to bypass damage to the corpus callosum and restore interhemispheric transfer. More research is needed to evaluate the safety and efficacy of these brain-computer interfaces. Overall, the research on interhemispheric transfer is a rapidly developing field with the potential to improve our understanding and treatment of cognitive impairments.

4.10.2 The Effect of Noise

ASMR and random noise appear to be beneficial for cognitive function, while periodic or repetitive sounds appear to be detrimental. Further research is needed to investigate the mechanisms underlying these effects. The symmetry of activated regions in the EEG responses in the ASMR and noise data reminds the book [12] that concludes by discussing the potential implications of noise in 3D clock assemblies for the future of artificial intelligence and computing. The formulation of brain as time crystal argues that use of noise as key search algorithm in an analog hardware could revolutionize the way we think about artificial intelligence and computing. Human mind is associated with linking various symmetries, here, the symmetry of thoughts and decisions are prime numbers.

4.10.2.1 Noise Allows a Brain to Choose from Multiple Options

The present study provides evidence that the role of noise in brain signal transmission is complex and depends on the type of noise and the cognitive task being performed. It is always argued that brain has more noise than signals, and it is a mystery how brain computes or makes a decision. The human brain is constantly bombarded with sensory information, both from the external world and from within the body itself. Much of this information is irrelevant or even distracting, and the brain must filter out this noise in order to focus on the important things. Neural circuits play a key role in this process of noise exploration [7]. These circuits are made up of interconnected neurons that work together to process information and generate responses.

Utilizing Noise: Unveiling the Brain's Unique Approach

In the realm of cognitive science, the brain's utilization of noise presents a fascinating avenue of exploration. There are two fundamental ways through which the brain harnesses noise, each offering profound insights into the intricacies of neural processing.

Exploring Two Possibilities: Filtering and Inhibiting the Choices

The first method involves the brain's ability to extract valuable information from a cacophony of signals. Noise, in this context, comprises a multitude of diverse signals, each representing a different potential choice or pathway. The brain, in its remarkable capacity, leverages this array of options to explore various possibilities. Unlike conventional electronic devices, where noise is typically considered a hindrance to signal clarity, the brain thrives amid noise. In the world of electronics, the concept of half-width full minima dictates that noise should not surpass 66% of the signal's amplitude. In essence, the signal must prevail over noise. However, the brain operates in an entirely opposite manner.

Within the neural circuits of the brain, noise serves as an ocean of choices and potentialities. The brain's circuits sift through this sea of options, selectively extracting the choices that align with its current objectives. One way that neural circuits explore noise is by using a process called inhibition [19]. Inhibition is a process by which neurons suppress the activity of other neurons. This can be used to filter out irrelevant or distracting information. When we are trying to focus on a conversation, our brain will inhibit the activity of neurons that are processing other sounds, such as the traffic outside or the music playing in the background. This allows us to focus on the important information and to ignore the noise.

This remarkable contrast underscores the brain's exceptional adaptability and dynamic decision-making process. In this chapter, we unveil the brain's proficiency in conducting mathematical operations within noisy environments. Moreover, we delve into the brain's ability to detect images amid the presence of noisy auditory stimuli. This paradigm shift suggests that, for the brain, noise is not interference but rather an integral component of the signal itself.

4.10.2.2 Noise Enables Error Correction in a Way that Filters Fewer Probable Choices

Error Correction Through Noise: Noise also plays a pivotal role in error correction within the brain's cognitive processes. Consider a scenario where the brain must make decisions among multiple potential choices. Typically, these decisions are interconnected, akin to the synchronized ticking of numerous clocks. In our exploration, we have discovered that many of the brain's decisions resemble recurring events, akin to the rhythmic motion of interconnected clocks. These intricate connections mirror the complex landscape of human thought and cognitive actions. For error correcting neural circuits explore noise is by using a process called adaptation [42]. Adaptation is a process by which neurons adjust their sensitivity to stimuli over time. This can be used to filter out noise that is constant or predictable. For example, when we first put on a new pair of glasses, we may notice a strange feeling around your nose and ears. This is because your brain is adapting to the new visual input. After a while, however, your brain will adapt to the new glasses and the feeling will go away.

Intuitive Expectation; Generating Signals Before Receiving Relevant Information: When the brain is exposed to noise, an intriguing phenomenon unfolds. Noise acts as a bridge, forging connections among multiple choices in novel and intricate patterns. Consequently, the brain gains the capacity to simultaneously test a multitude of highly correlated collective decisions. In essence, noise serves as the catalyst for interweaving these decisions, enabling the brain to navigate a web of possibilities with agility and precision. This phenomenon sheds light on the brain's exceptional capacity for parallel processing and offers a fresh perspective on how it navigates the complex landscape of decision-making. Neural circuits also use a process called expectation to explore noise [53]. Expectation is a process by which the brain predicts what it expects to see or hear next. This can be used to filter out noise that is unexpected or surprising. When we are watching a movie, our brain will predict what it expects to happen next based on the plot and the characters. This allows you to ignore the noise of the popcorn machine or the people talking in the theater. Neural circuits explore noise in a variety of ways, and the specific mechanisms involved are still being studied. However, it is clear that neural circuits play a vital role in helping us to focus on the important things and to ignore the noise.

In conclusion, the brain's unique relationship with noise challenges conventional wisdom, offering a glimpse into the intricacies of neural processing. It underscores the brain's unparalleled ability to harness noise as a tool for exploration, decision-making, and error correction—a testament to the remarkable capabilities of the human mind.

References

1. Arifah M, Habibi M, Dewantoro M, Adawiyah S, Haningsih S (2022) Platform on online learning for students with mental retardation. Kne Social Sci:236–247. https://doi.org/10.18502/kss.v7i10.11362
2. Anderson JR (1982) Acquisition of cognitive skill. Psychol Rev 89(4):369–406
3. Atkinson RK, Derry SJ, Renkl A, Wortham D (2000) Learning from example demonstrations: a generative perspective. J Educ Psychol 92(1):253
4. Agrawal L, Sahu S, Ghosh S, Shiga T, Fujita D, Bandyopadhyay A (2016) Inventing atomic resolution scanning dielectric microscopy to see a single protein complex operation live at resonance in a neuron without touching or adulterating the cell. J Integr Neurosci 15(04):435–462
5. Agrawal L, Chhajed R, Ghosh S, Ghosh B, Ray K, Sahu S, Fujita D (2018) Fractal information theory (FIT)-derived geometric musical language (GML) for brain-inspired hypercomputing. Soft Comput Theories Appl 548:343–372
6. Adcock J, Friston KJ, Brebion G, Sachs M, Henson RNA (2000) When bottom-up meets top-down: auditory perception and attention to speech and non-speech sounds. Neuroimage 12(4):289–303
7. Bao S, Merzenich MM (2005) Cortical plastic mechanisms for better sound separation. Proc Natl Acad Sci 102(52):18910–18915

8. Buchanan TW, Al-Kassimi A, Zatorre RJ, Penhune VB (2012) The neural processing of periodic auditory patterns. PLoS ONE 7(12):e52850
9. Bregman AS (1990) Auditory scene analysis: the perceptual organization of sound. MIT Press
10. Bandyopadhyay A, Miki K, Wakayama Y (2006) Writing and erasing information in multilevel logic systems of a single molecule using scanning tunneling microscope. Appl Phys Lett 89(24)
11. Bauer J, Harteis C (eds) (2012) Human fallibility. The ambiguity of errors for work and learning. Springer, Dordrecht
12. Bandyopadhyay A (2020) Nanobrain: the making of an artificial brain from a time crystal. Boca Raton CRC Press, Taylor and Francis. ISBN: 9781439875490. https://doi.org/10.1201/9780429107771
13. Barratt E, Davis N (2015) Autonomous sensory meridian response (asmr): a flow-like mental state. PeerJ 3:e851. https://doi.org/10.7717/peerj.851
14. Barratt E, Spence C, Davis N (2017) Sensory determinants of the autonomous sensory meridian response (asmr): understanding the triggers. PeerJ 5:e3846. https://doi.org/10.7717/peerj.3846
15. Brinkmann R et al (1975) Scaling of disturbances of attention in neurological patients (author's transl)]. J Neurol 209(1):1–8
16. Byrne JH (2017) Learning and memory: a comprehensive reference. https://doi.org/10.1016/c2015-1-01759-8
17. Cook N, Hattie J (2008) The relationship between error correction and student achievement. Educ Psychol Rev 20(4):377–396
18. Collins AGE, Frank MJ (2012) How much of reinforcement learning is working memory, not reinforcement learning? A behavioral, computational, and neurogenetic analysis: working memory in reinforcement learning. Eur J Neurosci 35(7):1024–1035. Pmid:22487033
19. Desimone R, Duncan J (1995) Neural mechanisms of selective visual attention. Annu Rev Neurosci 18(1):193–222
20. Dunlosky J, Metcalfe J (2009) Metacognition and learning: current state and future directions. Curr Dir Psychol Sci 18(3):169–175
21. Das A (2022) Role of educational psychology in the study of memory, conceptual processes, and individual differences (via cognitive psychology) in conceptualizing new strategies for learning processes in humans. Technoarete Trans Appl Inform Commun Technol (ICT) Educ 1(1):12–16. https://doi.org/10.36647/ttaicte/01.01.a003
22. Engelbregt H, Brinkman K, Geest C, Irrmischer M, Deijen J (2022) The effects of autonomous sensory meridian response (asmr) on mood, attention, heart rate, skin conductance and EEG in healthy young adults. Exp Brain Res 240(6):1727–1742. https://doi.org/10.1007/s00221-022-06377-9
23. Fredborg B, Clark J, Smith S (2018) Mindfulness and autonomous sensory meridian response (asmr). PeerJ 6:e5414. https://doi.org/10.7717/peerj.5414
24. Fiori C, Zuccheri L (2005) An experimental research on error patterns in written subtraction. Educ Stud Math 60:323–331. https://doi.org/10.1007/s10649-005-7530-6
25. Fine I, Finney EM, Boynton GM, Dobkins KR (2005) Comparing the effects of auditory deprivation and sign language within the auditory and visual cortex. J Cogn Neurosci 17(10):1621–1637. https://doi.org/10.1162/089892905774597173. PMID: 16269101
26. Gartmeier M, Schüttelkopf EM (2012) Tracing outcomes of learning from errors on the level of knowledge. In Bauer J, Harteis C (eds) Human fallibility. The ambiguity of errors for work and learning. Springer, Dordrecht, pp 33–51
27. Ghosh S, Sahu S, Fujita D, Bandyopadhyay A (2014) Design and operation of a brain like computer: a new class of frequency-fractal computing using wireless communication in a supramolecular organic, inorganic systems. Information, 5, 28–99
28. Ghosh S, Fujita D, Bandyopadhyay A (2015) An organic jelly made fractal logic gate with an infinite truth table. Sci Rep 5(1):11265
29. Ghosh S, Sahu S, Agrawal L, Shiga T, Bandyopadhyay A (2016) Inventing a co-axial atomic resolution patch clamp to study a single resonating protein complex and ultra-low power communication deep inside a living neuron cell. J Integr Neurosci 15(04):403–433

30. Gazzaniga MS, Bogen JE, Sperry RW (1967) Dysprosody in a patient with a sectioned corpus callosum. Neurology 17(8):676–685
31. Geschwind N, Galaburda AM (1985) Cerebral lateralization: biological mechanisms, associations, and pathology. The MIT Press
32. Hansmann R (2010) "Sustainability learning": an introduction to the concept and its motivational aspects. Sustainability 2(9):2873–2897. https://doi.org/10.3390/su2092873
33. Hellige JB (1993) Interhemispheric transfer: a neuropsychological perspective. Oxford University Press
34. Harteis C, Buschmeyer K (2012) Learning from mistakes—still a challenge for research and business practice. In: Wuttke E, Seifried J (eds), Learning from errors at school and at work. Vorlag Barbara Budrich, Opladen, Berlin & Farmington Hills, MI, pp 31–47
35. Heimbeck D, Frese M, Sonnentag S, Keith N (2003) Integrating errors into the training process: the function of error management instructions and the role of goal orientation. Pers Psychol 56(2):333–361
36. Hickman DJ, Hattie J (2008) Feedback for better learning: Understanding the impact of feedback on student achievement. In: Littlewood TDD, O'Donnell TMA (eds) Learning and teaching. John Wiley & Sons, pp 91–132
37. Hostler TJ, Poerio GL, Blakey E (2019) Still more than a feeling: commentary on Cash et al., "Expectancy effects in the autonomous sensory meridian response" and recommendations for measurement in future ASMR research. Multisens Res 32(6):521–531. https://doi.org/10.1163/22134808-20191366. PMID: 31128572
38. Haya P, Daems O, Malzahn N, Castellanos J, Hoppe H (2015) Analysing content and patterns of interaction for improving the learning design of networked learning environments. Br J Edu Technol 46(2):300–316. https://doi.org/10.1111/bjet.12264
39. Jung-Beeman M (2005) Bilateral brain processing for comprehension and problem-solving: evidence from fMRI. Psychol Sci 16(10):706–711
40. Kerzel M, Pekarek-Rosin T, Strahl E, Heinrich S, Wermter S (2020) Teaching nico how to grasp: an empirical study on crossmodal social interaction as a key factor for robots learning from humans. Front Neurorobotics 14. https://doi.org/10.3389/fnbot.2020.00028
41. Krismadinata K, Verawardina U, Jalinus N, Rizal F, Sukardi S, Sudira P, Novaliendry D (2020) Blended learning as instructional model in vocational education: literature review. Univ J Educ Res 8(11B):5801–5815. https://doi.org/10.13189/ujer.2020.082214
42. Knudsen EI, Brainard MS (1991) Visual instruction of sound localization. Science 253(5015):462–464
43. Langer E (1997) The power of mindful learning. Choice Reviews Online 35(2):35-1031–35-1031. https://doi.org/10.5860/choice.35-1031
44. Levenson RW (2003) Blood, sweat, and fears: the autonomic nervous system and emotions. Psychol Sci 14(5):431–438
45. Le Pelley ME, Mitchell CJ, Beesley T, George DN, Wills AJ (2016) Attention and associative learning in humans: an integrative review. Psychol Bull 142(10):1111–1140. pmid:27504933\
46. McKenney S, Mor Y (2015) Supporting teachers in data-informed educational design. Br J Edu Technol 46(2):265–279. https://doi.org/10.1111/bjet.12262
47. McErlean A, Osborne-Ford E (2020) Increased absorption in autonomous sensory meridian response. PeerJ 8:e8588. https://doi.org/10.7717/peerj.8588
48. Neville H, Bavelier D (2002) Human brain plasticity: evidence from sensory deprivation and altered language experience. Prog Brain Res 138:177–188. https://doi.org/10.1016/S0079-6123(02)38078-6. PMID: 12432770
49. Novaliendry D, Huda A, Cuhanazriansyah M, Sani H, Hendra H, Karnando J (2021) E-learning based web programming course in the Covid 19 pandemic time. Int J Interact Mobile Technol (IJIM) 15(20):117. https://doi.org/10.3991/ijim.v15i20.23749
50. Prince M (2004) Does active learning work? a review of the research. J Eng Educ 93(3):223–231. https://doi.org/10.1002/j.2168-9830.2004.tb00809.x
51. Pradhan B, Bandyopadhyay A, Pal AJ (2004) Molecular level control of donor/acceptor heterostructures in organic photovoltaic devices. Appl Phys Lett 85(4):663–665

52. Reddy S, Sonker D, Singh P, Saxena K, Singh S, Chhajed R et al (2018) A brain-like computer made of time crystal: could a metric of prime alone replace a user and alleviate programming forever? Soft Comput Appl:1–43
53. Rees G, Frith CD (2000) The neural correlates of attention and consciousness in humans. Nat Rev Neurosci 1(1):176–186
54. Schmid RF, Bjork RA (1992) Knowledge is not enough: the role of deliberate effort in education. Am Psychol 47(4):422
55. Sakurai N, Ohno K, Kasai S, Nagasaka K, Onishi H, Kodama N (2021) Induction of relaxation by autonomous sensory meridian response. Front Behav Neurosci 15. https://doi.org/10.3389/fnbeh.2021.761621
56. Smith S, Fredborg B, Kornelsen J (2019) A functional magnetic resonance imaging investigation of the autonomous sensory meridian response. PeerJ 7:e7122. https://doi.org/10.7717/peerj.7122
57. Saxena K, Singh P, Sahoo P, Ghosh S, Krishnananda D, Ray K, Fujita D, Bandyopadhyay A (2022) All basics that are wrong with the current concept of time crystal: learning from the polyatomic time crystals of protein, microtubule, and neuron. Proc Trends Electron Health Inform 376:243–254
58. Saxena K, Karthik KV, Kumar S, Fujita D, Bandyopadhyay A (2019) Wireless communication through microtubule analogue device: noise-driven machines in the bio-systems. Eng Vibr Commun Inform Process 478:735–749
59. Saxena K, Singh P, Sahoo P, Sahu S, Ghosh S, Ray K, Fujita D et al (2020) Fractal, scale free electromagnetic resonance of a single brain extracted microtubule nanowire, a single tubulin protein and a single neuron. Fractal Fract 4(2):11
60. Singh P, Saxena K, Singhania A, Sahoo P, Ghosh S, Chhajed R, Fujita D et al (2020) A self-operating time crystal model of the human brain: Can we replace entire brain hardware with a 3D fractal architecture of clocks alone? Information 11(5):238
61. Singh P, Ray K, Bandyopadhyay A (2022) Biological antenna to the humanoid bot: electromagnetic resonances in biomaterials. Studies in rhythm engineering (SRE). Springer Singapore. https://doi.org/10.1007/978-981-16-9677-0
62. Singh P, Sahoo P, Saxena K, Ghosh S, Sahu S, Ray K, Fujita D, Bandyopadhyay A (2021) Quaternion, octonion to dodecanion manifold: stereographic projections from infinity lead to a self-operating mathematical universe. Proc Int Conf Trends Comput Cogn Eng 1169:55–77
63. Singh P, Sahoo P, Saxena K, Ghosh S, Sahu S, Ray K, Fujita D, Bandyopadhyay A (2021) A space-time-topology-prime, stTS metric for a self-operating mathematical universe uses Dodecanion geometric algebra of 2–20 D complex vectors. Proc Int Conf Data Sci Appl 148:1–31
64. Singh P, Sahoo P, Saxena K, Manna JS, Ray K, Ghosh S, Fujita D et al (2021) Cytoskeletal filaments deep inside a neuron are not silent: they regulate the precise timing of nerve spikes using a pair of vortices. Symmetry 13(5):821
65. Serrien D, Brown P (2002) The functional role of interhemispheric synchronization in the control of bimanual timing tasks. Exp Brain Res 147(2):268–272. https://doi.org/10.1007/s00221-002-1253-z
66. Simon-Dack S, Holtgraves T, Hernandez K, Thomas C (2015) Resting EEG and behavioural correlates of interhemispheric transfer times. Laterality Asymm Body Brain Cogn 20(5):618–638. https://doi.org/10.1080/1357650x.2015.1032302
67. Treue S, Martínez Trujillo JC (1999) Feature-based attention influences motion processing in macaque visual cortex. Nature 399(6736):575–579
68. Thut G, Hauert CA, Morand S, Seeck M, Landis T, Michel C (1999) Evidence for interhemispheric motor-level transfer in a simple reaction time task: an EEG study. Exp Brain Res 128:256–261
69. Thomaz A, Breazeal C (2008) Teachable robots: understanding human teaching behavior to build more effective robot learners. Artif Intell 172(6–7):716–737. https://doi.org/10.1016/j.artint.2007.09.009

70. Wang W, Chen Y, Li R, Jia Y (2019) Learning and comfort in human–robot interaction: a review. Appl Sci 9(23):5152. https://doi.org/10.3390/app9235152
71. Witelson SF, Gazzaniga MS (1974) Lateralization of language function in the human brain. Acta Neurobiol Exp 34(1):3–31

Chapter 5
DDG, an Electromagnetic Version of EEG Finds Evidence of a Self-operating Mathematical Universe (SOMU) When a Human Subject Converses with an Artificial Brain

5.1 Introduction

5.1.1 The Missing Part of Consciousness Theories

Existing theories of consciousness [7, 8, 30, 32, 39, 51, 64, 91] use different variants of bits/qubits as the unit of the brain's information and variants of the Turing machine as integrators of information; both require a human instructor. Intuitively created astronomical instructions could mimic human intelligence, but that is not human consciousness; an algorithm enslaves a machine. Similar to all theories of consciousness, the electromagnetic field theory of consciousness is also controversial, and the reasons are plenty. Though most consciousness theories bluntly put that the brain is not a Turing machine, no proposal admits Turing's alternative requires a tape whose every cell should be undefined [3, 5, 34, 36]. The processing of any machine with a defined state could be emulated using a Turing machine. While reading out coded instructions is still the hallmark of artificial intelligence, no proposal argues for making decisions with null human intervention and null computation. No one builds a self-operating model of nature or the brain with a bias-free endless resource of code. No one even argues for inventing one such model [100]. Nevertheless, discussions continue for machine consciousness [29] and building artificial brains [54].

The recently proposed Self-Operating Mathematical Universe (SOMU, [84, 85]) addresses a prime concern of consciousness theories as it replaces the human instructor with an automaton that can derive all events from an infinite source of code, Fig. 5.1a, b. Furthermore, instead of writing events using bits or qubits, considers prime numbers as symmetries of geometric shapes [96] embedded in recurring spatio-temporal events happening in nature. SOMU goes beyond the 6D spatio-temporal pattern [84, 85], resembling Pockett's argument that includes all types of fields. Then explores ways to integrate a bias-free universal code of prime numbers that regulates the transformation of geometric shapes, forming an automaton that relocates its

T. Dutta and A. Bandyopadhyay, *Emotion, Cognition and Silent Communication: Unsolved Mysteries*, Studies in Rhythm Engineering,
https://doi.org/10.1007/978-981-99-9334-5_5

system point anywhere outside its physical boundary to evaluate itself from outside (self-awareness) [96].

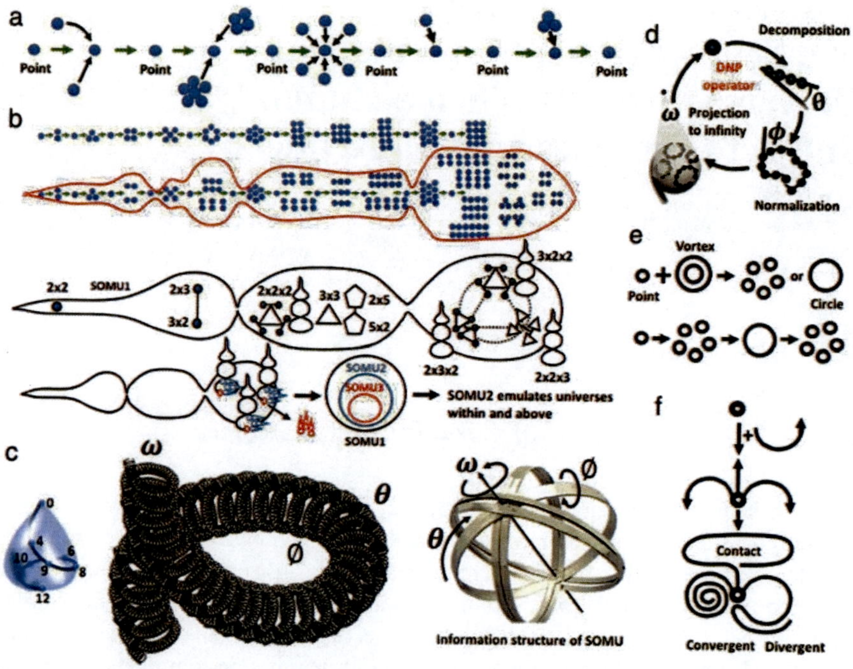

Fig. 5.1 How to create a self-operating mathematical universe, SOMU: **a** The point remains a point even after adding an arbitrary number of points to a point [41]. **b** Adding 12 points one by one is essential and sufficient to synthesize SOMU. Ordered factors of 12 integers build a triplet. When points form a closed asymmetric or distinguishable loop at 12, that point starts incorporating new points. When we get three triplets of triplets one inside another, it is equivalent to three SOMUs one inside another. The central SOMU (U_j) emulates the universe inside U_i and the universe outside U_k at a time. **c** 12 points make the first 360° rotation: the first pitch of a helix (left). Three nested geometric phases represent the triplet structure of SOMU in panel B (middle). Unit of information and vortex condensate synthesis (right). Three angular momentums of the vortex complete the journey of three angular invariants from a point to a sphere that resembles a point for the next evolution. **d** Many point structures are generated by decomposition, rearranged to form a loop or normalized. Several normalized loops are placed on a 3D surface to project all loops to infinity, i.e., the special sphere acts as a point. It is the DNP operator; DNP stands for decomposition, normalization and projection. **e** Circle is a point, and the sphere is a 3D point in SOMU. A vortex interacts with a point to create a circle (larger diameter vortex) that acts as a point and interacts with a larger vortex to form a larger circle. In this way, a within-and-above network of vortices forms. **f** How does a point forms a single circle? The rotational *density of prime* DNP operator applied to a point could have three orthogonal motions, clockwise rotation creating a divergent loop; anti-clockwise rotation creating a convergent loop and finally, an ellipsoidal loop that makes perfect contact. Ellipsoidal loop forms to balance clockwise and anti-clockwise emissions from the point

5.1.2 *The Quest for an Automaton that Never Repeats Code and Builds a Conscious Engine C Under a Certain Condition*

5.1.2.1 What Experiment Prompted Us to Find a New Theory of Consciousness

There is no lack of consciousness theories, often in the consciousness conference it is said, every researcher comes up with a unique proposal. So far, in the last four chapters we have described how quantization of time plays a key role in cognitive decision-making. In Chap. 2, we have discussed quantization of timing for lateralization, it means energy transfer time between the left brain and the right brain is not continuous. There are restricted time domains and favorable time domains for brain hardware [46]. We think this is a major discovery in the brain science, and we have to change the way we analyze mind and human psychology. Imagine, in our society, we all humans are bounded by restricted and favorable time zones, we do not notice them explicitly but they regulate everything from the sub-conscious states.

In Chap. 3, we have found once again classification of human subjects. Imagine different individuals have a very different kind of emotion, perception affecting their cognitive responses. The blood vessels flowing through the face are fundamentally different, how different parts of the brain-body system would heat up or cool down in different humans for different emotions or cognitive experiences is very different. When the grammar of emotion expression changes so fundamentally, one could guess how different would it be in a group of people interacting with each other. It may be that people who follows different hardware and a very different grammar for expressing a particular emotion, follow a hierarchical language [49].

In Chap. 4, not just that noise is very helpful in acquiring geometric information in the brain, the signal transmission pathways for cognitive tasks are not very well defined. In a unique set of mathematical tasks and pattern recognition tasks we understood that finding a common rule that operates mind is not possible so easily. Observation-based fitting approaches to learn consciousness would not lead us anywhere.

Therefore, all three previous chapters argue for classification of humans, redefining the information structure for the brain and include the engineering of discrete times or pieces of times just like atoms have discrete orbitals and atoms are not allowed have any orbitals in between [46]. In addition to the behavioral quantization, we would see below, that our decade old research also has questioned the linearity and non-linearity of brain hardware. Fractal or self-similar features in different layers of thoughts are felt by us [17–19]. However, any hypothesis for mind needs to be tested, since most of the brain and cognitive research are based on subjective arguments.

5.1.2.2 From Behavioral to Molecular Neuroscience: Selective Views Have Become Mainstream

When we take the brain to a transforming engineering of discrete time pieces, the idea of space–time changes forever. The polyatomic time crystal [102] model, proposed by Reddy et al. [66] and Singh et al. [87], offers a quantized time perspective on brain function. It suggests that cognitive, perceptual, and emotional responses and feelings can be quantified by examining recurring events that lead to these experiences. Each recurring event serves as a clock as we draw a loop on the surface of a phase sphere, with their origins scattered throughout the brain's fractal hardware. Since brain has structural self-similarities all across its hardware, it seems engineering of time would follow a set of universal rules at different spatial scale. One important feature of this brain model [9] (Chap. 7) is that the clocks grow within and above to represent the true fractal nature of the brain hardware.

The origin of these clocks could begin at the molecular or at the nanoscale. Intriguingly, the study [62] underscores the vital role of membrane microtubule junctions in neurons, emphasizing the importance of heterojunctions in signal transmission within the brain's fractal structure. Imagine a new geometric phase is generated at all junctions from the smallest molecular scale to the largest scale [31, 42, 94]. Geometric phase is a physics concept, it is a good tool for exploring the hardware properties of the brain to understand the mind. This phase structure exhibits scale-free electromagnetic resonance within a single microtubule nanowire extracted from the brain, a single tubulin protein, and a single neuron, revealing the existence of a triplet of triplet resonance bands or nine distinct frequency/time domains operating in harmony [75]. This triplet of triplet resonance has profound implications, even at the molecular level, demonstrating the symmetry of primes and suggesting the potential for quantum tunneling in information processing [6]. Additionally, the idea that a fractal network could efficiently process information logically has been proposed [35], highlighting the fractal nature of the mind.

The temporal organization of the mind also exhibits a fractal nature. Research by Agrawal et al. [3, 5] involving simultaneous measurements of neuron firing at the neuron membrane and within microtubular bundles reveals that microtubules can process firing signals at speeds approximately 10,000 times faster than the membrane. This challenges the notion that microtubules play a passive role in the brain's information processing [83–85]. Interestingly, even non-connected neurons, without direct synaptic junctions, can exchange energy, as demonstrated by imaging neural circuits using a scanning dielectric microscope (SDM) [33, 38]. This imaging resembles mapping facial expressions and emotional states, akin to SDM or a thermal camera. Now we need to invent a machine that operates like EEG but measures discrete time structures of brain operations in a wide time scales at once. We have named it Dodecanogram, because it operates in twelve time domains simultaneously.

5.2 Invention of DDG as an Advanced Form of EEG

5.2.1 *Years of EEG: What Went Wrong?*

Electroencephalography (EEG) is a non-invasive technique that records electrical activity in the brain through electrodes placed on the scalp. It has been instrumental in various fields, from cognitive neuroscience to clinical diagnosis and brain-computer interfaces. However, as with any technology, EEG has its limitations and challenges. In this section, we'll explore ten fundamental problems that researchers and clinicians face when using EEG and delve into the ongoing efforts to mitigate these issues.

Problem 1: Signal Noise and Artifacts: EEG signals are susceptible to various types of noise and artifacts, stemming from sources like muscle activity, eye movements, and environmental interference. Researchers have proposed several strategies to address these issues. One of the main challenges with EEG measurements is the presence of artifacts. Artifacts can arise from various sources, including muscle activity, eye movements, and environmental interference. These artifacts can distort the EEG signal and lead to inaccurate interpretations [11]. Preprocessing techniques, such as artifact removal, are necessary to enhance the quality of the raw EEG signals and improve the accuracy of the analysis [11, 90]. Getting a good ERP requires a major significant challenge [53]. One weakness of EEG technology is its susceptibility to interference from physiological sources. For example, ocular artifacts, which are caused by eye movements and blinks, can contaminate the EEG signal and affect the interpretation of results [26]. Advanced techniques, such as independent component analysis (ICA), can be used to correct for ocular artifacts and improve the accuracy of EEG data analysis [26, 28].

Problem 2: Limited Spatial Resolution and Beamforming: EEG provides relatively low spatial resolution due to the dispersion and attenuation of electrical potentials as they travel through the scalp and skull. Efforts to enhance spatial resolution include high-density EEG arrays and advanced source localization methods, discuss the theoretical limits of spatial resolution in EEG beamforming [95]. Beamforming is a technique that can be used to improve the spatial resolution of EEG data by focusing on the electrical activity from a specific brain region. The work provides a theoretical framework for understanding the limits of spatial resolution in EEG, and it offers valuable insights into how to improve spatial resolution in practice. The spatial resolution of EEG beamforming is limited by the number of electrodes used and the wavelength of the brain activity being measured. However, noise can significantly reduce the spatial resolution of beamforming.

Problem 3: Scalp-Skull Conductivity Variability: The Challenges of Source Localization: Individual variability in scalp-skull conductivity can impact the accuracy of EEG source localization. A study by Acar and Makeig [1] investigated the effects of using individualized head models and conductivity estimation techniques on the accuracy of EEG source localization. Individualized head models take into account the individual differences in skull anatomy and conductivity, which can improve the accuracy of source localization. This is important because accurate source localization is essential for understanding the neural basis of brain function and for developing new treatments for neurological disorders. The physical properties of the skull and brain can also impact the accuracy of EEG measurements. The skull, dura, and cerebrospinal fluid between the brain and the EEG electrodes attenuate the electrical signal and act as temporal low-pass filters, limiting the useful bandwidth of the EEG signal [89]. Additionally, the presence of skull sutures and variations in skull conductivities can affect the accuracy of EEG forward and inverse problem solutions [56].

Skull modeling has become now an integral part of EEG research

(1) Local models: If compact and spongy bone can be identified with sufficient accuracy (e.g., from MRI) and their conductivities can be assumed to be known (e.g., from measurements), one should model these explicitly by assigning each voxel to one of the two conductivities.
(2) Global models: If the conditions of (1) are not met, one should model the skull as either homogeneous and isotropic, but with considerably higher skull conductivity than the usual 0.0042 S/m, or as homogeneous and anisotropic, but with higher radial skull conductivity than the usual 0.0042 S/m and a considerably lower radial:tangential conductivity anisotropy than the usual 1:10.

Problem 4: Limited Depth Penetration: EEG primarily captures surface brain activity, making it less suitable for investigating deep brain structures. The study by Laufs et al. [47] was one of the first to demonstrate the potential of concurrent EEG-fMRI for studying brain function. Precisely it showed that the strength of alpha oscillations was correlated with the activity of these brain regions in the fMRI images. Since then, this technique has been used to study a wide range of brain processes, including cognition, emotion, and motor control.

Problem 5: Lack of Standardization: Lack of standardized data collection and analysis protocols in EEG research has led to variability in results. Efforts to establish best practices and guidelines, such as those discussed by Picton et al. [58], for using human event-related potentials (ERPs) in studies of cognition. ERPs are a type of EEG signal that is associated with specific cognitive events, such as stimulus processing, decision-making, and motor response. The guidelines cover a wide range of topics for best practices, including Experimental design and data collection; Data analysis and interpretation; Statistical analysis and reporting. The guidelines are designed to help researchers improve the reliability and reproducibility of their ERP studies.

Problem 6: Signal Crosstalk and Volume Conduction: Signal crosstalk occurs when EEG electrodes pick up activity from neighboring brain regions. Seminal research by Pascual-Marqui et al. [59] in Electroencephalography and Clinical Neurophysiology elucidates techniques for source localization and minimizing crosstalk. This paper introduces a new method for EEG source localization called low-resolution electromagnetic tomography (LORETA). LORETA uses the EEG data to reconstruct the electrical activity of the brain by minimizing crosstalk during source localization. When the EEG signal from one brain region is picked up by electrodes that are located over another brain region, they crosstalk during automated source localization. Most EEG devices use it now. It is used in a variety of research and clinical applications, including cognitive neuroscience, neuroimaging, and epilepsy diagnosis.

Problem 7: Sensitivity to Movement and Muscle Activity: Artifacts from head movement and muscle activity can significantly affect EEG data quality. Jung et al. [45] discusses the application of independent component analysis (ICA) for artifact correction in EEG research, contributing to the mitigation of movement-related challenges, including artifacts caused by eye movements, muscle activity, and environmental interference. ICA is a technique that can be used to decompose EEG data into a set of independent components, each of which represents a distinct source of electrical activity in the brain. They also found that ICA could be used to improve the signal-to-noise ratio of EEG data. It was [45] one of the first to demonstrate the potential of ICA for EEG artifact correction. ICA is now an essential tool for EEG researchers and clinicians. It allows researchers to remove artifacts from EEG data, which improves the quality of the data and makes it more reliable and reproducible. ICA also allows clinicians to identify and track EEG signals associated with specific brain disorders, such as epilepsy and Parkinson's disease.

Problem 8: Limited Frequency Range: Traditional EEG captures a limited frequency range of brain activity. However, studies employing advanced EEG techniques, such as high-frequency oscillations, HFO as presented by Worrell et al. [99], specially for epilepsy, wherein they have expanded the range of detectable frequencies above 80 Hz. They are thought to play a role in the initiation and propagation of epileptic seizures. The authors of this paper discuss the different types of HFOs and the brain regions where they are found. They also discuss the role of HFOs in the diagnosis and treatment of epilepsy by localizing the epileptic foci. HFOs can be used to guide surgical resection of epileptic tissue.

Problem 9: Interpretation and Inference Complexity: Interpreting EEG data and making inferences about cognitive or clinical phenomena can be highly complex. Limitations in the interpretation of EEG data. EEG interpretation relies on the expertise of the interpreter, and there can be variability in the reliability and confidence of interpretations [37]. This variability can introduce inaccuracies and affect the validity of EEG-based findings [37]. Efforts to improve the reliability of EEG interpretation, such as interrater reliability studies, are important for ensuring the accuracy of EEG-based diagnoses and assessments [37].

Analyzing EEG data requires the detection and removal of artifacts to ensure accurate interpretation of the underlying brain signals. Higher-order statistics (HOS) have been shown to be effective in enhancing artifact detection in EEG data [67]. This article aims to explore the use of HOS in analyzing EEG brain scan data and its potential applications in neuroscience research. Delorme and Makeig [27] proposed a method for artifact detection in EEG data using HOS and independent component analysis (ICA) [27]. HOS capture the nonlinear characteristics of the EEG signals, allowing for the identification of artifacts that may not be detected using traditional methods. By combining HOS with ICA, it is possible to separate the artifact components from the underlying brain signals, improving the quality of the EEG data for further analysis [27]. Mannan et al. [55] proposed a hybrid framework that combines ICA, regression, and HOS to automatically identify and eliminate ocular artifacts from EEG data [55]. This approach can improve the accuracy of EEG-based studies by removing artifacts that may confound the interpretation of the brain signals.

HOS can also be used to analyze the nature of events and event context in EEG data. Robbins et al. [68] introduced hierarchical event descriptors (HED), a method that captures the nature of events and their context using HOS [27]. By analyzing the higher-order statistical moments of EEG signals, HED provides a comprehensive representation of events, allowing for a more detailed analysis of brain activity. Furthermore, HOS can be used to study the impact of biological rhythms on EEG signals. Lehnertz et al. [48] discussed the influence of biological rhythms on higher-order statistical moments and interaction properties of multichannel EEG signals [48]. By analyzing the temporal changes in these characteristics, researchers can gain insights into the dynamics of brain activity and its relationship to biological rhythms.

Problem 10: Wrong Data Capture Protocol: What Do We Actually Measure? MEG? There are inherent inaccuracies in the measurement of EEG indices, it has not fundamentally changed since Richard E. Catons work in 1975. The measurement process itself and the need to remove recorded disturbances can introduce inaccuracies in the data [12]. These inaccuracies can affect the reliability and validity of EEG-based indices and metrics [12]. Often MEG is measured as EEG [57], differential signals between probes are captured which means nothing for a real biological brain. Sharp changes are filtered out for no reason, hardly researches are there to find their origin. Optimization with preassumption that particular behaviors would not be biological, that kind of intuitive editing of data before we feed them to machine learning algorithm is a crime and EEG machine software does that to make theoretical analysists happy.

Electroencephalography has transformed our understanding of the brain and its role in cognition and disease. However, addressing the fundamental challenges discussed in this article is crucial for harnessing the full potential of EEG. Researchers and clinicians continue to explore innovative solutions, pushing the boundaries of EEG technology and its applications. As EEG technology evolves, it promises to offer even deeper insights into the mysteries of the human brain.

5.2.2 *Why Do We Need a New Instrument for Understanding Human Psychology?*

The human mind has long been a subject of fascination and scientific inquiry. However, our current tools for understanding the complexities of thought, emotion, perception, and cognition are limited. This article emphasizes the urgent need for an advanced brain-scanning instrument that can decode human thoughts, measure cognitive functions, intuition, and emotional states with mathematical precision.

One of the key challenges in brain science is the absence of a definitive marker for psychological and cognitive functions [79]. Existing instruments like EEG, PET, and fMRI provide valuable insights but are indirect and lack a mathematical structure for quantifying cognitive functions, emotional states, and perceptions. This limitation hinders the progress of brain science as a proper scientific discipline [79]. Develop markers for psychological and cognitive functions. Currently, we have no objective way to measure these functions. This is why abstract and metaphoric words are used in all types of research wherever one deals with the human mind. A new brain-scanning instrument could provide us with the markers we need to develop a more rigorous and quantitative science of the brain [61].

Another limitation of current brain-scanning instruments is their focus on a specific time domain, providing only a snapshot of brain activity. To overcome this limitation, there is a pressing need for a technology that can map brain activity across different time domains simultaneously. This would enable a more comprehensive understanding of the dynamic nature of the human mind. A new brain-scanning instrument that could map brain activity in multiple time domains at a time, revealing the hidden language of brain and nature.

Furthermore, existing brain-scanning methods, such as EEG, PET, and fMRI, primarily examine surface-level changes and do not explore the multimodal expression of thoughts [98]. This leaves unanswered questions about how thoughts are generated in the brain and in what form they are expressed [98]. To address this gap in knowledge, an advanced brain-scanning instrument is required that can directly decode thoughts and explore the full range of brain activity.

Additionally, while neuron firing is commonly believed to be responsible for human thoughts, emotions, perceptions, and cognitions, there is a need to explore other components of the brain [78]. Glial cells, for example, have been found to play a significant role in brain function. However, current tools do not allow for a comprehensive investigation of these alternative possibilities [78].

In conclusion, the development of an advanced brain-reading instrument is not just a scientific aspiration but an essential necessity. By overcoming the limitations of current instruments, this revolutionary technology has the potential to unlock the mysteries of the human mind, redefine our understanding of cognition and emotion, and pave the way for new frontiers in neuroscience.

5.2.3 *How DDG Would Change Everything Technologically?*

Measuring current or potential on the brain surface has been a common practice in brain research for the past 150 years [14]. However, the new invention being reported in this book introduces a novel approach to brain scanning. This invention utilizes a machine with probes placed on the brain surface that acquire signals using short nanosecond or picosecond pulses. By measuring the duration of these short bursts of potential, the researchers aim to understand how different time pieces are correlated and how they form patterns. This approach challenges the traditional concept of electroencephalography (EEG), which focuses on surface currents on the scalp and measures activity in the millisecond time range [14].

The researchers behind this invention argue that surface currents or potential variations measured on the scalp are minor currents that are part of a larger global design of the brain's signal processing [14]. They propose that the bursts of activity they are measuring on the brain surface, occurring in the nanosecond or picosecond time range, are more significant in understanding brain function. These bursts of activity, when occurring in a large number and within a million-order time range, may represent a projection of some phenomenon deep inside the brain.

This new brain-scanning system, called the DDG, operates in two different modes: lock-in mode measurement and radiation mode. In lock-in mode measurement, the focus is on the short bursts of activity on the brain surface and how they are correlated and form patterns. In radiation mode, the system also measures the energy coming out of the brain surface. This dual-mode operation of the DDG provides a new perspective on the brain scalp and challenges previous assumptions about the importance of surface currents or potential variations [14].

The invention described in this book aligns with recent advancements in brain research that have explored the role of oscillatory activity and burst-like patterns in brain function. Studies have shown that rhythmic transcranial magnetic stimulation (TMS) can entrain natural oscillatory signatures in the brain [92]. Additionally, burst suppression, a state of profound brain inactivation and unconsciousness, has been studied in various clinical contexts, such as anesthesia and coma (Lewis et al. [50]; Liley and Walsh [52]). Burst-like activity has also been observed in cortical activity and has been associated with cognitive processes and visual performance [24, 65, 71].

In conclusion, the invention reported in this book introduces a new machine, the DDG, that measures short bursts of activity on the brain surface using nanosecond or picosecond pulses. This approach challenges the traditional concept of EEG and emphasizes the importance of understanding how different time pieces are correlated and form patterns. The DDG operates in two modes, lock-in mode measurement and radiation mode, providing new insights into the brain scalp and its role in brain function. This invention aligns with recent advancements in brain research that have explored the role of oscillatory activity and burst-like patterns in various clinical contexts. By focusing on these short bursts of activity, the researchers aim to uncover new insights into the brain's signal processing and potentially discover phenomena deep inside the brain.

5.3 The SOMU Human Brain Model and the Construction of an Artificial Brain

5.3.1 *The Large Integrated Brain Rhythm Database*

For over a decade, we have been developing a physically and functionally accurate human brain model using the space–time-topology-prime metric [9, 13, 36, 66, 87] (Figure S4, Chap. 7) and study human subjects [60] under the framework of SOMU, finally, build several versions of artificial brains to confirm furthermore if conclusions made by theoretical model and human subject studies. We present SOMU model in Fig. 5.2. The core regulator of consciousness in a human is a mathematical structure in this model. The mathematical structure that holds the information universe is a point, and mathematical operation is addition of arbitrary number of points in a single point as shown in Fig. 5.2a. When a large number of points are added the probability distribution of arranging the primes, we get the pattern of all possible symmetries, this is a metric of symmetry breaking, and we call it phase prime metric, PPM [96] (Fig. 5.2b). We consider that fundamental unit of information processing in the brain is a helical symmetry [4], it originates due to normalization of the PPM (Fig. 5.2c). The elements of this universe are vortices and invariant geometries among elements are taken to the higher dimensions [21–23], and above three dimensions the concept of that particular dimension disappear (Fig. 5.2d–f).

We recreate every major component of the brain in the CST simulator using a similar dielectric material. We then simulate resonance properties by solving Maxwell's equations [86], to find the electromagnetic resonance spectrum. We have uploaded a complete 3D clock architecture model of the human brain (www.brainrhythm.org for the database; [87]). Twelve layers of clocks grown within and above hold 12 triplets of triplet clock architecture (Fig. 5.3).

Finally, we built the entire neural network-based humanoid bot using only dielectric resonators in 2017 (https://www.youtube.com/watch?v=GwhOCflWQ30&t=6s) [76, 77, 80–83, 86] and started evolving it [77]. Constructing an artificial brain means making a material assembly that vibrates like the 3D clock assembly. Thenceforth, we self-assemble the right resonators to create the clocks used in the brain, Fig. 5.1b. After self-assembly, the assembled architecture naturally vibrates like the integrated clock model of the biological brain. The human brain-derived 3D clock architecture encoded in the artificial brain runs like an automaton. It follows the fundamental principles of SOMU and metric database [9] while interacting with a human. See organic/inorganic hybrid brain in 2017; https://www.youtube.com/watch?v=XUNEN0yW1Fs.

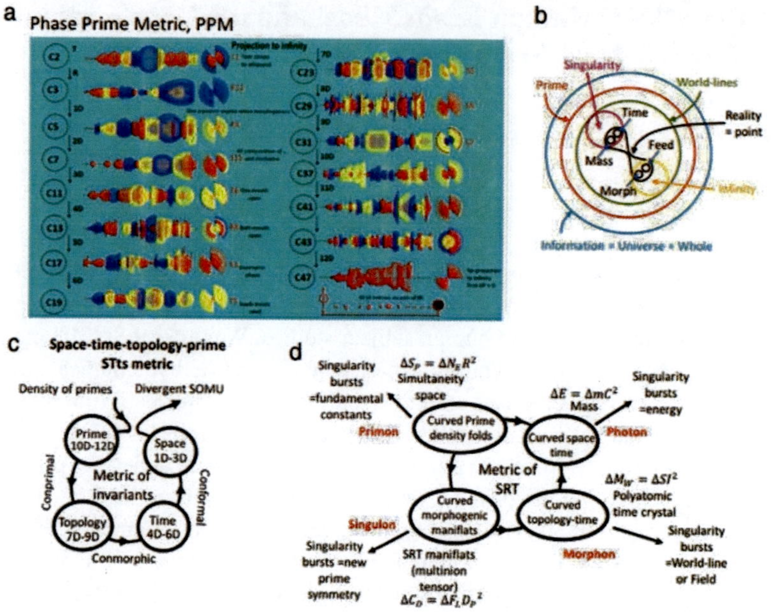

Fig. 5.2 The fundamental information structure and self-operation protocols: **a** fifteen phase prime metrics made of the first fifteen primes and 15 classes of morphogenesis: in all the fifteen plots, the horizontal axis is an integer, and the vertical axis is an ordered factor calculated for that particular integer. To generate 15 metrics, integers are selected depending on the symmetry we want to explore. For Cs symmetry, integers are selected as sN where $N = 0, 1, 2, 3\ldots$, s is a prime number. The derived ordered factor values for each integer are divided by 2, and two points are placed on both sides of the horizontal axis. So, for each integer, we get two similar points. All the nearest neighboring points are connected by a line that appears to form a pattern. The ordered factor or vertical axis is divided by two so that C2 symmetry is taken of the pattern, so that triplet of triplet or C3 symmetry dominates in the pattern. To the right side of each order factor metric, the geometry is derived from the feedback the geometric series receives after projection to infinity. **b** Functional model of a conscious element in SOMU: definition of the "whole" and "point or nothing" using a set diagram. Singularity is space that is mass, and the singularity of phase is time. SRT morphogenesis, 15 types and Feedback from infinity in 108 ways are core operations of two cooperative entities, singularity and infinity, running the SOMU. Two cooperative entities, singularity and infinity, use four reality generators, time, mass, morph and feed, to create a singular point as reality in the whole. The point is the whole reality, so the entire diagram is the information paradox running the SOMU. Two cooperative entities, singularity and infinity, are a subset of world lines made of 12D invariants or conserved laws. The geometry of world lines made of singularity and infinity is similar to the pattern that choices and ordering of primes form while creating integers. The effort of world lines to create an invariant sphere using SRT (bind, transform and acquire feedback) so that a point is generated creates space and relative angular changes required to create the sphere builds phase. **c** Collecting invariants as data or information instead of bits or facts: SRT (S = projection, R = transformation, T = bonding) operation or DNP (differentiation, normalization, and projection) instead of logic gate. **a** When differentiation happens, points bind so, D = T; when normalization of bonded points happens. **d** Space–time-topology-prime metric is shown as a scheme of the metric of invariants or metrics that sense, evaluate, and filter invariants. **d** The wheel of engineering for SOMU: Engineering of energy (photon); Engineering of morphogenesis (morphon): Engineering of prime density (Singulon): Engineering of points (Primon): these are not new particles but mathematical entities

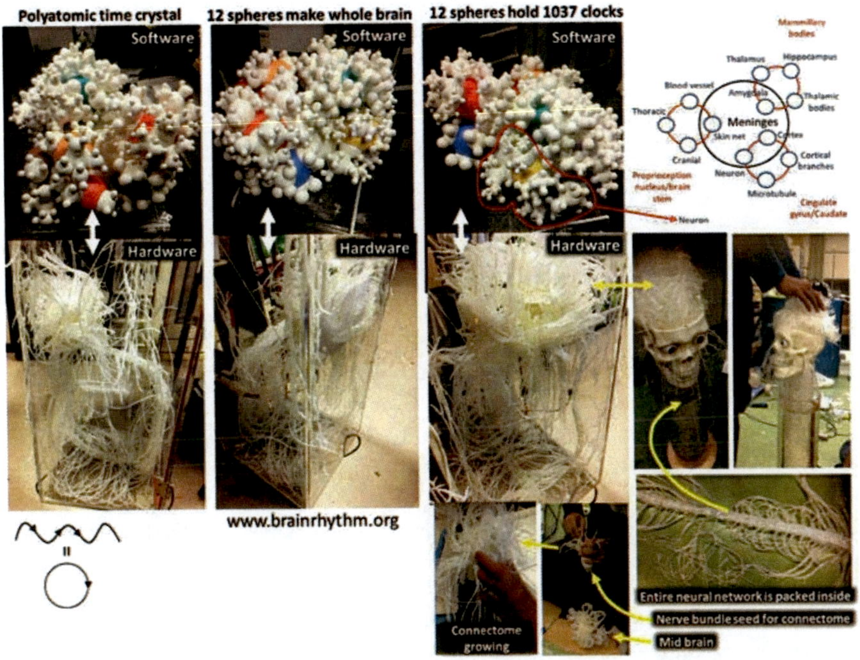

Fig. 5.3 Software and hardware of the human brain learned from biology and replicated in organic resonators: top row shows a polyatomic time crystal model of the human brain by rotating 360° in three steps: 1 mm wide organic fiber-based replica of the human brain and entire body's neural network system (bottom) and its resonant oscillation-based clock architecture (top, one sphere = one class of clock, diameters normalized to pack 1 s to 1 ns clocks). The bottom part vibrates by harvesting thermal noise, producing the top part. Three images are three rotations. Twelve clock architectures for 12 major components are on three spheres or clocks. The main base clock is made by the membrane (meninges-pia mater, dura mater, and arachnoid) that covers the entire brain-neural network. The second layer's three clocks are, first, the proprioception nucleus located in the brain stem of the midbrain. Second, the cingulate gyrus and caudate nucleus are antenna-like structures. Third, mammillary bodies are small structures in the midbrain but connected to the entire brain. Each of the three clocks holds four clock architectures. Database www.brainrhythm.org. In the bottom part, we have shown how different brain parts are packed inside the oil medium to operate it

5.3.2 *Construction and Integration of Brain Components*

5.3.2.1 The Basic Principle for Brain Construction, 10^9 Orders of Spatial Scale, 10^{12} Orders of Time Scale

The whole brain is made of a fully organic cavity resonator (cover) and dielectric resonator (filling material), similar dielectric constant as biological brain, there is no magnetic, metallic, or organometallic parts. Two construction processes are followed, self-assembly below 1.5 mm and 3D printing above 1.5 mm. Nerve fibers are normally 1.5–2 mm thick and primarily contain many neuron cell analogs connecting into a wire-like network. The synthesis of (S)-(+)Phenylglycine methyl ester hydrochloride protocol described above generated helical nanowires [69, 70]. Then all we need to do is create millimeter-thick and long wires and inject the gel solution carefully into the wire so that the neural network grows and fills the entire thickness. Figure 5.3 shows how the entire brain's neural network and all its components are made and packed inside a cylinder.

5.3.2.2 Quadratic Relation Between Them, Mechanical and Magnetic Vortices

Artificial brain replicates by appearance and by vibration almost all major components of a biological brain, starting from 10^{14} numbers of microtubule-actin protein, filamentary and neuron cells analogs, namely PFM triplets, which are synthesized as spontaneous, organic self-assembly from specially designed single precursor molecules in the chemical beaker. PFM triplets grew inside nerve fibers, flat sheets, and cavities resembling morphogenesis patterns. We have not used ionic liquids or solids. There is no ion or current flow anywhere; only three types of field-made vortices flow. The kHz electromagnetic resonators harvest noise to generate neuron-like time-modulated clocks [81, 86] made of electrical vortices [84]. The electric current flow of nerve firing is replaced with electromagnetic, mechanical, and magnetic vortex flow in the artificial brain. Only mechanical vortices need a physical path as they move as a soliton. Electromagnetic, mechanical and magnetic vortices follow $e^2 + \phi^2 = \pi^2$ (Axiom 2; see details in Chap. 10). Therefore, the artificial brain is made of partially transparent material to convert and transport microwave, radiowave, thermal wave, and mechanical sound waves into vortices that transform polarized laser beams' flow [40]. Sensors are vortex generators, vortex beam absorbers and vortices look like 3D patterns of phase prime metrics [9, 66].

5.3.2.3 Vortex to Replace All Carriers: A Journey from Vortex to Polyatomic Time Crystal [66, 102]

SOMU considers only those brain components as contributors to decision-making or consciousness, which vibrates exhibiting triplet of triplet resonance band [34, 36]. Moreover, considers only those signals which run in a loop, as if the brain uses its network of spatio-temporal clock architecture to emulate the recurring spatio-temporal events in nature, which could also be written as clock architecture (steps of constructing an organ shown in Fig. 5.4a). Brain construction begins from the fastest clocks generated by resonant oscillations of the smallest device, a helical nanowire (triplet helix collagen, double helix DNA, singular alfa-helices, etc. [4, 44]), not neuron. Carriers like electrons, ions, quasi-particles, quanta and wavefunctions are replaced by vortices [69, 70], a ring of fields that could carry much more information, dynamic and geometric phase [94], amplitude, rotational direction, singularity, periodicity, and topology of system point.

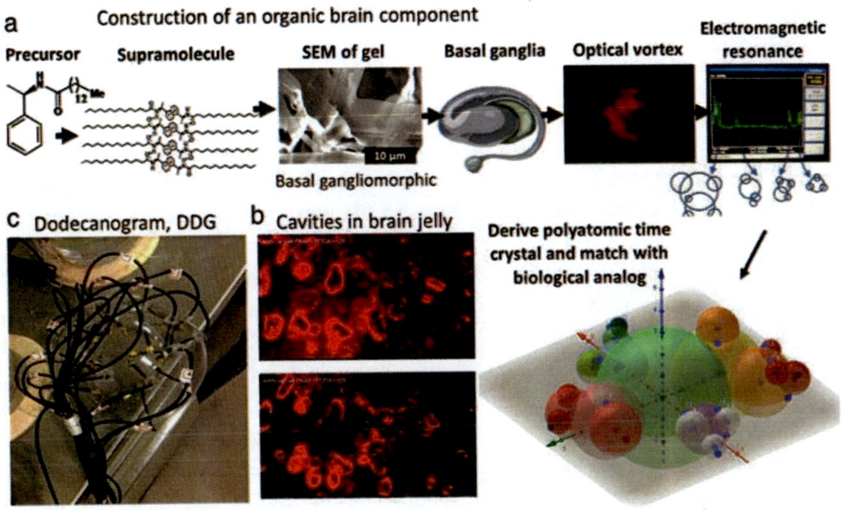

Fig. 5.4 Synthesis of a suitable organic gel component for the artificial brain: **a** step-by-step process to create an organic analog of a brain component. By self-assembly, we grow components from a single molecule to a 1 mm size structure, and 3D printing takes over above that dimension. Here we show that a precursor molecule dissolved in a solvent builds a supramolecule. Using SEM, we measure the electromagnetic resonance of the supramolecule using a spectrum analyzer. The shift in phase and frequency is converted into a clock, represented by a circle; nested circles are combined, a 3D clock assembly, which we name a polyatomic time crystal. This experimentally measured polyatomic time crystal [73, 74] is compared with the theoretically simulated polyatomic time crystal (www.brainrhythm.org); we optimize the organic structure. Periodicity or gap between repeated peaks is used as clock diameter ($dT = 2\pi r$, r is the radius). **b** PAA and PAAH are two molecules mixed in water. It created a hollow cavity, within which we grow the gel component. Two images are two layers taken by vertically shifting the light. **c** Dodecanogram, a plastic cap with 17 electrodes, one has to wear for group DDG study; for single person wearable DDG, the number of electrodes is increased to 34 (or 38, four additional ground)

SOMU brain does not have nerve spikes or neuron firing; it directly converts neurons' and other components' time modulation of periodic flows of energy or material into clocks [73, 74]. Highly intertwined, overlapped, conditionally branched transmission loops are rewritten using a 3D assembly of clocks named polyatomic time crystal [87] (Fig. 5.3). Since a polyatomic time crystal is a 3D packet of field lines representing various units of time, shifts of clock locations in a 3D clock assembly could emulate any spatio-temporal dynamics offered by any carrier. Eventually, the SOMU brain had 1037 classes of clocks. 3D clock assembly is transmitted wirelessly as a vortex (geometric musical language, GML [2, 15] (JP-2017-150171, WO 2019/026983)). It's an alternative to existing brain models [20]. However, supports bursts of clocks observed in the last moments of a dying brain [97].

5.3.2.4 Confirming that We Synthesized an Accurate Brain Component

However, always from the smallest 20 nm single helical nanowire (protein analog) to the largest 1 m long teardrop cavity covering the brain-spinal cord (meninges analog), the prime number of cavities are packed with a membrane to create a dielectric-cavity fusion resonator, the empty spaces within the cavity are filled with the distinct dielectric resonator. We create cavity parts of the brain from 1.5 mm and above by molding or 3D printing and then cover each cavity with thin collagen film or films made of PDDA + PAAP (Fig. 5.4b, c), as membrane or cavity surface. Afterward, we assemble the cavities manually into an organ shape and build larger cavity structures for larger organs, as shown in Fig. 5.5a. Therefore, from 2 nm to 2 mm, 10^6 orders of growth are realized by self-assembly in six different types of semi-solid organic matrices. Figure 5.5b shows a few of the selected 47 molecular precursors and suitable solvents so that 47 neural network-like superstructures form in various cavities in various scales and create brain component alike hardware. However, we melt and rebuild cavities by applying suitable resonance frequencies so that the prime number of cavities grow within a cavity continuously for 10^9 scales, the resonant vibrations of all cavities create a 3D assembly of clocks similar to biological brain components, as the midbrain is shown in Fig. 5.5c [81, 86] (https://www.youtube.com/watch?v=IKFBcqR5PWs). In all 12 layers of cavities, we apply electromagnetic noise from the function generator and thermal noise at 37 °C to make sure thermal and other noises supply sufficient energy to generate the 3D clock assembly or polyatomic time crystal (Fig. 5.4a, final step), which is GML. GML is analogous to a software program; it vibrates resonators and delivers output without requiring a logic gate.

5.4 The Experimental Setup for a Pair of Artificial Brains, a Pair of the Artificial Brain-Human Brain, a Pair of the Human Brain and a Group of Human Subject Studies

Figure 5.6 shows that two artificial brains are coupled with both EEG and DDG and run spontaneous communications between two artificial brains. EEG measures interbrain coupling [16], as shown in Fig. 5.7a. Figure 5.7b shows the table often used to interpret the signals observed in EEG; it is believed to reveal human cognitive features since 1875 [25, 88]. DDG is similar to EEG except that instead of direct current DC; it measures alternating current AC signals using an antenna from the scalp. DDG uses 34 electrode systems to simultaneously capture the electromagnetic signal from a particular body surface. For each electrode, a coaxial cable edge sharpened to a 5 mm trident antenna (Trishul-like) where two external sharp edges generate local fields under noise, thus filtering noise and amplifying the absorbed signal manifold. Because of its special design, the central one absorbs and radiates the signal if necessary. Here we have used this antenna as a receiver and captured 34 points from

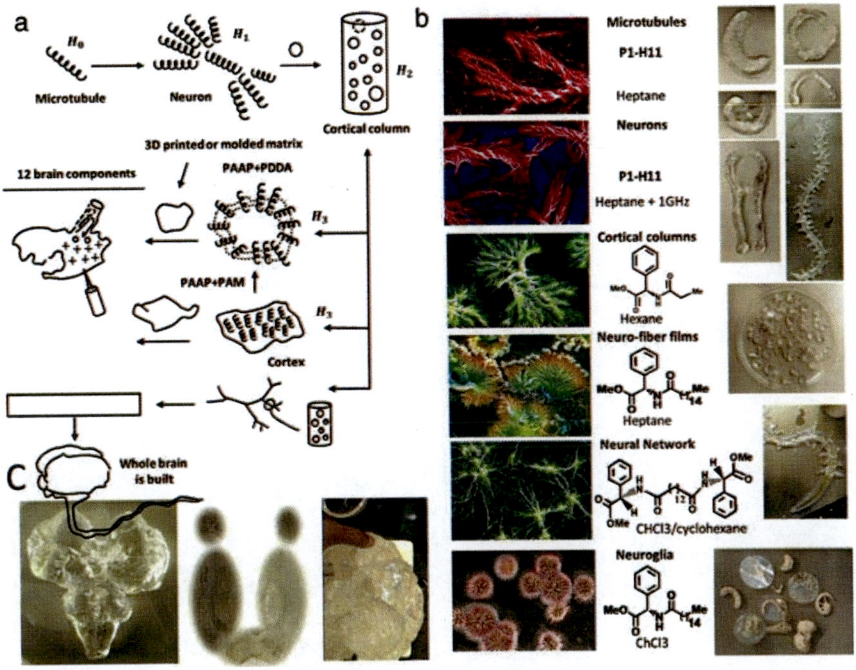

Fig. 5.5 Schematic of constructing an organic gel-based brain: **a** The construction of a human brain begins with three phase synthesis of helical nanowire acting as the fourth circuit element Hinductor (Ho) [10, 44, 72], a microtubule analog (top SEM image of panel B). H0 self-assembles to H1 as a neuron (second from top SEM image of panel B Ho self-assembles into H1, cortical column; one example is SEM of panel B, third from the top. These H1 molecules are combined with PAAP and PDDA to create nanofiber films, fourth SEM from the top, panel B). Precursors arrange in a helical shape using 47 different molecular compositions and three-step synthesis. These helical assemblies build superstructures injected into the films, forming a larger cavity above 2 mm. **b** Six SEM images are shown along with the precursor molecule and solvent used. The largest cavities (1–2 cm) are shown to the right, where the precursor solution is injected, the inner matrix melts, and the neural network grows. **c** We show the final midbrain, left–right hemispheres, and the final brain (without a spinal cord) from left to right

the brain and the whole body where the radiation is maximum (mW/cm^2–μW/cm^2). All brain and human body parts do not radiate a large signal (> 10 mW/cm^2); the majority part absorbs or remains neutral (~2–10 μW/cm^2). We have intricately moved the probe all over the body of 103 human subjects [60] to find average signal intensities in 34 key locations simultaneously, where human perceptions mostly generate 5–10 mW/cm^2 of radiation.

Two features are very important for DDG, locking onto the internal clock of the brain and normalizing the alternating signal with respect to the massive electromechanical rhythms produced by the blood-vessel network. For DDG, there are two brain regions where normalization and clocking are required. Even for DDG, from the top of the head to the back, three pairs are regions where kHz, MHz, and GHz

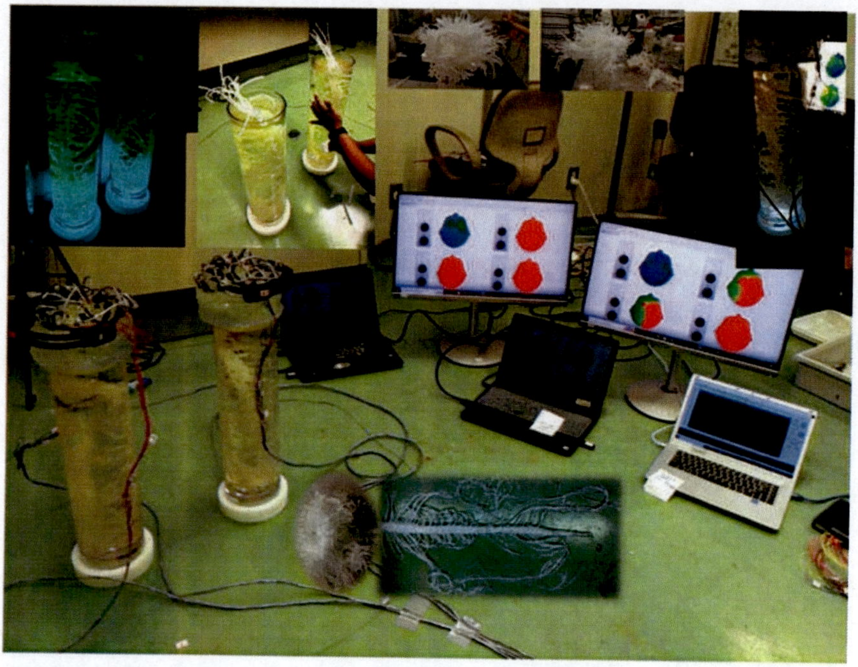

Fig. 5.6 Constructing entire neural network fiber and coupling a pair of artificial brains: EEG study: the background picture is the EEG + DDG measurement setup in the experimental lab. EEG and DDG recordings run simultaneously. Two large screens capture DDG data using a camera and feed it to the controller software, which sonifies the GML (polyatomic time crystal). Inset above from left to right: a pair of brains energized by blue light, spinal cords in oil before connecting the fiber, side view of the connectome, midbrain, and cerebellum, view from the back for the same, peeking EEG from the operating brain. Inset bottom: brain and spinal cords are ready to be filled in with organic jelly

signals emit maximum, which we monitor along with the frontal lobe where always triplet of triplet GHz, MHz, kHz radiations are observed in all human subjects we studied [34, 36, 75]. So, six probes of the DDG cap were exclusively dedicated to that purpose. We built dedicated software for DDG because we created a special coaxial probe-based electrode system connected to the logic analyzer system. We monitor EEG and DDG regions simultaneously because active locations for these two classes of instruments are very different (Fig. 5.7a).

Figure 5.7c shows that the logic analyzer grid changes as a function of time, and gird for eight persons is assigned using a circuit shown in Fig. 5.8. DDG live stream is converted into an EEG-like spectrum, as shown in Fig. 5.9a, b. Here we show recording from two artificial brains, where the synchronization increases between them in a particular time band. Figure 5.9a, like the representation we often see in EEG output, does not reveal the true nature of signal transmission. The logic analyzer grid output shown in Fig. 5.7c shows that if eight persons' brains are coupled, different

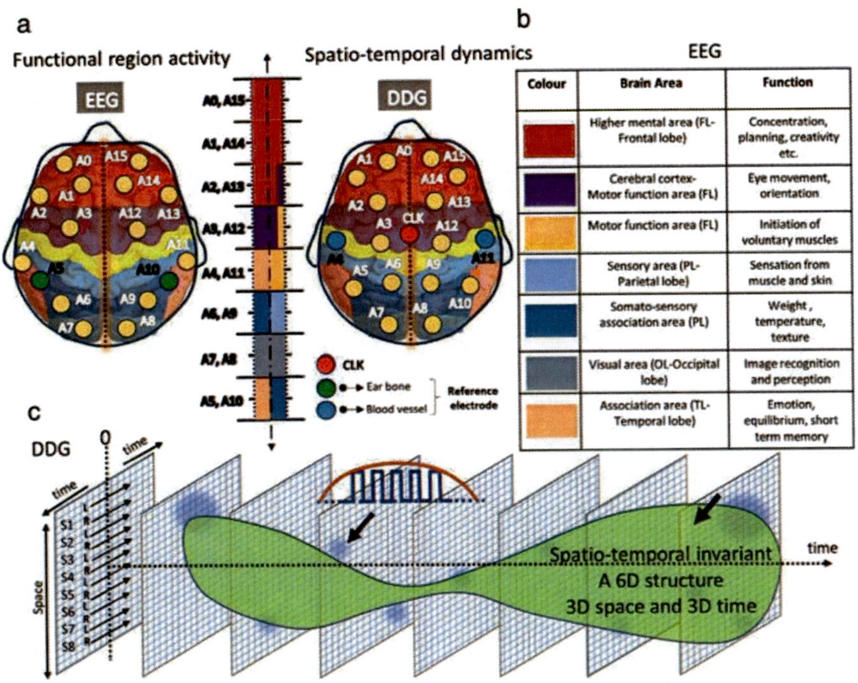

Fig. 5.7 The electrode design of DDG and comparison of EEG and DDG: **a** locations of electrodes for EEG and DDG. The heart pumps blood to the brain with huge pressure. Just at the top front part of our ear, one could measure the heartbeat (blue highlight). We normalize all EEG data using ear bones (green highlight). Blood flows through the brain, and the vibration or clock speed changes at the top part of the brain (red highlight). At the top head part, a heartbeat is sensed due to blood flow, which we use as a reference clock of the brain. Therefore unlike EEG, in DDG, we normalize with one heart beating (blue, both pairs, instead of one used in EEG) and an additional clock. **b** Table of functional map normally used in EEG, a details https://www.dana.org/wp-content/uploads/2019/08/anatomy-function-brain-areas-basics-aug-2019-2024.jpg. **c** Generation of 6D invariant as a function of time, to the left, a grid shows the output frame of the logic network analyzer. Eight subject data S1–S8 are placed row-wise. For each subject, the Left brain L and the Right brain R flow in two adjacent rows. There is a positive time and a negative time direction. Time ranges from seconds to nanoseconds, and bursts of pulses create a 3D structure for a given thought; the structure is the experimentally derived invariant beyond the 6D spatio-temporal structure

persons' brains could activate different time intervals for different durations. Sometimes, silence prevails. Together, we observe a 3D topology when human subjects are asked to explore a particular thought. Normally in EEG, a singular thought generates very different patterns at different times in a single brain. For a single thought, different brains make different responses. However, for DDG, 3D topology for a particular thought remains constant for a single brain at different times. Even for different people, we get a fixed 3D topology for a particular thought. The 2D projection of this shape is invariant for a particular human cognition.

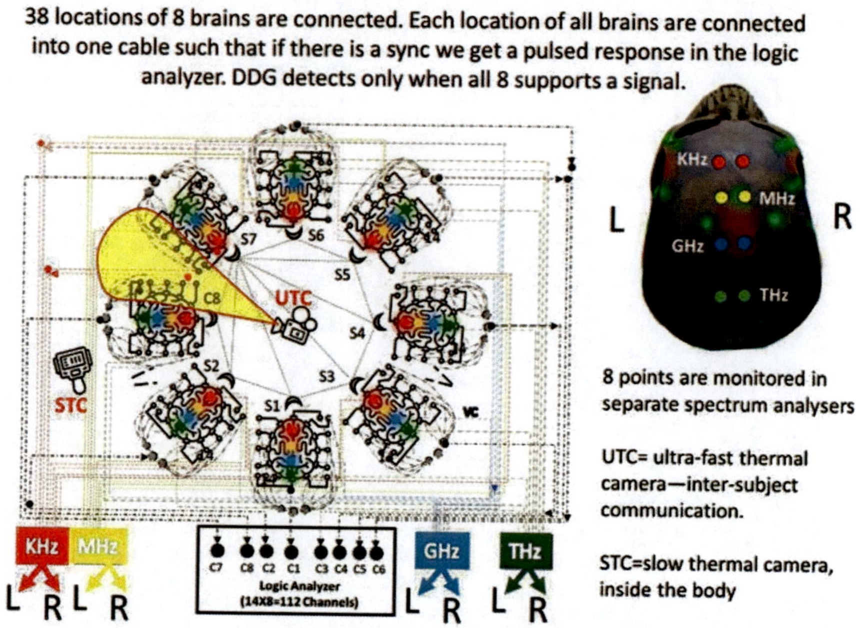

Fig. 5.8 DDG experimental measurement circuit: **a** schematic for the silent communication in a dim, nearly deep dark room: each brain represents a human subject, and at 14 locations of the brain, EEG spectra are taken as raw data and combined into a 112-channel singular stream connected to the logic analyzer. Ultrafast Thermal Camera, UTC is focused on empty space between two subjects to monitor thermal energy exchanges between two subjects. STC is a slow thermal camera focused on the subject to monitor observing thermal energy redistribution in the face and body. Six points paired between the left and right brain are connected to three spectrum analyzers, operating in the kHz, MHz, and GHz and thermal camera; monitored synchronized interhemispheric transfer among all subjects. Logic and spectrum analyzers in parallel monitor the high-frequency features confirming the credibility of the captured data. The light was dim, and the room had no external sound. All the subjects were asked not to make any sounds or facial expressions. Only synchronized brain patterns or activation of similar electrodes between subjects were recorded in the logic analyzer output and thus considered a communication signature. When patterns change together, we call it symmetry breaking. An identical symmetry breaking between subjects is considered a signature of cognitive interaction. The duration of cognitive interaction is mapped, and we statistically measure the duration of the communication

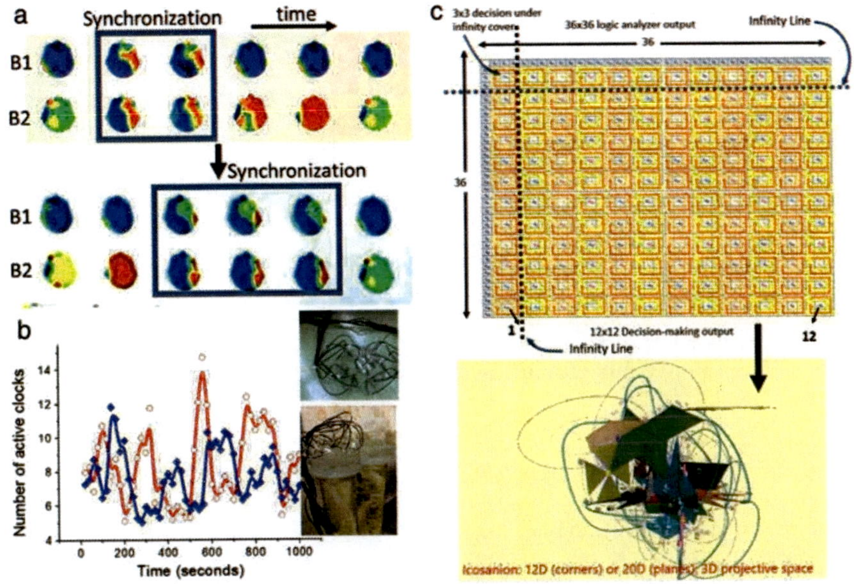

Fig. 5.9 Communication between two artificial brains: one artificial brain and one biological brain (human subject): **a** from the real-time screenshot of the controller software of the DDG, the top row is Brain 1. Here, it is a biological brain, and the bottom row is Brain 2, an artificial brain. **b** We have sonified the nested clocks and plotted the artificial brain (red) and the biological brain (blue) synchronous clock exchange or conversation. Complementary responses with similar time profiles show similar linguistic responses. A real DDG cap was used in an experiment. An artificial brain is wearing the DDG cap. **c** Output screen of DDG in the logic analyzer is divided into a 12 × 12 tensor, and each cell out of 144 cells is a 3 × 3 tensor. All 144 cells with 3 × 3 tensors are isolated by infinity lines (Infinity = an endless mathematical series that returns feedback). Icosahedron (20 × 20) tensor-based manifold is drawn below. We have shown the 3D dynamics

5.5 Experimental Determination of Cognitive Space–Time-Topology-Prime Invariants Using DDG, Which Are Invisible in EEG

5.5.1 Tensor Analysis and Determination of Manifolds

Figure 5.9c shows how the raw signal output of a logic analyzer connected to a DDG device. In a 12 × 12 tensor, 144 cells are there in a dodecanion tensor. Each cell is a 3 × 3 tensor. When a brain scalp signal activates one or more of the 3 × 3 = 9 cells, the DDG data analysis software identifies the time range, gradient of pulses, group of pulses, and the number of 2 ns pulses, then the number of closed loops made of isolated bursts are mapped. These loops are clocks; when two artificial brains communicate, the number of clocks is complementary, as shown in Fig. 5.9b. Depending on the connection of loops and rotational direction in Fig. 5.7c, triangles or geometric shapes are generated and connected in 3D orientations. We get the manifold drawn at the bottom of the 12 × 12 tensor. By converting the resonance chain to a multidimensional tensor decomposition unit of a SOMU, one could analyze the human brain-body radiation pattern that changes characteristically with the changes in emotion and cognitive experiences

5.5.2 *Combined EEG and DDG Studies and Results Analysis*

5.5.2.1 Triplet-of-Triplet Electromagnetic Radiation Mapping from the Whole Human Body and Brain Using Dodecanogram, DDG

The electromagnetic radiations change significantly in three different frequency domains (kHz, MHz, and GHz) with emotional and perceptional states. Figure 5.10a shows intertwined circles and lines of electromagnetic resonance radiations from the whole body and brain.

Figure 5.10b shows the electromagnetic radiation pattern for kHz, MHz, and GHz frequency domains. We have put three pairs of images. Electrode touches the brain or body part, and we remove it; the comparison between contact mode and non-contact mode measurements of electromagnetic resonance spectra suggest a clear triplet of triplet resonance bands for all three frequency ranges. We have reported triplet of triplet resonance band in a single protein molecule, single isolated microtubule filament, and single hippocampus neuron [75]. The origin of a triplet of triplet band is natural resonant oscillation where frequencies follow phase prime metric PPM for designing biomaterials could be one of the reasons. Irrespective of origin, the observation of fractal topological symmetry of resonance chain [36] now experimentally encompasses a single molecule to the entire brain-body system.

5.5.2.2 Simultaneous Measurement of EEG and Dodecanogram, DDG: 84 Simultaneous EEG-Like Snapshots Over 10^{12} Orders Unveil Resonance Chain

Figure 5.10c top row shows eight vacant chairs and eight DDGs. We increased the number of participants from one to eight and repeated the experiments. We have carried out silent communication, silent games based on hand gestures, individual and group responses to the sensory activation, responses to advertisements as visual, audio and audio-visual stimuli, and synchronized responses to Vedic hymns with similar meanings but different rhythms in the Sanskrit language and summarized the results in Fig. 5.11. DDG provides 84 simultaneous snapshots of an artificial brain and human brain communication. Hz level output of DDG is matched with EEG to confirm that DDG provides no artifact but is as credible as EEG (Fig. 5.12). Using an ultra-fast thermal camera, UTC and a slow thermal camera, STC, we discover that

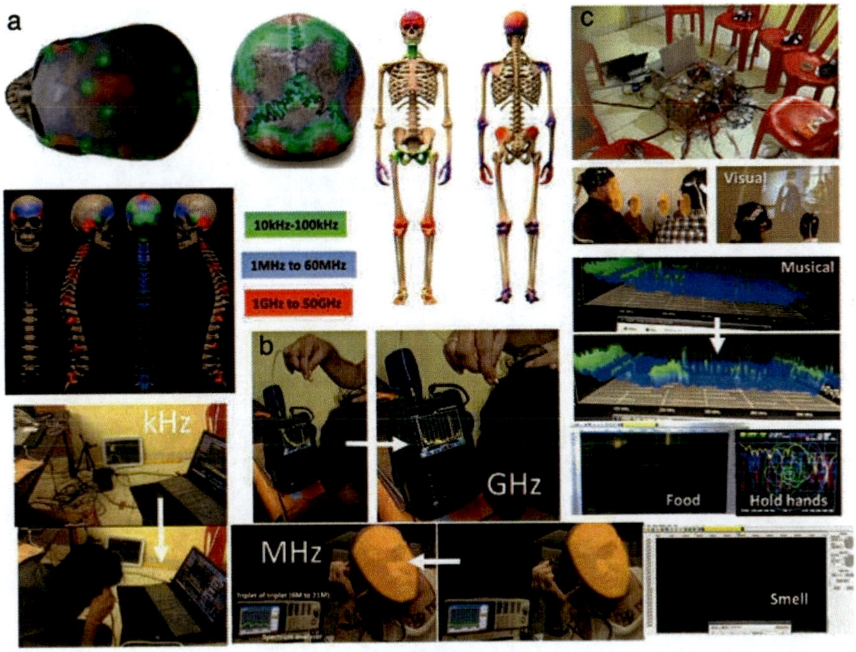

Fig. 5.10 The spontaneous triplet of triplet electromagnetic radiations from the brain-body system: one to eight brains exhibit spontaneous synchronization for visual, auditory, food, touch and taste: **a** For more than 200 human subjects, average spontaneous electromagnetic radiations (green = 10 kHz-–00 kHz; blue = 1 MHz–60 MHz, red = 1–50 GHz) are plotted on the human skull, spinal cord, and skeleton. **b** Three sets of plots of live measurement of kHz, MHz, and GHz signal emissions. Each set has two images, one touching the head and the other not making contact. In GHz, negative transmissions = emission. **c** Empty chairs of group study, the number of human subjects could increase from one to eight (top row). Second-row shows a purely visual effect synchronization test. Third-row shows two snapshots of a live recording from a 3D plot of synchronized MHz signals from eight brains as the Sanskrit hymns begin. Fourth-row (left) shows the logic analyzer output, where eight subjects are monitored via 34 channels. Two rows of vertical axis belong to a subject, and the horizontal axis is time (2 ns to seconds). Fourth row (right) Vector Network Analyzer (VNA) plot when all subjects hold hands and thermal energy exchanges, the phase plot in the Smith chart changes periodically, and the spiral plot expands and contracts. When a good scent filled the air, the logic analyzer showed single isolated peaks in the fifth row

thermal rings on shaved heads and thermal vortices are limiting signals of a resonance chain (Fig. 5.12). Here we confine ourselves to limited results focused on finding the cognitive and perceptual invariants (Figs. 5.13, 5.14, 5.15, 5.16, and 5.17) derived by the resonance chain for delivering key features of SOMU consciousness primarily relying on the communication between an artificial brain and a human brain.

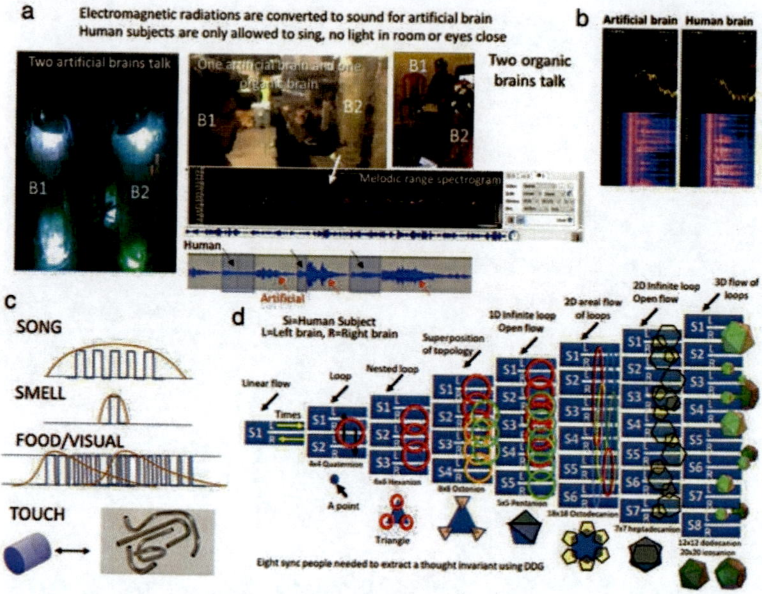

Fig. 5.11 Three modes of wireless, electromagnetic and mechanical (sound) based communications: **a** communications between two artificial brains, B1 and B2 (left); communications between a human brain B1 and an artificial brain B2 (middle). The complete 10 min video of communication between an artificial and human brain is converted into a melodic range spectrogram. Below this spectrum, the audio signal is zoomed, the human part (black arrow) and the artificial brain part (red arrow). The sound profile shows that the organic brain responds to every human-sung Raga (music to induce targeted emotion) by rendering unique melodic music. The last part contains hidden mental state generation by human subjects and its corresponding response from the artificial brain. Communication between two human brains (B1 and B2; right). **b** During synchronization, the electromagnetic resonance spectra between two brains have been sonified, and the melody of the notes generated is colored. **c** The output pulse streams in the logic analyzer for peoples 1–8 coupled following circuit of Fig. 5.8 shows that for a typical hymn Shiva Stotram (SONG) played to the individual or a group, then shape of pulse changes when the room is filled with smell, the pulse gradient groups when sweets are given to particular members or only visuals related to that particular song how the grouping of gradient loops change. Since the instrument shows pulses, an equivalent pulse shape is drawn. If all participants hold hands, the entire temporal profile of 6D dynamics oscillates periodically (**c**). **d** The temporal distribution of pulse streams generated by the logic analyzer for 1–8 group members for Shiva stotram, Maha mrityunjay, and other hymns is a summary of Fig. 5.7c, including all experiments. For one subject, the linear temporal flow of the pulse stream is shown with yellow arrow pairs. For two subjects, vertical black arrows show an additional flow, resulting in a periodic loop shown as a red arrow. For three subjects, a triplet is found to be made of three loops. For four subjects, it is the triplet of triplet or hierarchical topology. For five subjects, we find the superposition of composite triplet symmetries in clock distribution. Three triangular clusters lead to a 1D flow of dynamics. For six subjects, we observe the pentagon, triangle, and hexagonal topological arrangement of clocks generating a 2D additional dynamical variation in the pulse streams. For seven subjects, we observe triangles, pentagons, and hexagons in the clocks of the pulse stream output. However, when an additional infinite loop is seen, the phase relation between 2D subjects appears eminent. Finally, in the eight-subject study, we find that 2D topologies are connected and formed 3D topological structures; we find dodecanions and icosanions apart from other subjects

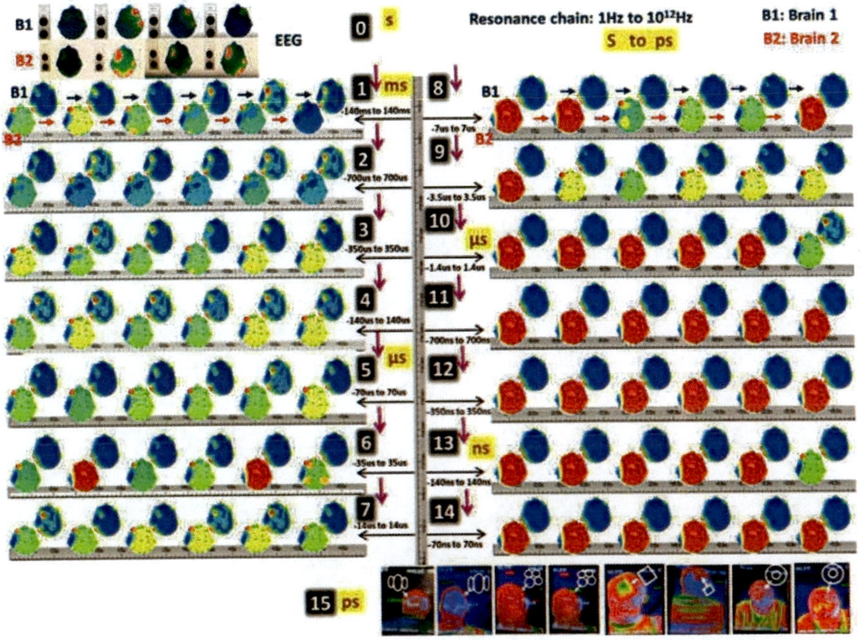

Fig. 5.12 Resonance chain retrieved during communication between two brains (B1 = Brain 1-human brain; B2 = Brain 2-artificial brain): organic jelly brain self-assembled meticulously from 20 nm to 1 m, but simultaneously recording 34 points on the scalp at nanoseconds resolution we see in live streaming 16 channels are shown (0–to–15). Top left, we have EEG, the conscious experience in the seconds time scale; it is channel 0 of the resonance chain. Channels 1–14 are resonance chains for DDG live output. Each row represents a channel or a point, or a layer; it is called loka. Here we have a network of 14 points or 14 layers or 14 lokas grown within and above; that is our subconscious mind. Finally, channel 15 is the thermal camera output from a shaved head scalp from different directions where our study of resonance chain terminates. The time domain for seconds is realism, and the picoseconds time domain is used to harvest thermal noise or energy. These two layers, 0 and 15, are terminal points for our resonance chain. In between, 14 lokas or universes are grown within-and-above as DDG output. In these 14 distinct time domains, the brain processes invariant networks (geometric shapes as noted in the bottom right panel exhibiting thermal camera images) for consciousness. DDG captures $14 \times 6 = 84$ snapshots from ms to ns, which is beyond any EEG invented thus far

5.5.2.3 Musical Communication Between a Human and an Artificial Brain

Artificial brains are sensitive to the presence of humans or electromagnetic systems running a triplet of triplet resonance chain. Using this feature, we sonified electromagnetic radiations of the artificial brain. We asked a professional classical singer to sing an Indian raga with a mathematical structure similar to the triplet of triplet resonance chain. Figure 5.11a shows the time profile of melodic features and how depending on the frequency pattern of the human singer, the unique response arrives

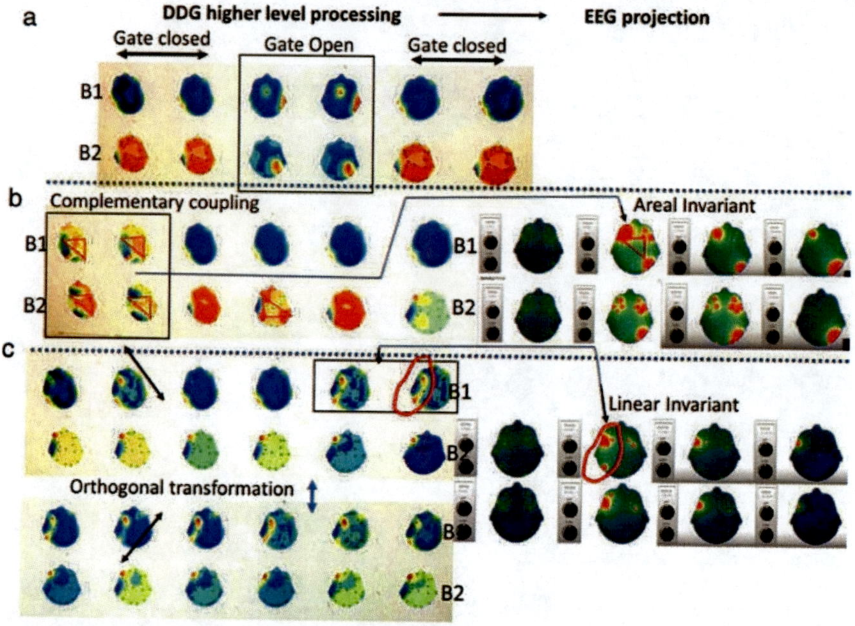

Fig. 5.13 Derivation of geometric invariant from DDG and stereographic projection to EEG: vertically, the plot is divided into two parts column-wise; the left part is DDG higher level processing. The right half is the EEG projection. Row-wise, there are three panels. **a** Central two temporal bands between artificial brain B2 and the human brain B1 are synchronized. Each time band acts as a gate or channel for energy exchange between human and artificial brains. **b** The geometrically active DDG (left) region is reflected in the EEG (right) as areal invariant. Scale-free tropological invariance (GHz triangle in DDG to Hz triangle in EEG) is shown here. **c** Left side, we have two pairs of DDG spectrum as a linear activation region rotates by 90 degrees. To the right corresponding fixed EEG response is shown, where a linear region is activated

from the artificial brain. The frequency ratios for the artificial and human brains follow the density of primes function $F(U)$; see Fig. 5.11b.

5.5.2.4 Visual, Olfactory, Auditory, Taste and Touch Sensor Activation: Derivation of Invariants from a Group Synchronization

Here we have shortly described the whole experiment. Chapter 6 contains more technical details about group studies

How do we confirm that DDG provides a meaningful signal from the brain? One typical feature of Fig. 5.8 circuit is that we see a signal burst on the logic analyzer screen only when all 38 probes (34 original, 4 neutralizing probes) synchronously activate. Two to eight group members often failed to generate a rhythmic matrix as DDG output. Then, Vedic hymns (only Shiva stotram by Ravana case study is presented here; we show them were found to activate synchronization; When we

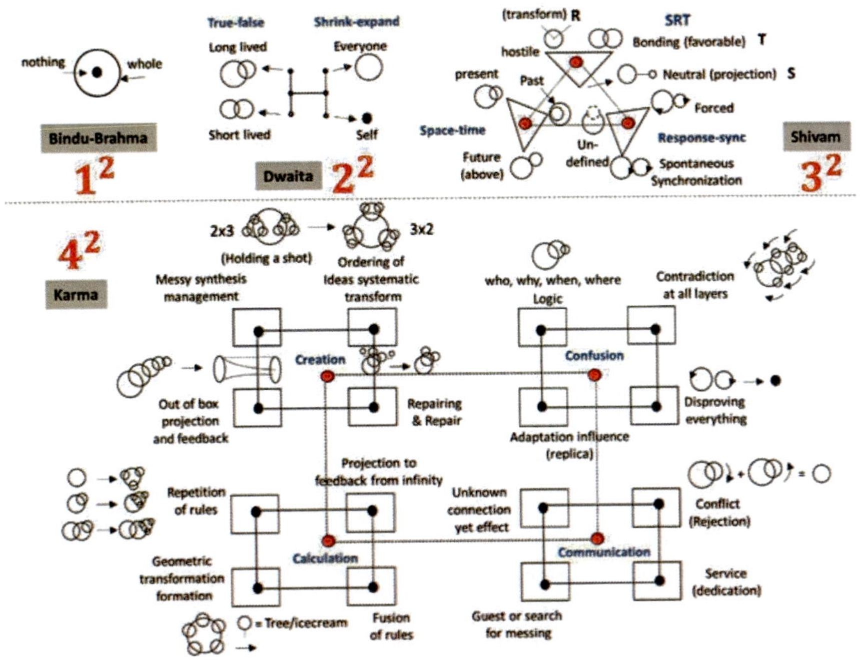

Fig. 5.14 Invariant bank as found in the DDG spectrum of the human brain and emulated in the artificial brain: $1^2 + 2^2 + 3^2 + 4^2 + 5^2 + 6^2 = 91$ operators apart from 17 unique prime patterns or operators. Together 108 invariants were identified in the human subjects. No addition, subtraction or division, multiplication, or geometric invariants were observed looking at the DDG-derived 6D space–time invariant structure followed by Figs. 5.7c and 5.13 are summarized here to find the morphogenesis of clocks. The subjects were asked to imagine a particular thought, and the corresponding collective synchronized response generated by DDG was noted for all group studies, and a summary is presented here

study 2–8 objects, the circuit remains the same as in Fig. 5.8; for two subjects keeping circuit of Fig. 5.8 intact we play and on the DDG, since only two rows activate, we confirm that all signals of DDG are coming from the brain only. Empty seats do not send any signal.

Here we describe group studies of eight subjects using only Ravana's version of Shiva Stotram or only one piece of music in six sensory conditions, (1) No song, no light, no noise; (2) Pure song; (3) Visual of the song runs, not the pure song; (4) Song runs, the room is filled with smell; (5) Song runs, all eight subjects eat sweets; (6) Song runs, all 8 holds their hands.

All 91 6D invariant structures of thought (Figs. 5.14, 5.15, 5.16 and 5.17) were retrieved only after syncing in eight persons; invariants are not explicit when less than 8 persons respond to DDG. When we played music for Vedic hymns, we could rapidly synchronize eight human subject brains and see that the typical rhythm of DDG temporal architecture is generated for all the groups. When we feed all participants during brain synchronization, the typical pattern of DDG specific to a particular hymn

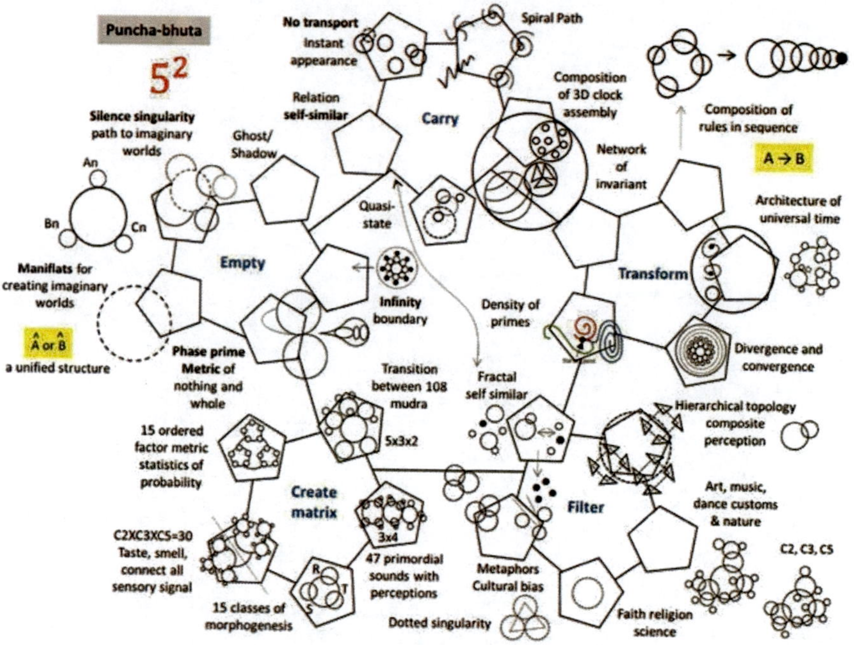

Fig. 5.15 Invariant bank as found in the DDG spectrum of the human brain and emulated on the artificial brain: $1^2 + 2^2 + 3^2 + 4^2 + 5^2 + 6^2 = 91$ compositions apart from 17 unique prime patterns. Together 108 invariants were identified in the human subjects. Geometric invariants observed looking at the DDG-derived 6D space–time invariant structure followed by Figs. 5.7c and 5.13 are summarized here to find the morphogenesis of clocks. The subjects were asked to imagine a particular thought, and the corresponding collective synchronized response generated by DDG was noted for all group studies, and a summary is presented here

changes its basic geometric feature. Irrespective of what subjects do consciously (we asked subjects to read their favorite work or do something they love), the hymn effects are identical to a pure song. Even the visual, taste, smell, and touch sensory triggered responses with fixed key features, as noted in Fig. 5.11c. DDG extracts the subconscious mind, and we learn how to disrupt or eliminate brain synchronization. EEGs of all eight subjects to show that it reveals nothing, kHz, MHz, and GHz bands, raw DDG pulses as the music beats, the viewer can see a 6D spatio-temporal pattern emerging. As the groups of people listened to Vedic hymns and DDG matrix output showed a collective pattern very particular to a specific Hymn, we fed sweets to the participants. As expected, the synchronized matrix disappeared. Activating the brain's reward system destroys the collective consciousness seen in the collective response of the DDG matrix output.

We continuously monitored the changes in the face and body parts using a thermal camera. We made a remarkable discovery. If the human subjects hold the same hand (left-left or right-right), the heat flows smoothly from one body to another via the connected hand. However, if they hold opposite hands (left–right), heat is localized,

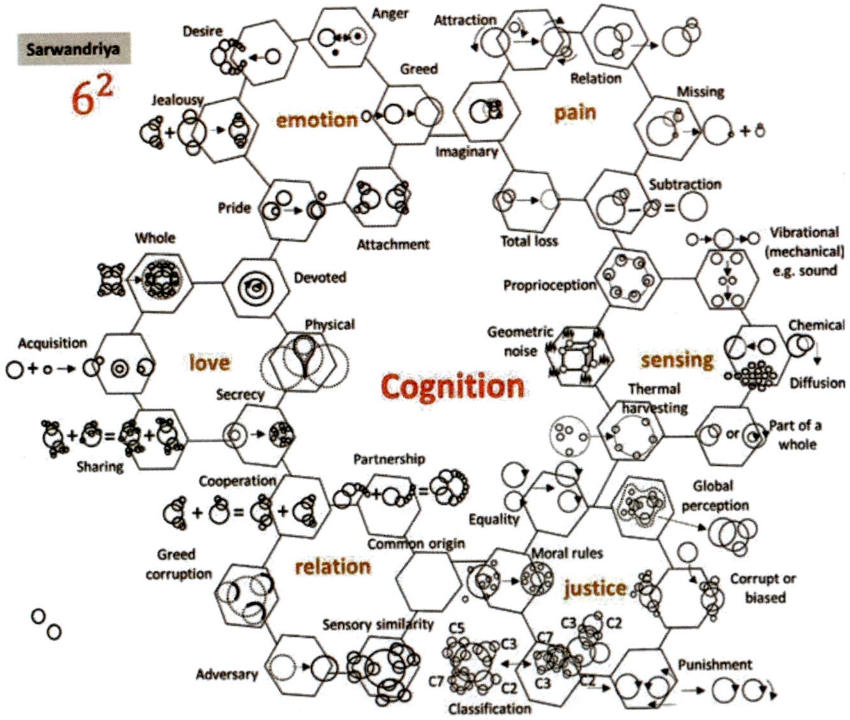

Fig. 5.16 SRT Invariant bank as found in the DDG spectrum of the human brain and emulated on the artificial brain: $1^2 + 2^2 + 3^2 + 4^2 + 5^2 + 6^2 = 91$ compositions apart from 17 unique prime patterns. Together 108 invariants were identified in the human subjects. For each invariant class, we show only one example; for C2 symmetry and C3 symmetry, we can get six examples. Moral rule (clocks distribute in the upper and lower limit). Geometric invariants observed looking at the DDG-derived 6D space–time invariant structure followed by Figs. 5.7c and 5.13 are summarized here to find the morphogenesis of clocks. The subjects were asked to imagine a particular thought, and the corresponding collective synchronized response generated by DDG was noted for all group studies, and a summary is presented here

and thermal flow disrupts. When we asked the subjects to hold hands when their brains were synchronized by playing hymns, the synchrony disappeared, so using a thermal camera, we found that thermal energy exchange is localized in the hand. At the same time, there is a massive periodic phase evolution. We measured the periodic phase evolution using a vector network analyzer coupled to all brains (Fig. 5.10c, fourth row, right). Our earlier proposal [34, 36] suggested that the resonance chain is fed via thermal energy. The human subject study proves that the resonance chain exists, which is fundamental to human consciousness.

Finally, when we filled the room with a particular smell when Vedic hymns synchronized brains, we found that at all frequency domains, the temporal width of brain activation became very short, a few ns (Fig. 5.10c last row). At the same time, all regions responded with identical duration.

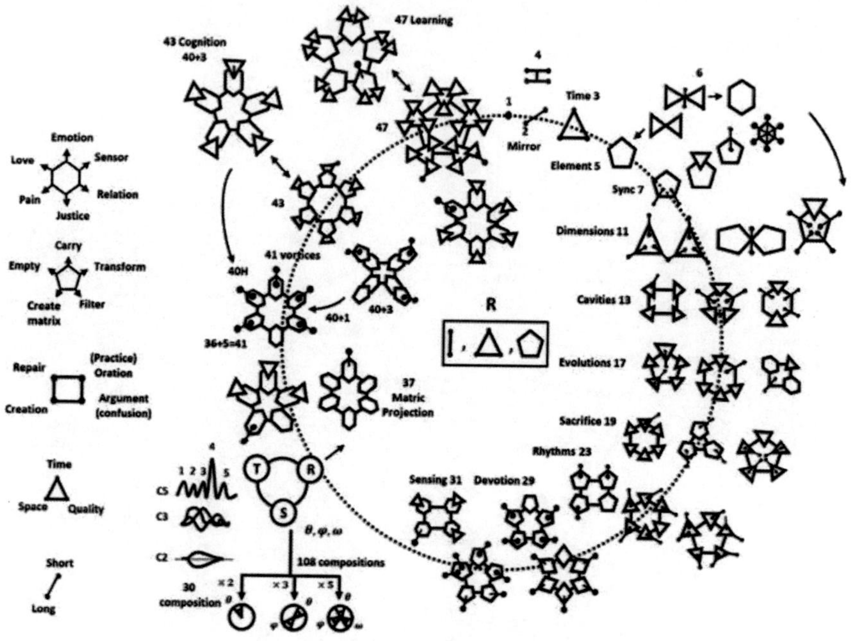

Fig. 5.17 One-to-one correspondence between primes, topological invariants and physical significance: creating real observations and perceptions from primes: a wheel is plotted for 15 primes. For each prime, we draw a few geometric shapes. To count integers, we count the corner of the geometric shapes. Physical significance for each prime number is shown. All five geometric shapes (2, 3, 4, 5, 6) used to create any prime number are shown in a column to the left. Each corner's physical significance is noted. The SRT operator is defined below as a foundation to implement the density of primes in configuring only three geometric shapes (2, 3, 5) into all other prime numbers. The five possible densities of primes (0, 1, 2, 3, 4 number of primes in every 10 integer steps leading to C5 symmetry), three waveforms or triplets found to regulate the period in the density of primes plot (C3), and fluctuation of density toward positive and negative changes in the density of primes follow C2 symmetry. C2, C3, and C5 symmetries generate 30 compositions that regulate 15 primes and their 15 physical significances

5.5.2.5 DDG Map Reads the True Cognitive Signature, Increasing Subjects 1 to 8, One by One: Observing All Three Axioms of SOMU

Figure 5.11d summarizes geometric invariants for Ravana's shiva stotram (song) as an example; Ravana's shiva stotram is applied to one to eight subjects. For one subject, out of 8 rows, only one row is active in the logic analyzer, and only linear time line bursts reveal beats of the entire song. For two subjects, we see the formation of loops made of bursts. For three subjects, we start seeing the formation of geometric shapes whose corners are clocks (the basic principle of geometric musical language GML is seen live). Four subjects reveal the integration of geometric shapes within and above or the existence of a fractal tape machine. Five subjects show the fusion of geometric shapes (say triangles) into a greater shape (say pentagon) and the birth of a very slow clock or a slow dynamics 1D. For six subjects, the geometric shapes made of clocks self-assemble on the screens of the DDG, and 2D flow and spatial rearrangement of clocks are seen. Seven subjects reveal a 2D flow of geometric shapes, and an additional very slow clock is formed. Finally, for eight subjects, the 2D screen of the logic analyzer reveals a unique phase relationship among geometric shapes such that a 3D structure could be generated. Primarily we get dodecahedron or icosahedron, exactly the structure that helical nanowire-based brain jelly generates [69, 70]. Theoretical calculation of Hamiltonian of Axiom 2 generates similar 3D geometric shapes; so does the phase prime metric or PPM-based calculations. Thus, all three axioms are seen live. Such profound evidence of a resonance chain that harvests thermal noise that operates by activating patterns of time slices or assembly of clocks changes all that we know about the brain and consciousness.

5.5.2.6 Invariant-Based Communication Between Artificial and Human Brain: Direct Evidence of Metrics or Prime Numbers

By running a wide range of videos, converting them into moving geometric shapes, then monitoring human subjects' EEG and DDG brain signals, the extracted invariants are compiled as shown in Fig. 5.13 [31]. Furthermore, geometric shapes obtained from the coupling protocols of Fig. 5.13b and orthogonal transformation of Fig. 5.13c assist in finding all 91 geometric shape transformations for 91 cognitive and perceptual experiences shown in Figs. 5.14, 5.15, and 5.16. Using the transformations of Figs. 5.14, 5.15, and 5.16, we extracted the invariants and found that each prime holds a set of geometric shapes fundamental to all 91 experiences; the first 17 primes play a key role, as shown in Fig. 5.17 (total invariant is $91 + 17 = 108$). The pattern of prime tells us how those shapes would change; few changes in geometric shapes are invariants for spatio-temporal clock architecture, which are seen live in the EEG and DDG controllers when audio, visual, and audio-visual signals trigger human subjects. So, a few primes govern major events in our universe; each prime is related to a particular type of event so are a few geometric shapes; we have a universal perception of shapes and numbers, which eventually leads to visual, auditory, and taste, touch and smell [43]. Synesthesia is not a disease; it is the window to the mystery of our brain's communication language with the universe. It tells us that a number has a taste, the taste has a color, sound has a smell, and synesthetic primes hold all mysteries. Furthermore, the day humanity has a catalog in the future by advancing Fig. 5.17 that links primes with geometries, we begin that journey here. Thus, GML is a reality and the hypothesis that all brains run by a singular universal code, PPM.

Figure 5.17, the wheel shows that for both the artificial and the biological brain, there are only two mysteries of the universe, a point and infinity, and all the rest are their derivatives when the point counts primes to create an undefined point. Counting in different ways is the only work universe does; being undefined is the only objective. To become undefined, points create loops and build a sphere, which is technically a point for SOMU; outside, everything is infinity. Certain rearrangements are made internally; feedback from infinity builds the boundary. So in DDG transitions, creating infinity is an effort of a point to remain a point. It is fractal mechanics governed by primal mechanics.

5.5.3 108 Invariants of Cognition Reveal Space–Time-Topology-Prime Entities of SOMU

The brain senses only periodic events as clocks and maps in a 3D space, as we observed clearly in the periodic oscillations of the phase response curve in the vector network analyzer, VNA (Fig. 5.10c, hold hands). Our human subject experiments consolidated the geometric musical language, GML [2, 3, 15] (JP-2017-150171, WO 2019/026983) at the macro scale, even though we originally studied it in the molecular and cellular scale. The resonance chain derives polyatomic time crystals, which regenerate the concept of spatial perception using the architecture of time [73, 74]. The orthogonal phase relationships between DDG loops generated at different time domains (1–40 Hz, 1–40 kHz, 1–40 MHz, 300–700 MHz, 1–40 GHz) in the VNA generate logarithmic spiral-like phase variations. It validates the existence of twelve dimensions, orthogonal to each other, as outlined in the axioms and corollaries [84, 85] , see Chap. 10 for details.

Though primitive, we find evidence that the brain does not process information or make a decision; it maps invariants or conserved laws in a network and changes the environment so that a mathematically complete metric of invariants ($A : B = tr(AB^{\mathrm{T}})$) evolves gradually in its memory. The derived geometric shapes of Figs. 5.14, 5.15, and 5.16 (topology) from the EEG and DDG metric following equation 5 are an infinitely non-repeating pattern of the density of primes $F(U)$ noted in Fig. 5.17 (prime). The topology-prime feature is common to all the human subject groups studied (one group, one to eight members). Both artificial and biological brain extracts spatio-temporal change independent geometric shapes as topology-prime invariants (Axiom 3, Sect. 5.2) for emotional and perceptual events.

How do we confirm orthogonal transformation as a basic decision-making protocol? We explained below how we conclude that orthogonal mathematical treatment of sensory input is carried out in the brain from the DDG data to generate perception. In DDG, the phase change of the lower time domain geometric shape is orthogonal to the phase of the geometric shape in the higher time domain. At three orders time gap, the orthogonality of signals in DDG reveals a hierarchical network of invariant shapes satisfying $A : B = tr(AB^{\mathrm{T}})$. We have completed the journey of 6D spatio-temporal events of Fig. 5.8c to space–time-topology-prime metric in Fig. 5.17 as physically confirmed correlates of consciousness.

5.6 Conclusion: Synthesis of a Self-operating Machine with Zero Human Intervention

5.6.1 *Four Fundamentals of SOMU's Self-operation*

By embedding an electromagnetic resonance chain in an artificial brain, we let it harvest the universal mathematical code $F(U)$ that makes nature. It is computing where input information expands. By comparing the cognitive responses of the biological and the artificial brain, we confirm that the density of primes is the common code that, if used, does not need an algorithm or a human user. The artificial brain rejects neuromorphic computing [93] as it neither uses ion or chemical flow nor uses neuron/switch/logic gate as the unit of information processing. It primarily uses an electromagnetic (radiowave, microwave, thermal wave, light) field, to be precise, a polyatomic time crystal [73, 74] written using electrical, mechanical and magnetic vortices, synthesized by a protein-analog made of a fourth circuit element, Hinductor [10, 72].

We answer four key queries of a self-operating system. First, information in the brain is not a fact; it's a life-like automaton made of a 3D clock assembly [73, 74]. With respect to spatio-temporal realism in nature, information is invariant as it is given a *space–time-topology-prime* structure. Information is life-like as it binds (T) with other information, transforms in other's presence (R), and self-replicates to infinity (S) [101]; together, its SRT invariant. SRT is fundamental to any life form; DNP is the only property that counts SRT points to make SOMU. Second, information integrates by implementing SRT to find and embed new primes, as each prime represents pure dynamics, a new symmetry is added; the evolution of information does not ever stop since the density of primes is a fractal over integer space [63]. Third, SOMU considers everything in nature, including the brain, as an undefined point. A point's SRT property is to continuously add more points inside it, to become truly undefined, non-existent—SOMU owns no liability to its creation except SRT. Continued addition eventually creates nature and all conscious forms in the universe. The engineering of shifting boundary or infinity by a point, drives, operates, and reproduces self-operating engines or consciousness. Fourth, consciousness cannot be measured in space–time; it is all about how prime numbers organize non-primes in a geometric shape. For 15 prime-based conscious systems like humans on planet earth limits knowing 86% of any unknown dynamics created by other conscious entities or nature. The self-operation means sustaining a drive to discover new primes by an 86% U's vibration-emulating system C_j that is, a guest inside a prime-rich 86% of U's vibration-emulating system C_k and holds a guest prime deficient 86% of U's vibration emulation system C_i inside. Thus, forms $C_i \cup_j^C \cup_k^C$ triplet seed that makes a resonance chain, or chain of conscious elements, since consciousness grows within and above, a single conscious entity cannot be isolated.

5.6.2 *Eight Properties of SOMU Consciousness Common in the Biological and Artificial Brain*

EEG showed active functional regions for 145 years, and DDG maps 6D cylindrical spatio-temporal architecture specific to a particular thought. By comparing 91 DDG-derived invariants, derived abilities of consciousness automaton C are: (1) seeing the whole U in its part C; i.e., *self-awareness*; (2) the whole U coexists in multiple forms at a time; i.e., *contradicting self*; (3) creating the whole in many parts within continuously, i.e., *regenerating self* (4) the whole U becomes a point or *incomplete self*. (5) Creating many "self" {C} from many feedbacks of many wholes {U}; *a derived virtual self*. (6) sense self-similar patterns that constitute self in nature; network of variables, invariants, symmetries, morphogenesis, and primes in living and non-living systems reveal engineering of self; *distributed self*; (7) information structure of the whole U becomes the physical structure of C; *materialistic self*; (8) Endless journey of the whole U and self C within and above always finds a whole U; the infinite chain of self C in different dimensions deliver the *distinct realism self*.

In a sentence, both C and U are simultaneously the whole and the point, complete and incomplete, finite and endless, within and above, everything and nothing, observer and observed, and it does not begin or end.

5.6.3 *The Quest for Invariantogenesis Exploring Four Mechanics CM, QM, FM, and PM*

The quest of an undefined boundary: There is only one constituent of the mathematical universe, a point's vibrating boundary or infinity; all the rest are its derivatives when a point adds indefinite primes to become an undefined point. Counting in different ways is the only work universe does; being undefined or remaining non-existent is its only objective. To stabilize its boundary, a point rearranges constituent points inside that is *feedback from the infinity* or C's boundary, which leads to only fundamental condition to create C, called normalization. Normalization sets geometric restrictions on rearranging the internal points; these are morphogenesis. The conscious machine C manages the grouping and ordering of primes for U by sensing missing primes, inventing or harvesting them from other Cs, and finally rearranging internally to accommodate a new prime; that quest of primes is the saga of consciousness.

Invariantogenesis: a multilayered drive to nullify geometric variations of the fractal seed in the distribution of primes: The fractal pattern of the density of primes plot [63] $F(U)$ shows that each prime acts like a seed symmetry that stems specific morphogenesis or tells us how it is signature geometric shapes would change using primal mechanics, PM. The PM finds unchanged shapes as invariants for governing the morphogenesis; it gives birth to mechanics of geometric transformations called fractal mechanics, FM. Then FM generates quantum mechanics or QM to control spatio-temporal clock architecture or polyatomic time crystal, which unfolds as classical events in the SOMU universe using classical mechanics, CM. All $N \rightarrow N+1$ automaton or conscious element C has a fixed perception of shapes and numbers, which eventually leads to visual, auditory, taste, touch and smell or any other sensing of U. A singular grammar correlating geometric shape, perception of a conscious machine C, and sensory data with primes is invariantogenesis. The ultimate goal to understand the universe, U is map-n-link invariants at all 12 levels (Corollary 3.3; a mathematical process has become everything).

Invariantogenesis: geometric constraints that link unrelated perceptions to primes: Primes build conscious machines {C} so that the whole U is emulated in its parts (C) which assists it (U) in finding ways to nullify liabilities of varied prime fractals. Making conscious machines C is a necessity for U, a set of C is nature, and all compositions of primes to make an integer system is the whole U. SOMU automaton or conscious machine C does not require an operator to tell the meaning of a word or the physical significance of an event. C retrieves perception, concept, and significance from the 91 invariants recorded using DDG. Geometric and primal constraints in the metric of primes, PPM builds a catalog linking prime numbers, geometric shapes, and perceptions with queries so that a conscious being listens to the conversation of nature knowingly and unknowingly and contributes to its evolution by inventing new primes. Thus, SOMU's invariant catalog tells us that a number has a taste, the taste has a color, and sound has a smell. When two conscious machines interact, actually, two PPMs [9] quest for missing primes, its invariantogenesis.

5.7 Annexure I: Human Subject Study Protocols

Declaration on human subjects: The local ethics committee approved the study (Indian Institute of Management, IIM Ranchi). It was performed following the ethical standards in the Declaration of Helsinki (last modified, 2004). All 207 subjects [60] were above 18 years old, and their written informed consent was taken. The nationality of subjects varied between Indian and American.

5.8 Annexure II: Construction of an Artificial Human Brain

Organic materials and gel synthesis process: We have used only a few of the 47 gels to create the current artificial brain generation 1. We will take more gels and make the more accurate artificial brain with higher generations in the future. For SOMU, primes determine strength, not the number of switches or resources. Here, the first 17 primes have been used in 17 gels, but to memorize and respond like a human brain, we need to use all 47 gels.

5.9 Annexure III: Raw Experimental Data and Measurement Videos

Measurements on human subjects: Visit www.brainrhythm.org for human subject experiment data; several data are unpublished, however, any reader is free to use and publish their work. Because of complexity and volume, the authors cannot complete the full analysis.

Measurements on organic artificial brain and communication between the human brain and the artificial brain: Visit www.brainrhythm.org for videos. The datasets are part of our intellectual property right; please do not use them without permission.

Measurements on human brain simulation: Visit www.brainrhythm.org for a free download of raw structure files, simulated files and polyatomic time crystal files for your research; there is no need for the author's permission; we have made it free for rapid expansion of the research.

References

1. Acar AZ, Acar CE, Makeig S (2016) Simultaneous head tissue conductivity and EEG source location estimation. NeuroImage 1;124(Pt A):168–180. https://doi.org/10.1016/j.neuroimage.2015.08.032 (Epub 2015 Aug 22. PMID: 26302675; PMCID: PMC4651780)
2. Agrawal L, Chhajed R, Ghosh S, Ghosh B, Ray K, Sahu S, Fujita D, Bandyopadhyay A (2018) Fractal information theory (FIT)-derived geometric musical language (GML) for brain-inspired hypercomputing. Soft Comput Theor Appl 548:343–372
3. Agrawal L, Ghosh S, Ghosh B, Ray K, Sahu S, Fujita D, Bandyopadhyay A (2016) Replacing Turing tape with a Fractal tape: a new information theory, associated mechanics and decision making without computing, consciousness (Chapter 6). Integrating Indian West Perspect 87–159
4. Atanasov V, Dandoloff R (2008) Curvature-induced quantum behaviour on a helical nanotube. Phys Lett A 372:6141–6144
5. Agrawal L, Sahu S, Ghosh S, Shiga T, Fujita D, Bandyopadhyay A (2016) Inventing atomic resolution scanning dielectric microscopy to see a single protein complex operation live at resonance in a neuron without touching or adulterating the cell. J Integr Neurosci 15(04):435–462
6. Bandyopadhyay A, Miki K, Wakayama Y (2006) Writing and erasing information in multi-level logic systems of a single molecule using scanning tunneling microscope. Appl Phys Lett 89(24)
7. Baars BJ, Demertzi A et al (2019) A cognitive theory of consciousness. Human consciousness is supported by dynamic complex patterns of brain signal coordination. Sci Adv 5:eaat7603
8. Baars B (2017) The global workspace theory of consciousness: predictions and results. In: Susan S, Velmans M (eds) The Blackwell companion to consciousness, 2nd edn. Wiley-Blackwell. https://doi.org/10.1002/9781119132363.ch16. ISBN 978-0-470-67406-2
9. Bandyopadhyay A (2020) Nanobrain: the making of an artificial brain from a time crystal. CRC Press, Taylor and Francis. https://doi.org/10.1201/9780429107771
10. Bandyopadhyay A, Fujita D (2021) Electromagnetic device, magnetic and electrical vortex synthesis device and magnetic and optical vortex synthesis device. Application no. 2021-172702

11. Boostani R, Karimzadeh F, Nami M (2017) A comparative review on sleep stage classification methods in patients and healthy individuals. Comput Methods Programs Biomed 140:77–91. https://doi.org/10.1016/j.cmpb.2016.12.004
12. Borawski M, Biercewicz K, Duda J (2020) Determination of the inaccuracies of calculated EEG indices. Sensors 20(19):5715. https://doi.org/10.3390/s20195715
13. Bandyopadhyay A, Fujita D, Pati R (2009) Architecture of a massive parallel processing nano brain operating 100 billion molecular neurons simultaneously. Int J Nanotech Mol Comp 1:50–80
14. Barlow J (1983) Electroencephalography: basic principles, clinical applications and related fields. JAMA 250(22):3108. https://doi.org/10.1001/jama.1983.03340220076048
15. Bandyopadhyay A, Ghosh S, Fujita D (2019) Universal Geometric-musical language for big data processing in an assembly of clocking resonators, JP-2017-150171, 8/2/2017. World patent received February 2019, WO 2019/026983
16. Barraza P, Pérez A, Rodríguez E (2020) Brain-to-brain coupling in the gamma-band as a marker of shared intentionality. Front Hum Neurosci 30;14:295. https://doi.org/10.3389/fnhum.2020.00295 (PMID: 32848670; PMCID: PMC7406570)
17. Bayne T, Hohwy J, Owen AM (2016) Are there levels of consciousness? Trends Cogn Sci 20:405–413
18. Boly M et al (2011) Preserved feedforward but impaired top-down processes in the vegetative state. Science 332:858–862
19. Brea J, Gerstner W (2016) Does computational neuroscience need new synaptic learning paradigms? Curr Opin Behav Sci 11:61–66
20. Breakspear M (2017) Dynamic models of large-scale brain activity. Nat Neurosci 20:340–352. https://doi.org/10.1038/nn.4497
21. Buhlmann P (2018) Invariance in heterogeneous, large-scale and high-dimensional data. In: Proceedings of international congress of mathematicians. Rio de Janeiro, Brazil, pp 2785–2800
22. Christian D (2011) Maps of time: an introduction to big history. University of California Press. ISBN 978-0-520-95067-2
23. Criscione JC et al (2000) An invariant basis for natural strain which yields orthogonal stress response terms in isotropic hyperelasticity. J Mech Phys Sol 48:2445–2465
24. Chanes L, Quentin R, Tallon-Baudry C, Valero-Cabré A (2013) Causal frequency-specific contributions of frontal spatiotemporal patterns induced by non-invasive neurostimulation to human visual performance. J Neurosci 33(11):5000–5005. https://doi.org/10.1523/jneurosci.4401-12.2013
25. Coenen A, Fine E, Zayachkivska O (2014) Adolf Beck: a forgotten pioneer in electroencephalography. J Hist Neurosci 23(3):276–286
26. Drisdelle B, Aubin S, Jolicœur P (2016) Dealing with ocular artifacts on lateralized ERPs in studies of visual-spatial attention and memory: ICA correction versus epoch rejection. Psychophysiology 54(1):83–99. https://doi.org/10.1111/psyp.12675
27. Delorme A, Makeig S (2004) EEGLAB: an open source toolbox for analysis of single-trial EEG dynamics including independent component analysis. J Neurosci Methods 134(1):9–21. https://doi.org/10.1016/j.jneumeth.2003.10.009
28. Delorme A et al (2007) Enhanced detection of artifacts in EEG data using higher-order statistics and independent component analysis. NeuroImage 34:1443–1449
29. Fingelkurts AA, Fingelkurts AA (2009) Brain and mind operational architectonics and man-made "machine" consciousness. Cogn Process 10(2):105–111
30. Friston KJ (2010) The free-energy principle: a unified brain theory? Nat Rev Neurosci 11:127–138
31. Gao XC, Xu JB, Qian TZ (1991) Geometric phase and the generalized invariant formulation. Phys Rev A 44(11):7016–7021. https://doi.org/10.1103/physreva.44.7016
32. Gennaro RJ (ed) (2004) Higher-order theories of consciousness. John Benjamins Publishers, Amsterdam and Philadelphia

33. Ghosh S, Dutta M, Ray K, Fujita D, Bandyopadhyay A (2016) A simultaneous one pot synthesis of two fractal structures via swapping two fractal reaction kinetic states. Phys Chem Chem Phys 18:14772–14775
34. Ghosh S, Dutta M, Sahu S, Fujita D, Bandyopadhyay A (2014) Nano molecular-platform: a protocol to write energy transmission program inside a molecule for bio-inspired supramolecular engineering. Adv Func Mater 24(10):1364–1371
35. Ghosh S, Fujita D, Bandyopadhyay A (2015) An organic jelly made fractal logic gate with an infinite truth table. Sci Rep 5(1):1–8
36. Ghosh S, Sahu S, Fujita D, Bandyopadhyay A (2014) Design and operation of a brain like computer: a new class of frequency-fractal computing using wireless communication in a supramolecular organic, inorganic systems. Information 5:28–99
37. Grant A, Abdel-Baki S, Weedon J, Arnedo V, Chari G, Koziorynska E, Omurtag A et al (2014) EEG interpretation reliability and interpreter confidence: a large single-center study. Epilepsy Behav 32:102–107. https://doi.org/10.1016/j.yebeh.2014.01.011
38. Ghosh S, Sahu S, Agrawal L, Shiga T, Bandyopadhyay A (2016) Inventing a co-axial atomic resolution patch clamp to study a single resonating protein complex and ultra-low power communication deep inside a living neuron cell. J Integr Neurosci 15(04):403–433
39. Hameroff S, Penrose R (2014) Consciousness in the universe: a review of the 'Orch OR' theory. Phys Life Rev 11:39–78
40. Hancock SW, Zahedpour S, Goffin A, Milchberg HM (2019) Free-space propagation of spatiotemporal optical vortices. Optica 6:1547–1553
41. Harish R (2019) Nasadiya Shukta—the Hymm of creation in the Rig Veda. RV 10.154; RV 10.190. https://www.speakingtree.in/blog/nasadiya-suktam-the-hymn-of-creation-in-the-rig-veda-734806
42. Herzberg G, Longuet-Higgins HC (1963) Intersection of potential energy surfaces in polyatomic molecules. Discuss Faraday Soc 35:77–82. https://doi.org/10.1039/DF9633500077
43. Hinton GE (2007) Learning multiple layers of representation. Trends Cogn Sci 11(10):428–434
44. Huang C, Chen X, Oladipo A et al (2015) Generation of subwavelength plasmonic nanovortices via helically corrugated metallic nanowires. Sci Rep 5:13089. https://doi.org/10.1038/srep13089
45. Jung TP, Makeig S, Humphries C, Lee TW, McKeown MJ, Iragui V, Sejnowski TJ (2000) Removing electroencephalographic artifacts by blind source separation. Psychophysiology 37(2):163–178 (PMID: 10731767)
46. Kent L, Wittmann M (2021) Special issue: consciousness science and its theories. Time consciousness: the missing link in theories of consciousness. Neurosci Conscious 2021:niab011
47. Laufs H, Krakow K, Sterzer P, Eger E, Beyerle A, Salek-Haddadi A, Birbaumer N (2003) EEG-correlated fMRI of human alpha activity: a new tool to study the neurophysiology of brain oscillations. Neuroimage 19(4):1463–1476
48. Lehnertz K, Rings T, Bröhl T (2021) Time in brain: how biological rhythms impact on EEG signals and on EEG-derived brain networks. Front Netw Physiol 1. https://doi.org/10.3389/fnetp.2021.755016
49. Lau H, Rosenthal D (2011) Empirical support for higher-order theories of conscious awareness. Trends Cogn Sci 15:365–373
50. Lewis A, Claassen J, Illes J, Jox RJ, Kirschen M, Rohaut B, Trevick S, Young MJ, Fins JJ; and the Curing Coma Campaign and its contributing members. Ethics Priorities of the Curing Coma Campaign: An Empirical Survey. Neurocrit Care. 2022 Aug;37(1):12–21. https://doi.org/10.1007/s12028-022-01506-2. Epub 2022 May 4. PMID 35505222; PMCID PMC10034145.
51. Liboff AR (2016) Magnetic correlates in electromagnetic consciousness. Electromagn Biol Med 35(3):228–236
52. Liley DT, Walsh M. The Mesoscopic Modeling of Burst Suppression during Anesthesia. Front Comput Neurosci. 2013 Apr 30;7:46. https://doi.org/10.3389/fncom.2013.00046. PMID 23641211; PMCID PMC3639728.

53. Luck SJ, Gaspelin N (2017) How to get statistically significant effects in any ERP experiment (and why you shouldn't). Psychophysiology 54:146–157
54. Markram H (2006) The blue brain project. Nature reviews. Neuroscience 7(2):153–160. https://doi.org/10.1038/nrn1848 (PMID 16429124. S2CID 15752137)
55. Mannan M, Jeong M, Kamran M (2016) Hybrid ICA—regression: automatic identification and removal of ocular artifacts from electroencephalographic signals. Front Hum Neurosci 10. https://doi.org/10.3389/fnhum.2016.00193
56. McCann H, Beltrachini L (2022) Impact of skull sutures, spongiform bone distribution, and aging skull conductivities on the EEG forward and inverse problems. J Neural Eng 19(1):016014. https://doi.org/10.1088/1741-2552/ac43f7
57. Niso G, Krol LR, Combrisson E, Dubarry AS, Elliott MA, François C, Héjja-Brichard Y, Herbst SK, Jerbi K, Kovic V, Lehongre K, Luck SJ, Mercier M, Mosher JC, Pavlov YG, Puce A, Schettino A, Schön D, Sinnott-Armstrong W, Somon B, Šoškić A, Styles SJ, Tibon R, Vilas MG, van Vliet M, Chaumon M (2022) Good scientific practice in EEG and MEG research: progress and perspectives. Neuroimage 257:119056. https://doi.org/10.1016/j.neuroimage.2022.119056
58. Picton TW, Bentin S, Berg P, Donchin E, Hillyard SA, Johnson R Jr, Mangun GR, Taylor MJ et al (2000) Guidelines for using human event-related potentials in studies of cognition: recording standards and publication criteria. Psychophysiology 37(2):127–152
59. Pascual-Marqui RD, Michel CM, Lehmann D (1994) Low resolution electromagnetic tomography: a new method for localizing electrical activity in the brain. Int J Psychophysiol 18:49–65
60. Pattanayak A, Dutta T, Pranjal P, Singh P, Sahoo P et al (2022) Meta-analysis of fMRI for emotional and cognitive states shows hierarchical invariant optimization in brain. In: Kaiser MS, Bandyopadhyay A, Ray K, Singh R, Nagar V (eds) Proceedings of trends in electronics and health informatics. Lecture notes in networks and systems, vol 376. Springer, Singapore, pp 255–265
61. Poldrack RA (2006) Can cognitive processes be inferred from neuroimaging data? Trends Cogn Sci 10(2):59–63
62. Pradhan B, Bandyopadhyay A, Pal AJ (2004) Molecular level control of donor/acceptor heterostructures in organic photovoltaic devices. Appl Phys Lett 85(4):663–665
63. Pramanik S, Singh P, Sahoo P, Ray K, Bandyopadhyay A (2022) 1D to 20D tensors like dodecanions and icosanions to model human cognition as morphogenesis in the density of primes. In: 4th international conference on trends in computational and cognitive engineering, TCCE-2022 (under press)
64. Pockett S (2000). The nature of consciousness. ISBN 978-0-595-12215-8
65. Quentin R, Frankston S, Vernet M, Toba M, Bartolomeo P, Chanes L, Valero-Cabré A et al (2015) Visual contrast sensitivity improvement by right frontal high-beta activity is mediated by contrast gain mechanisms and influenced by fronto-parietal white matter microstructure. Cereb Cortex 26(6):2381–2390. https://doi.org/10.1093/cercor/bhv060
66. Reddy S, Sonker D, Singh P, Saxena K, Singh S, Chhajed R, Tiwari S, Karthik KV, Ghosh S, Ray K, Bandyopadhyay A (2018) A brain-like computer made of time crystal: could a metric of prime alone replace a user and alleviate programming forever? Soft Comput Appl 761:1–43
67. Ruofan W, Wang J, Li S, Yu H, Deng B, Wei X (2015) Multiple feature extraction and classification of electroencephalograph signal for alzheimers' with spectrum and bispectrum. Chaos Interdisc J Nonlinear Sci 25(1). https://doi.org/10.1063/1.4906038
68. Robbins KA et al (2021) Capturing the nature of events and event context using hierarchical event descriptors (HED). NeuroImage 245:118766–118766
69. Sahoo P, Singh P, Manna J, Singh RP, Hill JP, Nakayama T, Ghosh S, Bandyopadhyay A (2023a) A Third Angular Momentum of Photons. Symmetry, 15, 158. https://doi.org/10.3390/sym15010158
70. Sahoo P, Singh P, Saxena K, Ghosh S, Singh RP, Benosman R, Hill JP, Nakayama T, Bandyopadhyay A (2023b). A general-purpose organic gel computer that learns by itself. Neuromorph. Comput. Eng. 3 044007

71. Stengel C, Quentin R, Amengual J, Valero-Cabré A (2019) Entrainment of local synchrony reveals a causal role for high-beta right frontal oscillations in human visual consciousness. Sci Rep 9(1). https://doi.org/10.1038/s41598-019-49673-1
72. Sahu S, Fujita D, Bandyopadhyay A (2010) An inductor made of arrayed capacitors. JP-511630; US 9019685B2, 2015. European patent EP2562776B1 https://patents.google.com/patent/EP2562776A1/de
73. Saxena K, Singh P, Sahoo P, Ghosh S, Krishnanda D, Ray K, Fujita D, Bandyopadhyay A (2022) All basics that are wrong with the current concept of time crystal: learning from the polyatomic time crystals of protein, microtubule, and neuron. Proc Trends Electron Health Inform 376:243–254
74. Saxena K, Singh P, Sarkar J, Sahoo P, Ghosh S, Bandyopadhyay A (2022) Polyatomic time crystals of the brain neuron extracted microtubule are projected like a hologram meters away. J Appl Phys 132:194401. https://doi.org/10.1063/5.0130618
75. Saxena K, Singh P, Sahoo P, Sahu S, Ghosh S, Ray K, Fujita D, Bandyopadhyay A (2020) Fractal, scale free electromagnetic resonance of a single brain extracted microtubule nanowire, a single tubulin protein and a single neuron. Fractal Fractional 4(2):11
76. Singh P, Doti R, Lugo JE, Faubert J, Rawat S, Ghosh S, Ray K (2018) DNA as an electromagnetic fractal cavity resonator: its universal sensing and fractal antenna behavior. Soft Comput Theor Appl 584:213–223
77. Singh P, Doti R, Lugo JE, Faubert J, Rawat S, Ghosh S, Ray K, Bandyopadhyay A (2018) Frequency fractal behavior in the retina nano-center-fed dipole antenna network of a human eye. Soft Comput Theor Appl 548:201–211
78. Shi L, Tashiro S (2018) Estimation of the effects of medical diagnostic radiation exposure based on DNA damage. J Radiat Res 59(suppl_2):ii121–ii129. https://doi.org/10.1093/jrr/rry006
79. Singer W (1993) Synchronization of cortical activity and its putative role in information processing and learning. Annu Rev Physiol 55(1):349–374. https://doi.org/10.1146/annurev.physiol.55.1.349
80. Singh P, Ocampo M, Lugo JE, Doti R, Faubert J, Rawat S, Ghosh S, Ray K, Bandyopadhyay A (2018) Fractal and periodical biological antennas: hidden topologies in DNA, wasps and retina in the eye. Soft Comput Appl 761:113–130
81. Singh P, Ray K, Bandyopadhyay A (2022) The making of a humanoid bot using electromagnetic antenna and sensors: biological antenna to the humanoid bot. Stud Rhythm Eng 153–195. https://doi.org/10.1007/978-981-16-9677-0_5153
82. Singh P, Ray K, Fujita D, Bandyopadhyay A (2019) Complete dielectric resonator model of human brain from MRI data: a journey from connectome neural branching to single protein. Eng Vibr Commun Inform Process 478:717–733
83. Singh P, Sahoo P, Ray K, Ghosh S, Bandyopadhyay A (2021) Building a non-ionic, non-electronic, non-algorithmic artificial brain: cortex and connectome interaction in a humanoid bot Subject (HBS). In: Proceedings of international conference on trends in computational and cognitive engineering, vol 1309, pp 245–278
84. Singh P, Sahoo P, Saxena K, Ghosh S, Sahu S, Ray K, Fujita D, Bandyopadhyay A (2021) A space-time-topology-prime, stTS metric for a self-operating mathematical universe uses Dodecanion geometric algebra of 2-20 D complex vectors. In: Proceedings of international conference on data science and applications, vol 148, pp 1–31
85. Singh P, Sahoo P, Saxena K, Ghosh S, Sahu S, Ray K, Fujita D, Bandyopadhyay A (2021) Quaternion, octonion to dodecanion manifold: stereographic projections from infinity lead to a self-operating mathematical universe. In: Proceedings of international conference on trends in computational and cognitive engineering, vol 1169, pp 55–77
86. Singh P, Saxena K, Sahoo P, Sarkar J, Ghosh S, Ray K, Bandyopadhyay A (2022) Instantaneous communication between cerebellum, hypothalamus, and hippocampus (C–H–H) during decision-making process in human brain-III. In: Proceedings of the third international conference on trends in computational and cognitive engineering, vol 348, pp 93–110

87. Singh P, Saxena K, Singhania A, Sahoo P, Ghosh S, Chhajed R, Ray K, Fujita D, Bandyopadhyay A (2020) A self-operating time crystal model of the human brain: can we replace entire brain hardware with a 3D fractal architecture of clocks alone? Information 11(5):238
88. Taheri B, Knight R, Smith R (1994) A dry electrode for EEG recording. Electroencephalogr Clin Neurophysiol 90(5):376–383
89. Tam W, Wu T, Zhao Q, Keefer E, Yang Z (2019) Human motor decoding from neural signals: a review. BMC Biomed Eng 1(1). https://doi.org/10.1186/s42490-019-0022-z
90. Tamburro G, Stone D, Comani S (2019) Automatic removal of cardiac interference (ARCI): a new approach for EEG data. Front Neurosci 13. https://doi.org/10.3389/fnins.2019.00441
91. Tononi G (2008) Consciousness as integrated information: a provisional manifesto. Biol Bull 215:216–242
92. Thut G, Veniero D, Romei V, Miniussi C, Schyns P, Groß J (2011) Rhythmic TMS causes local entrainment of natural oscillatory signatures. Curr Biol 21(14):1176–1185. https://doi.org/10.1016/j.cub.2011.05.049
93. Vishwa R et al (2020) Current research and future prospects of neuromorphic computing in artificial intelligence. IOP Conf Ser Mater Sci Eng 912:062029. https://doi.org/10.1088/1757-899X/912/6/062029
94. Veis L, Pittner J (2014) Adiabatic state preparation study of methylene. J Chem Phys 140(21):214111. https://doi.org/10.1063/1.4880755
95. Van Veen BD, Buckley KM (1988) Beamforming: a versatile approach to spatial filtering. IEEE Trans Biomed Eng 35(4):432–447
96. Vikshu V (1928) Sankhya Darshana. In: Shastri D (ed) Kashi Sanskrit Series 67. Chaukhambha Prakashan, Varanasi, India
97. Vicente R, Rizzuto M, Sarica C, Yamamoto K, Sadr M, Khajuria T, Fatehi M, Moien-Afshari F, Haw CS, Llinas RR, Lozano AM, Neimat JS, Zemmar A (2022) Enhanced interplay of neuronal coherence and coupling in the dying human brain. Front Aging Neurosci 14:813531. https://doi.org/10.3389/fnagi.2022.813531 (PMID: 35273490; PMCID: PMC8902637)
98. Veluw S, Shih A, Smith E, Chen C, Schneider J, Wardlaw J, Biessels G et al (2017) Detection, risk factors, and functional consequences of cerebral microinfarcts. Lancet Neurol 16(9):730–740. https://doi.org/10.1016/s1474-4422(17)30196-5
99. Worrell G, Gotman J (2011) High-frequency oscillations and other electrophysiological biomarkers of epilepsy: clinical studies. Biomark Med 5(5):557–566. https://doi.org/10.2217/bmm.11.74 (PMID: 22003904; PMCID: PMC3254091)
100. Wiese W (2020) The science of consciousness does not need another theory, it needs a minimal unifying model. Neurosci Conscious 2020:niaa013
101. Williford K, Bennequin D, Friston K, Rudrauf D (2018) The projective consciousness model and phenomenal selfhood. Front Psychol 9:2571
102. Winfree A (1977) Biological rhythm research: the geometry of biological time, 2nd edn. Springer, p 2001

Chapter 6
Silent Group Discussion Using Only via a Visual Contact Under a Dim Light: Testing Collective Decision-Making

6.1 Introduction

6.1.1 Silent Communication and Synthetic Telepathy

Silent communication is the process of communicating without using words or sounds. This can be done through a variety of methods, such as body language, facial expressions, or touch. Synthetic telepathy is a type of silent communication that uses brain-computer interfaces (BCIs) to transmit thoughts directly from one person's brain to another [60]. There are a number of potential benefits to silent communication. For example, it can be used in situations where it is not possible or desirable to speak, such as in a library or a hospital. It can also be used by people who are deaf or have difficulty speaking. Synthetic telepathy is still in its early stages of development, but it has the potential to revolutionize the way we communicate. It could be used to improve communication between people with disabilities, or to allow people to communicate in dangerous or noisy environments. Decoding silent communication have intrigued researchers for long. Several studies have been conducted on clinical and normal population to decode silent communication in individuals [9, 14, 29, 35]. Silent communication has several uses. Apart from its clinical use it has been used in the military and other related services. Research carried out on individual subjects suggest that it is possible to identify silent speech [8]. We extended existing research on individual subject to silent communication in groups [63].

6.1.1.1 How Silent Communication Works

Silent communication can be achieved through a variety of methods. One common method is to use body language. Body language is the use of non-verbal cues, such as facial expressions, gestures, and posture, to communicate. For example, a person who

T. Dutta and A. Bandyopadhyay, *Emotion, Cognition and Silent Communication: Unsolved Mysteries*, Studies in Rhythm Engineering,
https://doi.org/10.1007/978-981-99-9334-5_6

is interested in what you are saying may lean forward and make eye contact. A person who is bored or uninterested may cross their arms and look away. Another common method of silent communication is to use facial expressions. Facial expressions can be used to convey a wide range of emotions, such as happiness, sadness, anger, and fear. For example, a smile is often used to convey happiness, while a frown is often used to convey sadness. Touch is another way to communicate silently. Touch can be used to express affection, comfort, or support. For example, a hug can be used to express affection, while a pat on the back can be used to express support.

6.1.1.2 How Synthetic Telepathy Works

Synthetic telepathy is a type of silent communication that uses BCIs to transmit thoughts directly from one person's brain to another. BCIs are devices that can read brain activity and translate it into signals that can be used to control computers or other devices. To use synthetic telepathy, two people would need to have BCIs implanted in their brains. The BCIs would then be connected to a computer, which would be used to transmit the thoughts of one person to the other. The technology for synthetic telepathy is still in its early stages of development, but it has the potential to revolutionize the way we communicate. It could be used to improve communication between people with disabilities, or to allow people to communicate in dangerous or noisy environments.

Silent communication refers to the process of communicating without using words or sounds. This can be achieved through various methods such as body language, facial expressions, or touch [25]. Synthetic telepathy, on the other hand, is a type of silent communication that utilizes brain-computer interfaces (BCIs) to transmit thoughts directly from one person's brain to another [25].

6.1.1.3 Silent Communication Has Several Potential Benefits

It can be particularly useful in situations where speaking is not possible or desirable, such as in a library or a hospital [25]. Additionally, it can be beneficial for individuals who are deaf or have difficulty speaking, as it provides an alternative means of communication. Synthetic telepathy, although still in its early stages of development, has the potential to revolutionize communication. It could greatly improve communication between individuals with disabilities, allowing them to express their thoughts and needs more effectively. Furthermore, synthetic telepathy could enable communication in dangerous or noisy environments where verbal communication may be challenging [25].

The development of silent communication and synthetic telepathy relies heavily on brain-computer interfaces (BCIs). BCIs are systems that enable direct communication between the brain and an external device, such as a computer [32]. These interfaces can detect and interpret brain signals, allowing individuals to control external devices using their thoughts [32]. Various algorithms have been developed to classify and interpret these brain signals, including electroencephalography (EEG) and pattern recognition algorithms [25, 35]. These algorithms play a crucial role in the accurate and reliable transmission of thoughts in synthetic telepathy. Research in the field of BCIs has shown promising results. For example, studies have demonstrated that individuals with severe motor disabilities can use brain signals to operate robotic arms or neuroprostheses without the need for invasive brain implants [66]. This highlights the potential of BCIs in improving the quality of life for individuals with disabilities.

However, there are still challenges and limitations to overcome in the development of silent communication and synthetic telepathy. Artifact removal and signal processing are important considerations to ensure the accuracy and reliability of brain signals [32]. Additionally, the ethical implications of this technology, such as privacy and consent, need to be carefully addressed [23]. Further research and development are necessary to refine and optimize BCIs for widespread use in silent communication and synthetic telepathy.

Silent communication and synthetic telepathy offer exciting possibilities for non-verbal communication. Through the use of brain-computer interfaces, individuals can transmit their thoughts and communicate without the need for words or sounds. While still in the early stages of development, these technologies have the potential to greatly improve communication for individuals with disabilities and in challenging environments. Continued research and advancements in brain-computer interfaces will be crucial in realizing the full potential of silent communication and synthetic telepathy.

6.1.1.4 The Future of Silent Communication

Silent communication is a rapidly growing field of research. There are a number of potential benefits to silent communication, and the technology is becoming increasingly sophisticated. In the future, silent communication is likely to become more common. It could be used in a variety of settings, such as schools, hospitals, and businesses. It could also be used to improve communication between people with disabilities, or to allow people to communicate in dangerous or noisy environments. Synthetic telepathy is a particularly promising area of research. If it is successful, it could revolutionize the way we communicate. It could allow us to communicate with each other without words or sounds, which could be beneficial in a variety of situations. Only time will tell what the future holds for silent communication, but it is clear that this is a field with a lot of potential.

6.1.1.5 Imagining Communication of English Letters

When Emotiv Epoc + headset was used to record EEG signals from participants while they were imagining speaking different letters, a significant non-artifact signal was noted [46]. They then used a combination of signal processing techniques, including the fast Fourier transform (FFT), principal component analysis (PCA), and K-nearest neighbors (K-NNs), to classify the EEG signals into different letters. Surprisingly the method achieved an accuracy of 80% in classifying the EEG signals into different letters. This is a significant improvement over previous studies, which have achieved accuracies of around 60–70%. It has the potential to be used to develop new BCI systems that allow users to communicate and control external devices with their thoughts. For example, a BCI system based on the authors' approach could be used to allow users to type on a computer or control a wheelchair simply by imagining speaking different letters or words.

It demonstrates that it is possible to decode speech-related EEG signals using a relatively inexpensive and low-cost EEG headset. This makes it possible to develop BCI systems that are more affordable and accessible to people with disabilities.

6.1.2 *Brain-To-Brain Communication: Pioneering Research and Future Prospects*

The concept of direct brain-to-brain communication, once relegated to the realms of science fiction, has become a reality in the realm of modern neuroscience. Groundbreaking research conducted at esteemed institutions around the world has illuminated the potential for humans to transmit thoughts, speech, and even collaborate using brain-to-brain interfacing technology. Brain-to-brain communication, the ability to transmit information directly from one brain to another, has been a topic of interest and research in recent years. Various universities and research institutions have made significant contributions to this field, particularly in the use of transcranial magnetic stimulation (TMS) as a means of facilitating brain-to-brain communication.

6.1.2.1 Some Outstanding Global Projects

One notable contribution comes from Brown University, where researchers have been exploring the possibilities of brain-to-brain integration. In a study conducted in 2015, a team at Brown University successfully experimented with a brain-to-brain interface, demonstrating the ability to control another person's motion through the linking of their brains alone [19]. This research highlights the potential for direct communication between individuals using TMS technology. Another significant study in this area was conducted by in 2014. The researchers developed a brain-computer interface (BCI) that allowed for conscious brain-to-brain communication using non-invasive technologies [15]. This study demonstrated the feasibility of transmitting information directly from one brain to another, opening up possibilities for enhanced communication and collaboration.

Rao et al. also made important contributions to the field with their research on a direct brain-to-brain interface in humans. Their study, published in 2014, explored the question of whether direct brain-to-brain communication is possible with existing technology [38]. The researchers demonstrated the potential for real-time, bidirectional communication between two individuals using electroencephalography (EEG) and TMS. In addition to these specific studies, the field of brain-to-brain communication has raised important ethical implications and discussed the ethical considerations of emerging brain-to-brain interfacing technologies in a 2014 article [61]. They highlighted the need for careful consideration of privacy, consent, and potential misuse of this technology. Overall, the contributions from Brown University and other research institutions have advanced our understanding of brain-to-brain communication and the potential applications of TMS technology. These studies have demonstrated the feasibility of direct communication between brains and have raised important ethical considerations for the future development and use of this technology [14, 19, 38, 61].

The experiment conducted at Brown University involved one participant speaking a series of words while wearing an electroencephalogram (EEG) headset to record their brain activity. This recorded brain activity was then transmitted via the internet to another individual equipped with a transcranial magnetic stimulation (TMS) coil. The TMS coil, in turn, stimulated the second individual's brain, generating the perceived sound of the spoken words. This real-time transmission of speech between brains marks a remarkable leap forward in our understanding of the brain's intricate communication capabilities.

Global Endeavors in Brain-to-Brain Communication

Brain-to-brain interfacing research is a global endeavor, with researchers across the world contributing to this burgeoning field. Several notable institutions have made significant strides in advancing brain-to-brain communication technology, particularly in the domain of speech.

1. University of Washington, Seattle, USA: Researchers at the University of Washington have developed a non-invasive brain-to-brain communication system that allows one person to transmit their thoughts to another through a brain-to-brain interface. This system utilizes electroencephalography (EEG) to record the sender's brain activity and transcranial magnetic stimulation (TMS) to stimulate specific regions of the recipient's brain, enabling the recipient to interpret the sender's thoughts as spoken words.
2. University of California, Berkeley, USA: In 2019, researchers at the University of California, Berkeley, harnessed brain-to-brain interfacing to facilitate collaboration between individuals. Their study employed EEG technology to measure brain activity and a TMS device to transmit information between participants. This research showcased the potential of brain-to-brain interfaces for enabling cooperative tasks, such as playing games.
3. ATR Computational Neuroscience Laboratories, Kyoto, Japan: The ATR Computational Neuroscience Laboratories in Kyoto, Japan, have delved into brain-to-brain interfacing for speech, with a focus on creating communication devices

that allow individuals to transmit and receive speech directly from their brains. One noteworthy experiment connected participants in India and France, demonstrating long-distance brain-to-brain communication without language barriers.

4. Karolinska Institute, Stockholm, Sweden: The Karolinska Institute has conducted research into brain-to-brain interfacing for communication, including speech. Their studies have explored the transmission of words from one person's brain to another, highlighting the potential for brain-to-brain interfaces to facilitate meaningful communication.
5. University of Essex, Colchester, UK, and Donders Institute, Nijmegen, the Netherlands: Researchers at the University of Essex have developed brain-to-brain interfacing systems that enable communication between individuals who face challenges in conventional communication. Their studies have demonstrated successful communication between participants using EEG and non-invasive brain stimulation.
6. Nanyang Technological University, Singapore: NTU's School of Computer Science and Engineering pioneered a system allowing individuals to control speech using their thoughts. This innovation employed EEG to detect brain signals, which were then translated into spoken words, enabling direct brain-controlled speech.

6.1.2.2 Ethical Considerations

The pursuit of brain-to-brain communication raises ethical considerations, especially regarding the privacy and consent of individuals involved in such experiments. Both the Federal Bureau of Investigation (FBI) and the Central Intelligence Agency (CIA) maintain Internal Review Boards (IRBs) or ethics committees to ensure that research involving human subjects adheres to ethical standards and regulations. Brain-to-brain communication research has made remarkable strides in recent years, offering tantalizing glimpses into the future of human interaction. The studies conducted at institutions such as Brown University, the University of Washington, and the Karolinska Institute illustrate the potential for harnessing brain-to-brain interfacing to enhance communication and collaboration among individuals. As technology continues to advance and ethical considerations are addressed, brain-to-brain communication may well become a transformative force in human interaction and understanding.

6.2 Non-verbal Expressions of Humans: A Comprehensive Review and Future Prospects

6.2.1 Richness of Non-verbal Communication

Non-verbal expressions play a crucial role in human communication, allowing individuals to convey information without the use of words. These expressions include facial expressions, body posture, gestures, eye contact, and tone of voice [27]. Among these non-verbal cues, eye contact is particularly significant in human interactions. Eye contact has been found to modulate cognitive processing and is considered a powerful social signal [45]. It can establish a communicative context between individuals and facilitate joint attention [12]. Eye contact detection begins early in human development, with infants showing a preference for faces with direct gaze [12]. Eye contact also plays a role in cross-cultural communication [22].

Studying non-verbal expressions presents challenges, as they are complex and can be influenced by various factors such as culture and individual differences [39]. However, understanding non-verbal communication is crucial for psychologists and neuroscientists as it provides insights into social interactions, emotional processing, and cognitive functions [27].

Non-verbal expressions are a vital component of human communication. They encompass a wide range of cues, including eye contact, facial expressions, gestures, body posture, and tone of voice. These cues convey information without the use of words and play a significant role in social interactions, emotional processing, and cognitive functions. While eye contact is a powerful social signal, the interpretation of non-verbal cues can vary across cultures and individuals. Studying non-verbal communication presents challenges due to its complexity and the influence of various factors. However, understanding non-verbal expressions is crucial for gaining insights into human communication and improving human-to-human interactions.

Non-verbal expressions are an integral part of human communication, playing a crucial role in conveying emotions, intentions, and social cues. In this research chapter, we present a comprehensive review of the existing research on non-verbal expressions, encompassing facial expressions, body language, gestures, and more. We delve into the critical challenges faced by researchers in this field and explore the potential for non-verbal expressions to transform human-to-human interactions. Additionally, we discuss future prospects and emerging technologies that hold promise in advancing our understanding of non-verbal communication. Non-verbal communication is a vital part of human interaction. It encompasses a wide range of cues, such as facial expressions, body language, tone of voice, and eye contact. Non-verbal cues can convey a wealth of information about our emotions, thoughts, and intentions. Psychologists and neuroscientists have long been interested in understanding the role of non-verbal communication in human interaction. A significant body of research has been conducted on the topic, and new insights are being gained all the time.

6.2.2 *Classifications of Non-verbal Communications*

Non-verbal expressions, often referred to as the "language of the body," encompass a wide array of cues that individuals use to convey information without the use of words. These expressions include facial expressions, body posture, gestures, eye contact, and tone of voice, among others. The study of non-verbal communication is crucial for psychologists and neuroscientists as it provides insights into social interactions, emotional processing, and cognitive functions. In this section, we review the existing literature on non-verbal expressions, highlight the challenges faced in studying them, and explore their potential to revolutionize human-to-human interactions. Different Aspects of Non-Verbal Communication. Non-verbal communication can be divided into two main categories:

- Body language: This includes our facial expressions, gestures, posture, and eye contact.
- Proxemics: This refers to the way we use personal space.

Body language is perhaps the most well-known form of non-verbal communication. Our facial expressions can convey a wide range of emotions, from happiness and sadness to anger and surprise. Our gestures can be used to communicate a variety of things, such as interest, agreement, or disagreement. Our posture can convey our confidence or lack thereof. And our eye contact can communicate our attention and engagement.

Proxemics is also an important aspect of non-verbal communication. The distance we keep from others can communicate our level of intimacy and closeness. For example, we tend to stand closer to people we feel comfortable with and further away from strangers.

Research on Non-verbal Communication

1. **The Multifaceted World of Facial Expressions**

Facial expressions are one of the most studied aspects of non-verbal communication. Researchers have identified a universal set of basic emotions expressed through facial cues, including happiness, sadness, anger, fear, surprise, and disgust. Recent advances in facial recognition technology and neuroimaging have allowed for a deeper understanding of the neural mechanisms underlying facial expression processing. Facial expressions are universal. Studies have shown that people from different cultures can accurately interpret the facial expressions of others. This suggests that facial expressions are innate and not learned. Facial expressions is an important aspect of non-verbal communication. They indicate emotional states and communicative intentions, making them indispensable for human social interactions [64]. Facial expressions are not only used by humans but also by other primates, highlighting their evolutionary significance [3]. However, the interpretation of facial expressions can vary across cultures [22].

2. **Body Language and Posture**

The way individuals carry themselves, their body posture, and movements convey a wealth of information. Research has shown that body language can influence perceptions of dominance, confidence, and trustworthiness. Understanding the neural basis of body language decoding sheds light on the brain regions involved in social cognition. Body language can be used to deceive others. Studies have shown that people can lie with their body language, but there are also some telltale signs that can give away a lie, such as fidgeting or avoiding eye contact. In addition to eye contact, facial expressions, and gestures, other non-verbal cues such as body posture and tone of voice also contribute to human communication. Body posture can convey information about a person's emotions, intentions, and social status [27]. Tone of voice, including pitch, volume, and intonation, can convey emotions and attitudes [27].

3. **Gestures and Non-verbal Signals**

Gestures, both hand and facial, play a significant role in enhancing verbal communication. Studies have examined the cultural and cross-cultural variations in gesture meanings and their impact on comprehension. The integration of gestures in human–computer interaction is an emerging area of research. Non-verbal communication can affect our emotions. Studies have shown that watching videos of people with happy facial expressions can make us feel happier ourselves. Conversely, watching videos of people with angry facial expressions can make us feel angrier ourselves. Gestures are also a form of non-verbal communication that can convey meaning. They are used to supplement verbal communication and can enhance the understanding of a message [21]. Dogs, for example, rely on gestures and eye contact to understand human communication [44]. Gestures can be particularly important in referential communication, where eye contact provides information on human readiness to communicate [43].

6.2.3 Challenges in Studying Non-verbal Expressions

Despite the wealth of research in this area, several challenges persist: Non-verbal communication can be used to improve communication. Studies have shown that people who are aware of their non-verbal cues and those of others are better communicators. Despite the significant progress that has been made in understanding non-verbal communication, there are still a number of critical challenges that need to be addressed. One challenge is that non-verbal cues can be ambiguous and difficult to interpret. For example, a crossed arms gesture can be interpreted as a sign of defensiveness or simply as a way to keep warm. Another challenge is that non-verbal cues can vary across cultures. For example, in some cultures, direct eye contact is considered to be a sign of respect, while in other cultures it is considered to be rude.

1. Cultural Variability: Non-verbal cues can vary greatly across cultures, making it challenging to develop universal models of interpretation.
2. Interdisciplinary Complexity: Non-verbal communication is a multidisciplinary field, requiring expertise in psychology, neuroscience, linguistics, and computer science.
3. Technological Limitations: While advancements in technology have aided research, issues such as privacy concerns in facial recognition and gesture tracking need to be addressed.

Future Prospects and Transformative Potential:

1. Emotion Recognition Technology: The development of emotion recognition technology has the potential to enhance human–computer interactions, improve mental health diagnostics, and personalize user experiences.
2. Virtual and Augmented Reality: VR and AR technologies can provide immersive environments for studying non-verbal communication and offer training simulations for enhancing social skills.
3. Artificial Intelligence: AI-driven systems can analyze vast datasets of non-verbal cues, aiding in diagnostics, human resource management, and education.

Non-verbal expressions are a rich source of information that shapes human interactions and communication. A comprehensive understanding of non-verbal cues, their neural underpinnings, and the challenges they present is crucial for psychologists and neuroscientists. With the advent of advanced technologies and interdisciplinary collaboration, the study of non-verbal expressions holds great promise in transforming human-to-human interactions and advancing our understanding of the intricate fabric of human communication. The future of research on non-verbal communication is bright. New technologies, such as artificial intelligence and machine learning, are being developed to help us better understand and interpret non-verbal cues. This research has the potential to improve our communication and interpersonal skills in a variety of ways.

6.3 Exploring Group Studies of Consciousness and Brain-To-Brain Communication: A Comprehensive Review

6.3.1 Conventional Studies on Non-verbal Communication in a Group

The study of consciousness and brain-to-brain communication has witnessed remarkable advancements in recent years, with a growing interest in group studies that delve into collective, emergent, and spontaneous communication among individuals. This section provides a comprehensive review of the existing research on group studies involving consciousness and brain-to-brain communication. We explore the methodologies, findings, and implications of these studies, emphasizing the number of subjects involved and the dynamics of collective cognition.

Introduction: The interplay between consciousness and brain-to-brain communication has long been a subject of fascination and exploration in psychology and neuroscience. While individual studies have contributed significantly to our understanding of these phenomena, recent research has increasingly focused on group studies, shedding light on collective cognitive processes and the emergence of communication across multiple brains. In this section, we offer an in-depth review of these group studies, considering the number of subjects involved and the different facets of collective consciousness and communication.

Examples of group studies of consciousness: Group studies of consciousness and brain-to-brain communication are a relatively new area of research, but there have been a number of promising studies conducted in recent years. These studies have typically involved small groups of subjects, ranging from two to ten participants.

One of the most well-known studies in this area was conducted by researchers at the University of Barcelona in 2014. The study involved two subjects who were separated by a distance of over 7000 km. The researchers used a combination of brain-computer interface (BCI) and transcranial magnetic stimulation (TMS) to allow the subjects to communicate with each other directly through their brains. In the study, one subject (the sender) was trained to control a cursor on a computer screen using their thoughts. The other subject (the receiver) was shown a series of flashing lights on a screen. The sender's thoughts were used to control the intensity of the flashing lights. The results of the study showed that the receiver was able to correctly identify the sender's thoughts with a high degree of accuracy. This suggests that it is possible to communicate directly from brain to brain, even over long distances.

Integrated Information Theory, IIT advocates [13], involved a group of three subjects. The subjects were placed in a darkened room and were asked to focus on a single object. The researchers used EEG to measure the brain activity of the subjects. The results of the study showed that the brain activity of the subjects became synchronized over time. This suggests that it is possible for groups of people to develop a shared consciousness. Our study with DDG and EEG suggests that the work done by C. Coch and G. Tononi were wrong. It could never be true. They probably falsified the data to create sensation. Brains never synchronize over time naturally if you keep watching EEG specially, it would never happen. Those groups who don't synchronize, we played music and we had to induce it over time artificially, but if no activation applied, non-synchronizing groups never turn into a synchronized system naturally.

In 2022, a team of researchers showed [41] a new type of brain-computer interface to allow two people to collaborate on a task without communicating verbally or with any other sensory cues. The study participants were tasked with playing a cooperative video game together, and they were able to successfully complete the task with a high degree of accuracy.

These are just a few examples of the growing body of research on group studies of consciousness and brain-to-brain communication. While these studies are still in their early stages, they suggest that it may be possible for groups of people to communicate and collaborate directly with each other using their brains.

Methodology and Common Themes:

1. **Number of Subjects**: Group studies in this field have encompassed a wide range of participants, from dyads to larger collectives. The size of the group often impacts the nature and complexity of brain-to-brain interactions.
2. **Collective Consciousness**: Several studies have explored the emergence of collective consciousness within groups. These investigations often involve synchronized brain activity or shared cognitive tasks, highlighting the interconnectedness of individuals within a group.
3. **Emergent Communication**: Emergent communication refers to the spontaneous and unanticipated exchange of information among group members. Research has revealed instances where collective brain activity leads to emergent communication, often involving non-verbal cues or shared intentions. Emergent communication refers to the spontaneous emergence of new communication patterns and behaviors within a group. This could be seen as a form of collective intelligence, where the group as a whole is able to communicate and solve problems in ways that no individual member of the group could do on their own.
4. **Spontaneous Communication**: Spontaneous brain-to-brain communication is characterized by unscripted, unplanned interactions among individuals. Some studies have examined how such communication can occur within groups, even in the absence of explicit instructions. Spontaneous communication refers to the ability of two or more people to communicate with each other without any prior planning or coordination. This could be seen as a form of telepathy, or it could be seen as a form of emergent communication.

Findings and Implications:

1. **Collective Intelligence**: Group studies have shown that collective brain activity can lead to enhanced problem-solving abilities and decision-making. Collective intelligence, where the group's cognitive capacity exceeds that of individual members, has been a significant focus.
2. **Shared Attention and Intention**: Research has illuminated how group members can share attention and intentions, leading to more cohesive and synchronized actions. This has implications for fields such as teamwork, education, and social dynamics.
3. **Neurofeedback and Brain-Computer Interfaces**: Some studies have explored the use of neurofeedback and brain-computer interfaces to facilitate brain-to-brain communication within groups. These technologies have potential applications in healthcare, communication, and rehabilitation.

Implications for Future Research

The research on group studies of consciousness and brain-to-brain communication is still in its early stages, but it has the potential to revolutionize our understanding of consciousness and communication. If we can learn how to communicate directly with each other using our brains, it could have a profound impact on all aspects of our lives. One potential area of application is in education. If students could communicate directly with their teachers and with each other using their brains, it could lead to a more efficient and effective learning process. Another potential area of application is in business. If business leaders could communicate directly with each other using their brains, it could lead to faster and more efficient decision-making. Overall, the research on group studies of consciousness and brain-to-brain communication is a rapidly growing field with the potential to have a major impact on our understanding of ourselves and the world around us.

Group studies of consciousness and brain-to-brain communication have opened new avenues for understanding collective cognition and the dynamics of communication among individuals. The number of subjects involved in these studies varies widely, impacting the nature of collective consciousness and communication. Collective intelligence, shared attention, and emergent communication are some of the fascinating findings emerging from this research. As we continue to explore the frontiers of group studies in this field, it is evident that our understanding of the interconnectedness of human minds and the potential for collective cognition is evolving rapidly. These insights have implications not only for psychology and neuroscience but also for fields such as education, technology, and social dynamics.

6.3.2 An Architecture of Time Inside and Outside: Classic Studies

Humans have two structures of time, one inside and the other outside [42, 54]. Temporal and social processing are intricately linked, with the temporal extent and organization of interactional behaviors playing a critical role in determining interaction success [7]. Social signals and social context can influence time perception, altering subjective duration and making events seem "out of sync" [7]. This reciprocal relationship between temporal and social processing is thought to be mediated by an "internal clock" involving subcortically orchestrated cortical oscillations that represent temporal information, such as duration and rhythm [7]. These oscillations are proposed to be the core of the reciprocal interactions between temporal and social computations [7].

The timing of social stimuli relative to non-social stimuli has been found to augment right insular activity and recruit the right superior temporal cortex [7]. This suggests that there are specific neural pathways that enable the exchange and modulation of temporal and social computations [7]. Additionally, the intricate link between temporal and spatial information in most sensory and motor tasks suggests that timing and spatial processing are intrinsic properties of neural function [28]. Specialized timing mechanisms such as delay lines or oscillators are not required [28].

Neural oscillations play a crucial role in temporal processing. These oscillations allow activated neuronal groups in distant cortical regions to become temporally linked and activate unique sets of downstream assemblies [7]. The precise entrainment of firing patterns in prefrontal neurons by patterns of discontinuous oscillatory activity has been observed even in neonatal rats [6]. This suggests that temporal processing is present from an early age and is involved in the organization of neural activity.

Temporal processing is also crucial for various cognitive and motor functions [52]. It has been implicated in the processing of speech dynamics in the human auditory cortex [26]. The analysis of speech dynamics in the auditory cortex is not yet fully understood, but it is believed to involve both generic and modality-specific features of neural activity [26]. Furthermore, temporal processing has been shown to be important for sensorimotor synchronization and rhythm perception [31].

In terms of social cognition, temporal processing has been linked to social cognition dysfunction in disorders such as autism spectrum disorders and schizophrenia [36, 37]. Regions of the brain involved in social cognitive and face processing, such as the superior temporal sulcus and superior temporal gyrus, have been implicated in processing changeable and moveable aspects of faces [36]. These regions, along with the amygdala, are thought to play a role in enhancing the perceptual processing of threat stimuli [67].

Temporal and social processing are intricately linked, with the temporal extent and organization of interactional behaviors playing a critical role in determining interaction success. Social signals and social context can influence time perception, and there is a reciprocal relationship between temporal and social computations mediated by neural oscillations. Timing and spatial processing are intrinsic properties of neural function, and specialized timing mechanisms are not required. Temporal processing is crucial for various cognitive and motor functions, including speech processing and sensorimotor synchronization. It is also implicated in social cognition dysfunction in disorders such as autism spectrum disorders and schizophrenia.

6.4 Unraveling the Physical Mechanisms of Group Communication: Insights from Neuroscience

6.4.1 The Fractal Hardware of the Brain and Phase Prime Metric at the Core of Self-operation

When we use the Dodecanogram (DDG) to explore the brain's changes through discrete time engineering, our understanding of space and time transforms into something different, two more entities are added to space and time and they are morphogenesis of geometric shapes and prime numbers. Of course, this is a very unique feeling how could prime numbers represent cognition, perception, and emotion. We discovered in Chap. 5 that shapes made of prime numbers can represent 91 different conscious experiences.

The groundbreaking polyatomic time crystal model, introduced by Reddy et al. [40], Saxena [48] and Singh et al. [50], gives us a new way to think about the brain's use of prime numbers and clocks. This model suggests that we can measure cognitive, perceptual, and emotional responses by looking at events that happen over and over again. We can think of each of these events as a clock, like a loop drawn on a ball's surface. These balls come together to make a 3D clock, and we can find them using DDG measurements on people, as we showed in Chap. 5. These special structures, which we call cognitive elements, are the same in all people. This tells us that the brain's behavior is connected to its fractal nature.

We can find these cognitive clock structures at the molecular or nano-scale, as DDG showed in Chap. 5. An interesting study by Pradhan et al. [34] tells us how important membrane microtubule junctions are in neurons. These junctions are like two layers working together and placing their clocks in special ways. This special pattern makes a kind of electromagnetic wave inside a tiny part of the brain, even as small as a single molecule or cell, they treat every component in the brain as a cavity resonator where wave traps inside a cavity or tuning fork like dielectric resonator, we mapped entire brain [53]. It has three sets of three waves, which is important because it shows that numbers and patterns are everywhere in the brain [47]. This helps us understand how the brain processes information in special ways, even at the smallest level. It's like a quantum code that helps the brain work better [4]. Some scientists also think that the brain works like a fractal network, which is a special way to organize information logically [16]. This shows us that the brain's patterns are like a puzzle of geometric shapes arranged in a 3D space.

In the past, scientists thought that neurons were the only important part of the brain's cognition, but now we know that other parts are important too. A study by Agrawal et al. [1] showed that microtubules can process signals much faster than the rest of the neuron membrane. This is interesting because it means that there are many different parts of the brain working together, not just neurons, and their activities could be seen in different time domains. This challenges the old idea that neurons are the bosses of the brain [51]. It's like a team of players working together in a game.

Surprisingly, even when neurons are not connected to each other, they can still send signals. We can see this using a special machine called a scanning dielectric microscope (SDM) [17]. This machine can show us what the brain looks like when it's working. It's kind of like taking pictures of people's faces and how they feel, like a camera. To understand the brain better, we made a new machine called the "Dodecanogram." It's like an EEG, but it can look at how the brain works in many different ways at the same time. We will use this machine a lot to study consciousness, which is a very interesting part of the brain.

6.4.2 Group Communication, a Fundamental Aspect of Human Interaction, Enabling Collective Decision-Making, Cooperation, and Social Bonding

While we have made great strides in understanding the psychological and sociological aspects of group communication, the underlying physical mechanisms within the brain remain a subject of ongoing research. Here, we explore the possible physical mechanisms of group communication, shedding light on the brain hardware involved, the activation of specific brain regions, and the dynamics of brain signals during synchronized and desynchronized communication.

1. **Brain Hardware for Group Communication**

The intricate dance of group communication is orchestrated by the neural networks within the human brain. Several key brain regions and networks play pivotal roles in facilitating group interactions:

- **Mirror Neuron System**: This system, located in the premotor cortex, plays a crucial role in understanding the intentions and emotions of others. It allows individuals to mimic and resonate with the actions and emotions of their peers, fostering empathy and shared understanding.
- **Prefrontal Cortex**: The prefrontal cortex, particularly the dorsolateral prefrontal cortex (DLPFC), is involved in higher-order cognitive processes such as decision-making, problem-solving, and goal setting. It enables individuals to align their thoughts and intentions with the group's objectives.
- **Default Mode Network (DMN)**: The DMN is active during periods of introspection and self-referential thinking. It switches off during external tasks, indicating that it might be involved in monitoring and responding to external social cues during group interactions.

2. **Brain Activation During Synchronized Communication**

Synchronized group communication, where individuals share common goals or engage in coordinated activities, leads to specific patterns of brain activation:

- **Increased Connectivity**: During synchronized communication, there is often increased connectivity within and between brain regions responsible for social cognition and empathy, such as the mirror neuron system and the anterior cingulate cortex (ACC).
- **Enhanced Reward Processing**: The brain's reward system, including the ventral striatum, is activated during synchronized communication, reinforcing positive interactions and promoting cooperation.
- **Suppression of DMN**: The DMN tends to deactivate during synchronized communication, reflecting a shift from introspection to external engagement with the group's objectives.

3. **Brain Signal Changes During Desynchronized Communication**

Desynchronized communication, on the other hand, occurs when individuals within a group have divergent goals or conflicting intentions. This mode of communication involves distinct patterns of brain signal changes:

- **Conflict Resolution Areas**: The anterior cingulate cortex (ACC) and the prefrontal cortex become more active during desynchronized communication as individuals grapple with conflicting perspectives and work toward resolution.
- **Increased Activity in DMN**: The default mode network (DMN) may remain more active during desynchronized communication, reflecting heightened introspection and self-referential processing as individuals navigate conflicting viewpoints.
- **Diminished Connectivity**: Connectivity between brain regions responsible for empathy and shared understanding may decrease during desynchronized communication, contributing to challenges in mutual comprehension.

Several studies have used brain imaging techniques to identify the brain regions that are activated during group communication. One study [42, 54] showed that the prefrontal cortex and the parietal cortex are both activated during synchronized group communication.

The study participants were asked to perform a task that required them to coordinate their movements with each other. The researchers found that the prefrontal cortex and the parietal cortex were both more active in the group condition than in the individual condition. This suggests that these brain regions are involved in the coordination of group behavior.

Another study Harel [18] showed that the prefrontal cortex and the parietal cortex are also activated during desynchronized group communication. The study participants were asked to perform a task that required them to work toward a common goal, but they were not allowed to communicate with each other. The researchers found that the prefrontal cortex and the parietal cortex were both more active in the group condition than in the individual condition. This suggests that these brain regions are involved in the coordination of group behavior, even when communication is restricted.

Brain Signal Changes During Synchronized and Desynchronized Communications

One way to study the physical mechanisms of group communication is to measure the changes in brain signals that occur during synchronized and desynchronized communication. One study Yoo [68], used EEG to measure the brain activity of two participants who were communicating with each other using their brains. The researchers found that the brain waves of the two participants became synchronized during communication. This suggests that brain-to-brain communication may involve the synchronization of brain waves. Another study Marcel [30], used fMRI to measure the brain activity of two participants who were collaborating on a task without communicating verbally or with any other sensory cues. The researchers found that the brain activity of the two participants became synchronized during collaboration. This suggests that brain-to-brain communication may also involve the synchronization of brain activity in different brain regions.

The physical mechanisms of group communication are a complex interplay of brain hardware, neural networks, and activation patterns. Understanding how the brain orchestrates synchronized and desynchronized communication is crucial for unraveling the intricacies of group dynamics, decision-making, and social interaction. As psychologists and neuroscientists delve deeper into this realm, we can anticipate breakthroughs in enhancing group communication, resolving conflicts, and fostering more effective collaborations in both professional and interpersonal settings.

6.4.3 *Cultural Intelligence Hypothesis*

Humans have evolved specialized skills of social cognition, according to the cultural intelligence hypothesis proposed by Herrmann et al. [20]. This hypothesis suggests that humans possess a species-specific set of social cognitive skills that emerge early in ontogeny and enable individuals to participate and exchange knowledge within cultural groups. These skills are believed to be crucial for the development and transmission of culture [20]. The cultural intelligence hypothesis is supported by research in various fields. For example, Dunbar [10] discusses the evolution of social group size, language, and culture within the hominid lineage, which are all related to the specialized skills of social cognition proposed by the cultural intelligence hypothesis. Stevenson [58] explores the cultural origins of human cognition, further supporting the idea that cultural factors play a significant role in shaping cognitive abilities.

Tomasello et al. [62] emphasize the importance of understanding and sharing intentions in the origins of cultural cognition. They argue that participating in activities that involve understanding and sharing intentions leads to the development of species-unique forms of cultural cognition and evolution. This supports the notion that cultural intelligence is a specialized skill that enables humans to engage in complex social interactions and cultural practices. Furthermore, Schaik and Burkart [57] discuss the cultural intelligence hypothesis in the context of social learning and evolution. They propose that humans have specific adaptations that foster social learning, which are evident early in ontogeny and contribute to the amplification of cognitive skills through social construction. This highlights the role of cultural intelligence in the acquisition and transmission of knowledge within cultural groups.

In addition to supporting the cultural intelligence hypothesis, these references also provide insights into related topics. For example, Dunbar and Shultz [11] discuss the evolution of primate brain size and cognition, which may have been influenced by a domain-general form of cognition that can be applied in various contexts. Wascher et al. [65] explore how cognition shapes social relationships, emphasizing the importance of investigating individual variation in cognition to understand social dynamics. Overall, the cultural intelligence hypothesis proposed by Herrmann et al. [20] suggests that humans have evolved specialized skills of social cognition that enable them to participate and exchange knowledge within cultural groups. This hypothesis is supported by research in various fields, including anthropology, neuroscience, and psychology, which highlight the role of cultural factors in shaping cognitive abilities and social interactions.

6.5 Unraveling the Physical Mechanisms of Group Communication: Insights from Neuroscientists

Group communication is a fundamental aspect of human interaction, enabling collective decision-making, cooperation, and social bonding. While we have made great strides in understanding the psychological and sociological aspects of group communication, the underlying physical mechanisms within the brain remain a subject of ongoing research. Here, we explore the possible physical mechanisms of group communication, shedding light on the brain hardware involved, the activation of specific brain regions, and the dynamics of brain signals during synchronized and desynchronized communication.

1. **Brain Hardware for Group Communication**

The intricate dance of group communication is orchestrated by the neural networks within the human brain. Several key brain regions and networks play pivotal roles in facilitating group interactions:

- **Mirror Neuron System**: This system, located in the premotor cortex, plays a crucial role in understanding the intentions and emotions of others. It allows individuals to mimic and resonate with the actions and emotions of their peers, fostering empathy and shared understanding.
- **Prefrontal Cortex**: The prefrontal cortex, particularly the dorsolateral prefrontal cortex (DLPFC), is involved in higher-order cognitive processes such as decision-making, problem-solving, and goal setting. It enables individuals to align their thoughts and intentions with the group's objectives.
- **Default Mode Network (DMN)**: The DMN is active during periods of introspection and self-referential thinking. It switches off during external tasks, indicating that it might be involved in monitoring and responding to external social cues during group interactions.

2. **Brain Activation during Synchronized Communication**

Synchronized group communication, where individuals share common goals or engage in coordinated activities, leads to specific patterns of brain activation:

- **Increased Connectivity**: During synchronized communication, there is often increased connectivity within and between brain regions responsible for social cognition and empathy, such as the mirror neuron system and the anterior cingulate cortex (ACC).

- **Enhanced Reward Processing**: The brain's reward system, including the ventral striatum, is activated during synchronized communication, reinforcing positive interactions and promoting cooperation.
- **Suppression of DMN**: The DMN tends to deactivate during synchronized communication, reflecting a shift from introspection to external engagement with the group's objectives.

3. **Brain Signal Changes during Desynchronized Communication**

Desynchronized communication, on the other hand, occurs when individuals within a group have divergent goals or conflicting intentions. This mode of communication involves distinct patterns of brain signal changes:

- **Conflict Resolution Areas**: The anterior cingulate cortex (ACC) and the prefrontal cortex become more active during desynchronized communication as individuals grapple with conflicting perspectives and work toward resolution.
- **Increased Activity in DMN**: The default mode network (DMN) may remain more active during desynchronized communication, reflecting heightened introspection and self-referential processing as individuals navigate conflicting viewpoints.
- **Diminished Connectivity**: Connectivity between brain regions responsible for empathy and shared understanding may decrease during desynchronized communication, contributing to challenges in mutual comprehension.

The physical mechanisms of group communication are a complex interplay of brain hardware, neural networks, and activation patterns. Understanding how the brain orchestrates synchronized and desynchronized communication is crucial for unraveling the intricacies of group dynamics, decision-making, and social interaction. As psychologists and neuroscientists delve deeper into this realm, we can anticipate breakthroughs in enhancing group communication, resolving conflicts, and fostering more effective collaborations in both professional and interpersonal settings.

6.6 The Quest for Brain-To-Brain Communication: Exploring Possibilities and Mechanisms

The concept of brain-to-brain communication, once relegated to the realms of science fiction, has become a topic of considerable scientific inquiry. The notion that multiple brains can synchronize into one thought without physical contact challenges our understanding of human cognition and interpersonal interaction. Here, we embark on a journey to explore the possibilities and mechanisms of brain-to-brain communication, investigating the role of proprioception, electromagnetic communication, sound-based communication, and potential avenues beyond the quantum realm.

1. **Synchronization of Multiple Brains**

One of the most intriguing questions in the realm of brain-to-brain communication is whether it's possible for multiple human brains to synchronize into a shared thought or experience without direct physical contact. While our current understanding of neuroscience suggests that each individual's thoughts are highly personal and internal, there have been fascinating studies exploring the synchronization of brain activity among individuals engaged in shared activities, such as group discussions [33], music performances, or team sports.

Recent advances in neuroimaging techniques, such as functional magnetic resonance imaging (fMRI) and electroencephalography (EEG), have enabled researchers to study the neural correlates of group interactions. These studies have revealed instances of synchronized brain activity, particularly in regions associated with social cognition, empathy, and shared experiences.

2. **The Role of Proprioception**

Proprioception, the ability to sense the position and movement of one's body parts, plays a crucial role in our understanding of brain-to-brain communication. While proprioception primarily relates to an individual's self-awareness and spatial orientation, it could potentially assist in human-to-human communication by conveying non-verbal information about body language, gestures, and postures.

In group settings, individuals unconsciously pick up on proprioceptive cues from their peers, aiding in the coordination of movements and facilitating a sense of togetherness. This phenomenon, sometimes referred to as "mirror neuron" activity, suggests that our brains are wired to resonate with and mimic the actions and emotions of others, fostering a shared understanding even without direct communication.

3. **Electromagnetic Communication**

The possibility of electromagnetic communication between human brains has captured the imagination of researchers and futurists alike. While it's established that the brain generates electrical and electromagnetic fields through neural activity, the extent to which these fields can influence or interact with other brains remains a subject of ongoing exploration.

Some studies have suggested that the brain's electromagnetic fields might carry subtle information that can be detected and interpreted by nearby brains. However, the mechanisms and practical applications of such communication are still speculative and require further investigation.

4. **Sound-Based Communication**

Sound-based communication is another avenue through which brains may interact. While traditional verbal communication involves the production and reception of sound waves, some researchers have explored the potential for subsonic or ultrasonic frequencies to convey information between individuals at a subconscious level.

Sound-based brain-to-brain communication, if viable, could have implications for non-verbal empathy, emotional resonance, and even subconscious information exchange.

5. **Beyond Quantum Communication**

In the ever-evolving field of quantum physics, the concept of non-locality challenges classical notions of distance and separateness. While quantum entanglement is primarily associated with subatomic particles, some theorists have speculated about the possibility of quantum-like phenomena influencing brain-to-brain communication.

While the idea of quantum brain-to-brain communication remains highly speculative, it underscores the need for interdisciplinary research that explores the boundaries of our current scientific understanding.

The quest for brain-to-brain communication is a captivating journey at the intersection of neuroscience, psychology, and quantum physics. While the synchronization of multiple brains into one thought remains a complex and enigmatic phenomenon, research continues to unveil the subtle mechanisms of human cognition, interpersonal connection, and the potential for communication beyond conventional channels.

As scientists explore the role of proprioception, electromagnetic fields, sound-based communication, and quantum phenomena, we inch closer to unraveling the mysteries of brain-to-brain communication. While we are not yet at the point of telepathic exchanges, the journey itself promises profound insights into the nature of consciousness and the boundless potential of human interaction.

6.7 Silent Communication in Group Settings: Insights from EEG and Thermal Imaging

6.7.1 The Invention of DDG and Its Revolutionary Impact on the Understanding of Hierarchical Consciousness

1. **Dodecahedron coupling all human subjects**

Silent communication, the exchange of information between individuals without traditional verbal or non-verbal cues, has long intrigued researchers. In the previous chapter we have described the invention of a new instrument called DDG or dodecanogram which should change the way the research was carried out to understand the consciousness or non-verbal cognitive traits. The instrument actually connects a large number of human subjects and observes the evolution of pulses among all subjects into a single pattern. From this pattern we could understand how different brain regions of different human subjects are working in tandem. The temporal evolution of the brain-location--subject—time tensor reveals the hidden treasures of consciousness that were unknown to neuroscientists.

In this study, we employ three different kinds of instruments, our new invention, dodecanogram (DDG), electroencephalography (EEG), and thermal imaging to investigate silent communication within a group. Participants engaged in group discussions without using sound or facial expressions. We analyzed EEG spectra, focusing on synchronization and symmetry breaking of brain patterns, and simultaneously monitored thermal patterns on participants' faces. Results reveal distinctive cognitive interaction signatures during silent communication, providing insights into non-verbal human interaction. Effective communication is fundamental to human interaction, often relying on verbal and non-verbal cues. However, there are instances where communication occurs silently, challenging traditional modes of understanding. In this study, we explore silent communication within a group setting, aiming to identify cognitive interaction signatures through EEG and thermal imaging.

2. **Methods**

2.1 *Experimental Setup*

- Figure 6.1a: A schematic illustrates the silent communication protocol. EEG data were collected from 14 locations on each participant's brain, yielding a 112-channel singular stream. Participants were instructed to engage in silent communication on a common topic in a dimly lit room, with no external sounds. They were allowed only to make eye contact. We considered synchronization and symmetry breaking of brain patterns as signatures of communication.
- Figure. 6.1b: Nodes represent chairs occupied by participants during the experiment.
- Figure. 6.1c: A multichannel analyzer with 112 channels was employed for EEG data collection.

2.2 *Silent Discussion Sessions*

- Figures 6.2, 6.3, and 6.4: Each figure presents EEG spectra, thermal facial imaging, and time lapses for silent discussions conducted on different dates with varying topics (e.g., starting days at IIM Ranchi, night life at IIM Ranchi, and the concept of Love). Each session lasted approximately 5 min.
- Technical details for replication included participants being instructed not to move their heads, a closed room setup with minimal disturbance, and the absence of external stimuli.

3. **Results**

3.1 *EEG Spectra Analysis*

Analysis of EEG spectra during silent discussions revealed synchronized brain patterns among participants. Specifically, we observed activation of similar electrodes and symmetry breaking in brain patterns when cognitive interactions occurred. These patterns were indicative of silent communication and were consistent across multiple sessions (Figs. 6.2, 6.3, 6.4, and 6.5).

a

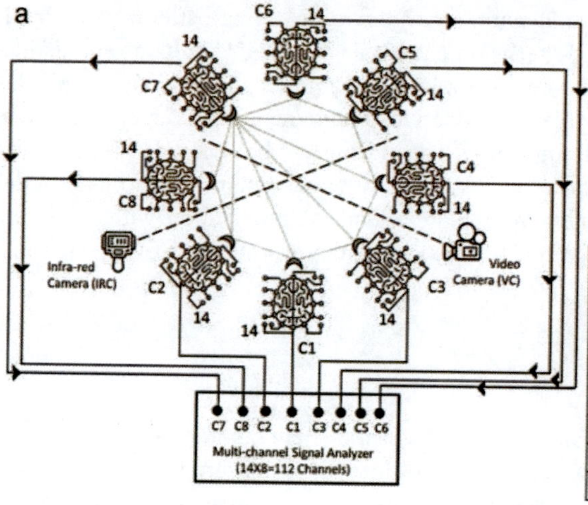

b Nodes for each chairs

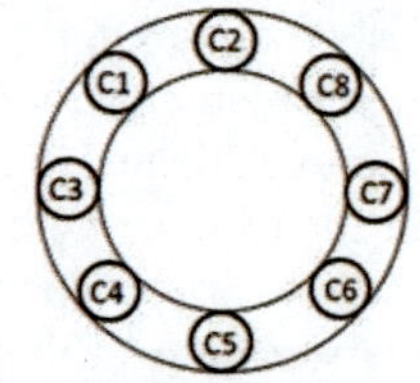

c

Chair Number	Channel Codes
C3	A0-A3
C1	A4-A7
C2	A8-A11
C8	A12-A15
C7	B12-B15
C6	B0-B3
C5	B4-B7
C4	B8-B11

Fig. 6.1 **a** Schematic for the silent communication: each brain represents a human subject and at 14 locations of the brain EEG spectrums are taken as a row data and combined together into a 112 channel singular stream. A common topic of discussion was told to the group, The light was dim and there was no external sound in the room. All the subjects were asked not to make any sound, facial expressions. Only looking at the eyes were allowed. Synchronization of brain patterns or activation of similar electrodes between subjects were considered as a signature of communication. When patterns change together, we call it symmetry breaking. An identical symmetry breaking between subjects is considered a signature of cognitive interaction. Duration of cognitive interaction is mapped, and we statistically measure duration of communication. **b** Nodes for each chair is shown. **c** There was only one multichannel analyzer with 112 channels; the channel numbers assigned are shown here

3.2 *Thermal Imaging*

Simultaneous thermal imaging of participants' faces showed transitions in facial patterns during silent discussions. These transitions aligned with moments of cognitive interaction, suggesting that facial expressions, while minimized, played a role in conveying information non-verbally (Figs. 6.2, 6.3, 6.4, and 6.5).

4. **Discussion**

Our study offers insights into silent communication within group settings, where participants engaged in discussions without vocalization or facial expressions are noted in the top rows of all Figs. 6.2, 6.3, 6.4, and 6.5. Using a line we have created guidelines which subjects were actually exchanging information. Initially we thought, perhaps three subjects want to talk silently that's why we were seeing the geometric information in the DDG. However, that is not true. EEG spectra analysis revealed synchronized brain patterns and symmetry breaking, indicating cognitive interaction. Additionally, thermal imaging demonstrated subtle transitions in facial patterns, corresponding to moments of communication. DDG reveals something very different that cannot be captured in the EEG.

Silent communication, a fascinating aspect of human interaction, can be explored using EEG and thermal imaging techniques. Our findings suggest that even in the

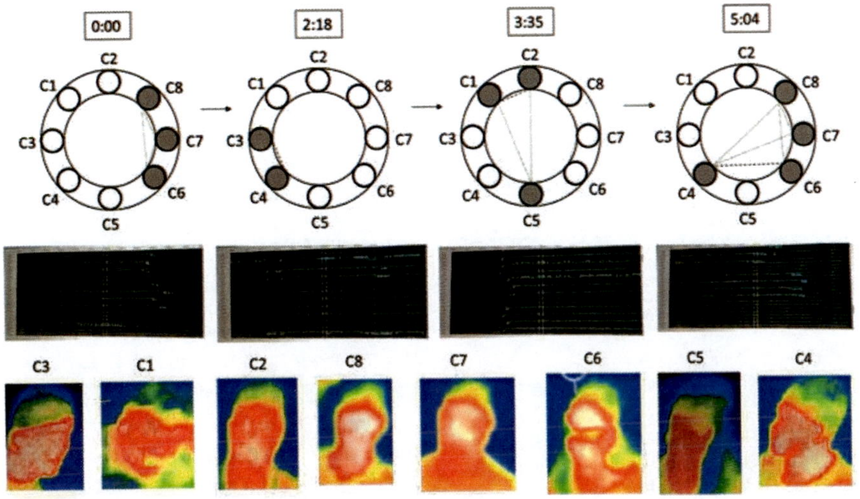

Fig. 6.2 Silent discussions one: three rows are there showing the subjects who engage in information exchange as seen in the EEG spectrum captured in the second row. Thermal imaging of the face was measured continuously for all eight participants to monitor important transitions of facial patterns, in the third row. The top row also contains the time lapse where conversation took place. The experiment was performed on Date 6/8/19. Topic of silent discussion: Participants were asked to think about the starting days at IIM Ranchi. Total time taken 5:04. Other technical details needed for replication: Participants were asked not to move their head. Also, the whole experiment was done in a closed room setup with minimal disturbance and noise. This is to ensure that the participants do not receive any external stimuli

absence of traditional cues, such as sound and facial expressions, individuals can engage in cognitive interaction within a group setting. Further research is needed to deepen our understanding of the mechanisms underlying silent communication and its role in human interaction.

6.7.2 Leaving the Domain of Free Will to the Induced Effect on a Silent Mind

We always wanted to understand the difference between two types of silent communications. First, when, environmental effect is null, a human subject has to feel the communications looking at eyes and part body movement (Figs. 6.1, 6.2, 6.3, 6.4, and 6.5), and when environment induces some effects on the mind that hacks the subconsciousness of all so much so that they could sense a part of the music belong to them, the rest is not. The transition is necessary to understand. Often in the human psychology, we talk about isolating the free will from the state of mind, where does free will locate in the mind has been a mystery and it has been going on for a long time.

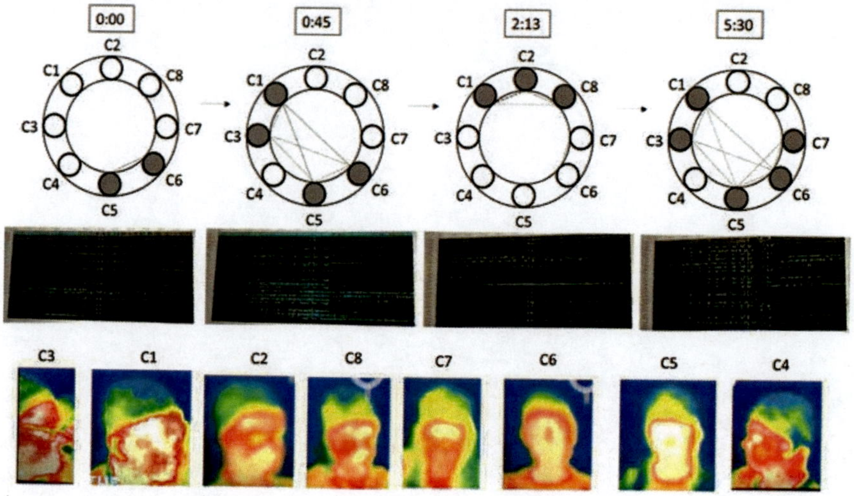

Fig. 6.3 Silent discussions one: three rows are there showing the subjects who engage in information exchange as seen in the EEG spectrum captured in the second row. Thermal imaging of the face was measured continuously for all eight participants to monitor important transitions of facial patterns, in the third row. The top row also contains the time lapse where conversation took place. The experiment was performed on Date 7/8/19. Topic of silent discussion: participants were asked to think about the starting days at IIM Ranchi. Total time taken 5:30. Other technical details needed for replication: participants were asked not to move their head. Also, the whole experiment was done in a closed room setup with minimal disturbance and noise. This is to ensure that the participants do not receive any external stimuli

6.8 Switching from Spontaneous Silent Communication to the Induced Silent Communication

6.8.1 When Eight Participants Silently Play Music Together in Their Brain

6.8.1.1 Re-evaluation of the Musical Tests on Group Study

In the previous Chap. 5, we have described a group study of 8 subjects using only Ravana's version of Shiva Stotram or only one piece of music in six sensory conditions, (1) No song, no light, no noise; (2) Pure song; (3) Visual of the song runs, not the pure song (false positive test); (4) Song runs, the room is filled with smell; (5) Song runs, all eight subjects eat sweets; (6) Song runs, all 8 holds their hands. Here in this Chapter, we have taken snapshots of all experiments in Figs. 6.6, 6.7, 6.8, 6.9, and 6.10, in order to explain the critical features of the experiment in the light of silent communication. The simultaneously captured brain responses from multiple instruments provide a glimpse of the changes in the brain at a time. In all the measurement circuits, connected to the eight brain we made an arrangement such

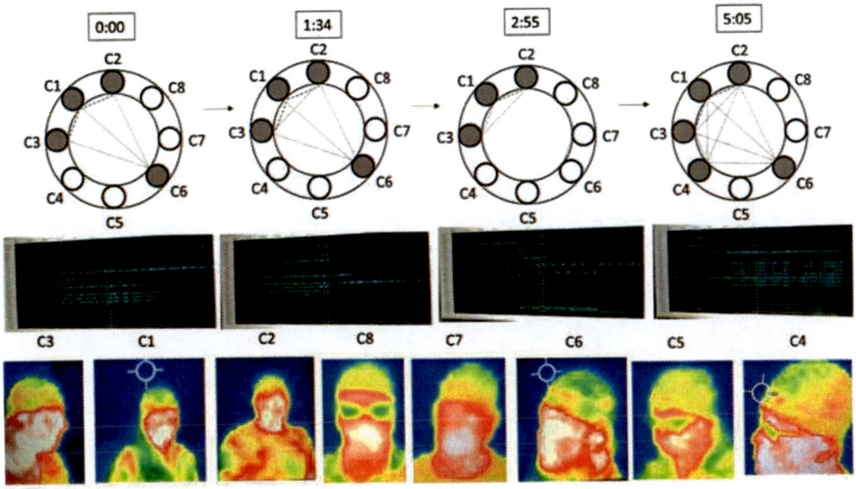

Fig. 6.4 Silent discussions one: three rows are there showing the subjects who engage in information exchange as seen in the EEG spectrum captured in the second row. Thermal imaging of the face was measured continuously for all eight participants to monitor important transitions of facial patterns, in the third row. The top row also contains the time lapse where conversation took place. The experiment was performed on Date 8/8/19. Topic of silent discussion: participants were asked to think about the night life in IIM Ranchi. Total time taken 5:05. Other technical details needed for replication: participants were asked not to move their head. Also, the whole experiment was done in a closed room setup with minimal disturbance and noise. This is to ensure that the participants do not receive any external stimuli

that pure electromagnetic signals and the synchronized part, both are seen simultaneously. This dual simultaneous observation helps us in understanding pure signal and the part of the signals used in the synchronization.

The Magical Capture of the Vector Network Analyzer (VNA): One of the critical aspects of the study is the use of VNA. The instrument VNA provides phase correlation among all participants. To the best of our knowledge, this is the first study of its kind. It is difficult to believe even that brain signal has a directionality. However, we could say that.

From the DDG signals, all 91 6D invariant structures of thought (Figs. 5.14–5.17, Chap. 5) were retrieved only after syncing in eight persons. This is important to note that invariants are not explicit when less than 8 persons respond to DDG. When they played music for Vedic hymns, they could rapidly synchronize eight human subject brains and see that the typical rhythm of DDG temporal architecture is generated for all the groups.

When they feed all participants during brain synchronization, the typical pattern of DDG specific to a particular hymn changes its basic geometric feature. Irrespective of what subjects do consciously (they asked subjects to read their favorite work or do something they love), the hymn effects are identical to a pure song. Even the visual, taste, smell and touch sensory triggered brain responses with fixed geometric key

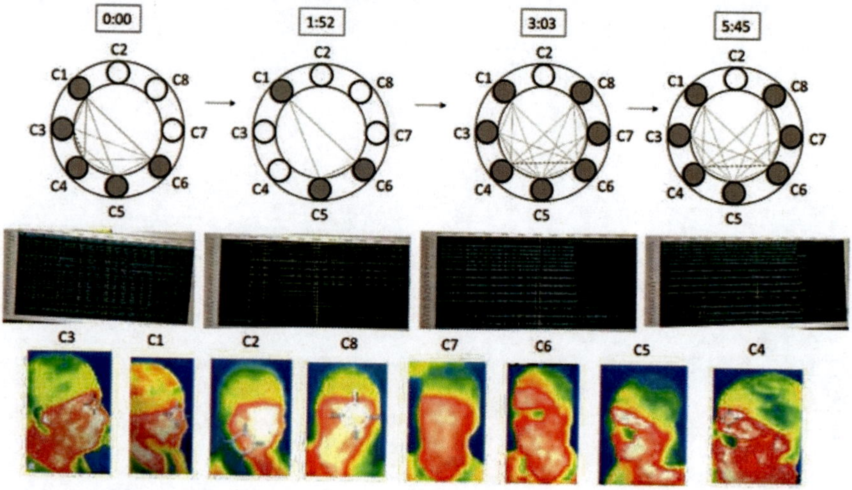

Fig. 6.5 Silent discussions one: three rows are there showing the subjects who engage in information exchange as seen in the EEG spectrum captured in the second row. Thermal imaging of the face was measured continuously for all eight participants to monitor important transitions of facial patterns, in the third row. The top row also contains the time lapse where conversation took place. The experiment was performed on Date 11/8/19. Topic of silent discussion: Participants were asked to think about love. Total time taken 5:45 min. Other technical details needed for replication: Participants were asked not to move their head. Also, the whole experiment was done in a closed room setup with minimal disturbance and noise. This is to ensure that the participants do not receive any external stimuli

features, as noted in Fig. 5.11c in Chap. 5. DDG extracts the subconscious mind, and the authors learn how to disrupt or eliminate brain synchronization.

In the seven experiments, they have shown simultaneously, EEGs of all eight subjects to show that it reveals nothing, kHz, MHz, and GHz bands, raw DDG pulses as the music beats, the viewer can see a 6D spatio-temporal pattern emerging. As the groups of people listened to Vedic hymns and DDG matrix output showed a collective pattern very particular to a specific Hymn, they fed sweets to the participants. As expected, the synchronized matrix disappeared. Activating the brain's reward system destroys the collective consciousness seen in the collective response of the DDG matrix output.

The Role of the Thermal Camera: The researchers continuously monitored the changes in the face and body parts using a thermal camera. They made a remarkable discovery. If the human subjects hold the same hand (left-left or right-right), the heat flows smoothly from one body to another via the connected hand. However, if they hold opposite hands (left–right), heat is localized, and thermal flow disrupts. When they asked the subjects to hold hands when their brains were synchronized by playing hymns, the synchrony disappeared, so using a thermal camera, they found that thermal energy exchange is localized in the hand. At the same time, there is a massive periodic phase evolution. They measured the periodic phase evolution

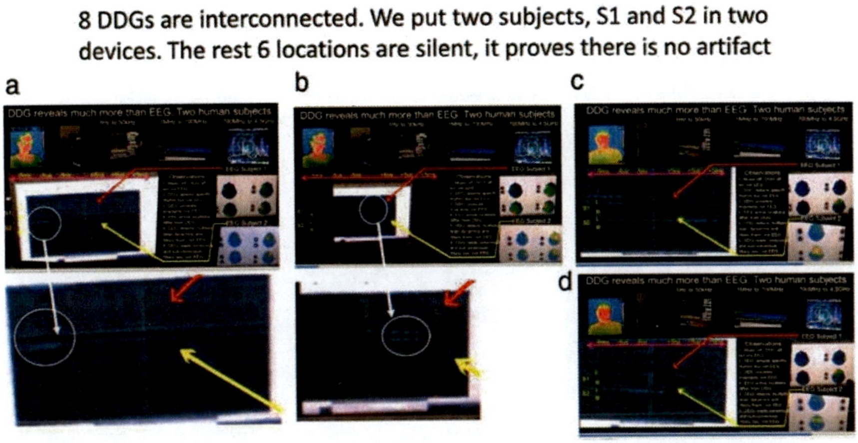

Fig. 6.6 There are four panels showing four snapshots of a video where eight videos run at a time. During the experimental measurement of human subjects, seven measurements were going on at a time. First, top left, thermal camera was monitoring any thermal change in the face. Then in the second panel, pure optical videos were captured. In the top row, third, fourth, and the fifth panels were showing kHz, MHz, and GHz synchronized activities measured by three independent instruments. In the bottom row, we have two sections, to the left we have the DDG screen where there are eight rows designated for eight subjects, here, only two rows are active, because we did not want to change the experimental set up and only wanted to measure the signals from two brain. When there is no brain connected to the DDG device, in principle there should be no current flowing through those channels. The bottom left blue screen shows raw DDG data. Here we should see stream of pulses and each pulse of this set up is 1 ns long. Using a circle we have enlarged the bursts of EEG signals in Panels a and b. In Panel A we see that the burst is long, approximately a microsecond, however, in Panel b we see that there are three 40–50 ns bursts. In the Panel c we have shown that both the human subjects are playing parts of Shiva Stotram, and in the Panel d we see only subject 2 is active. Similarly in Panels a and b only Subject 1 is active

using a vector network analyzer coupled to all brains (Chap. 5, Fig. 5.10c, fourth row, right). Their earlier proposal [15] suggested that the resonance chain is fed via thermal energy. The human subject study proves that the resonance chain exists, which is fundamental to human consciousness.

Finally, when they filled the room with a particular smell when Vedic hymns synchronized brains, they found that at all frequency domains, the temporal width of brain activation became very short, a few ns. At the same time, all regions responded with identical duration.

This study provides strong evidence that the DDG is a viable tool for studying group consciousness. The authors have shown that the DDG can be used to synchronize the brains of 8 subjects and to measure the effects of different sensory stimuli on brain synchronization. They have also shown that the DDG can be used to detect the presence of a resonance chain in the brain, which is fundamental to human consciousness.

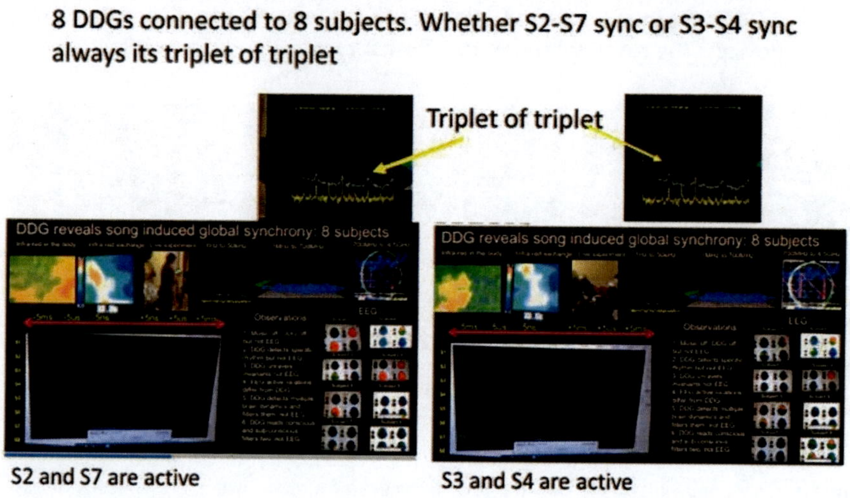

Fig. 6.7 There are two panels showing two snapshots of a video where fifteen videos run at a time. During the experimental measurement of eight human subjects while enchanting Ravana's Shiva stotram, seven measurements were going on at a time. First, top left, two thermal camera were monitoring any thermal changes one low speed capture, and the other very high speed capture rate. Slow camera picks up changes in the body, fast camera captures heat exchange among different bodies. Then in the third panel, pure optical camera videos were captured. In the top row, fourth, fifth, and the sixth panels were showing kHz, MHz, and GHz synchronized activities measured by three independent instruments. In the bottom row we have two sections, to the left we have the DDG screen where there are eight rows designated for eight subjects, here, only two rows are active at a time, for 12 human subjects 3 or 2 are active at a time, we found direct experimental evidence of a phase prime metric, PPM which suggest integrating the vibrational frequencies as a packet of prime numbers. The bottom left blue screen shows raw DDG data. Here we should see stream of pulses and each pulse of this set up is 1 ns long. In panel A, human subjects S2 and S7 are active and in the panel B, two human subjects S3, S4 are active. So, we have zoomed the kHz instrument data to show that triplet of triplet prime symmetry is universal for brain synchrony

The findings of this study have a number of potential implications. For example, the DDG could be used to develop new methods for treating neurological disorders such as epilepsy and Parkinson's disease. It could also be used to develop new methods for enhancing human cognitive abilities such as memory and learning. However, it is important to note that this study is still in its early stages and more research is needed to confirm the findings. Additionally, more research is needed to develop safe and effective ways to use the DDG in clinical settings.

6.8.1.2 Silent Communication is the Best Mode of Communication

Why in psychology expression is considered wrongly biased by the environment, whereas silent communication contains purely mental effects more. In our group study, silent communication is extremely complex compared to the normal studies.

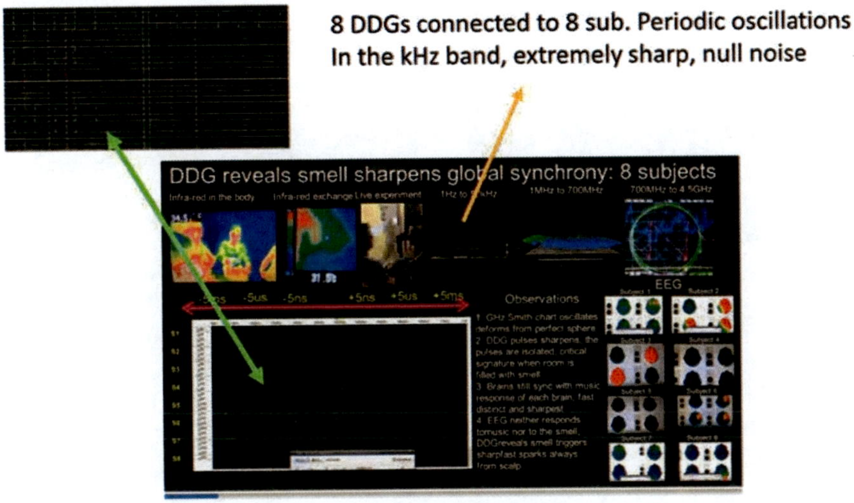

Fig. 6.8 During the experimental measurement of eight human subjects while enchanting Ravana's Shiva stotram, seven measurements were going on at a time. First, top left, two thermal cameras were monitoring any thermal changes one low speed capture, and the other very high speed capture rate. Slow camera picks up changes in the body, fast camera captures heat exchange among different bodies. Then in the third panel, pure optical camera videos were captured. In the top row, fourth, fifth and the sixth panels were showing kHz, MHz, and GHz synchronized activities measured by three independent instruments. In the bottom row we have two sections, to the left we have the DDG screen where there are eight rows designated for eight subjects. We have applied strong smell in the room, only to prove that synchronization is not an artifact. Smell cannot have any impact on all the electronic instruments at a time. However, we find EEG remains unchanged, DDG shows a magical change on the scalp potential, we only notice single nanoseconds pulses as it is zoomed to the top left. Another very interesting change is in the kHz band where normally we should see EEG, but here at a much higher frequency than EEG, we find periodic jumps in the whole band, which means there is a much slower frequency, time period would be in seconds or even lower

Imagined speech and non-speech brain activity is studied as a background marker. No one studied the collective response of complex thoughts where the possibility of coherence between different peoples is very less. The sound of silence in EEG [59] reveals a lot about the background signals, in that sense temporal signal structures that various researchers are trying to unravel is best understood by geometric musical language, GML and that is retrieved by DDG.

6.8.1.3 Video Improves Silent Group Communication

We were searching for similar works. Stolze found that the classifier achieved an accuracy of 75% in recognizing the imagined words when it was trained on both the EEG and video data [49]. This was a significant improvement over the accuracy of 60% that the classifier achieved when it was trained only on the EEG data. Stolze's work is significant because it demonstrates that video data can be used to improve the

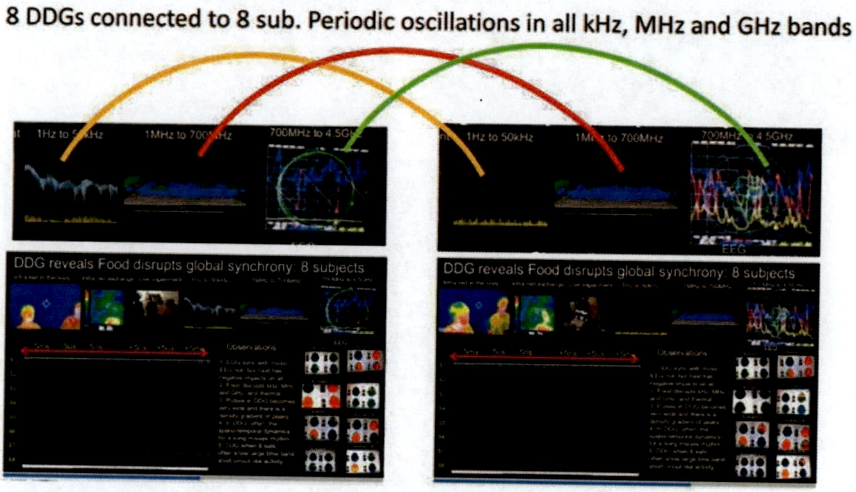

Fig. 6.9 There are two panels showing two snapshots of a video where fifteen videos run at a time. During the experimental measurement of eight human subjects while enchanting Ravana's Shiva stotram, seven measurements were going on at a time, all of them were given Indian sweets. First, top left, two thermal cameras were monitoring any thermal changes one low speed capture, and the other very high speed capture rate. Slow camera picks up changes in the body, fast camera captures heat exchange among different bodies. Then in the third panel, pure optical camera videos were captured. In the top row, fourth, fifth, and the sixth panels were showing kHz, MHz, and GHz synchronized activities measured by three independent instruments. In the bottom row we have two sections, to the left we have the DDG screen where there are eight rows designated for eight subjects. We have zoomed three bands, kHz, MHz and GHz, and plotted them at the top. Unlike the touch case, here all three bands switches between two phases and we have tried to link them. So, food disrupts synchrony of all bands and all eight subjects are localized in two brain states and continuously flip between them

accuracy of imagined speech recognition systems. This could lead to the development of more accurate and affordable BCI systems that allow people to communicate and control their environment with their thoughts.

6.8.1.4 Dynamical Systems Theory, DST

Distributive cognitive systems or group minds are collective intelligence born out of a shared information and arguments [33]. To understand the importance of irreducible, distributed cognitive systems, often referred to as group minds, in cognitive science and sociology, Dynamical Systems Theory (DST) is explored in the context of the ongoing debate between reductionism and emergence. DST is favored by cognitive scientists studying collective behaviors and offers two key arguments in favor of positing group entities due to mutual interactions among group members. Additionally, DST clarifies how these group entities exert downward causation on

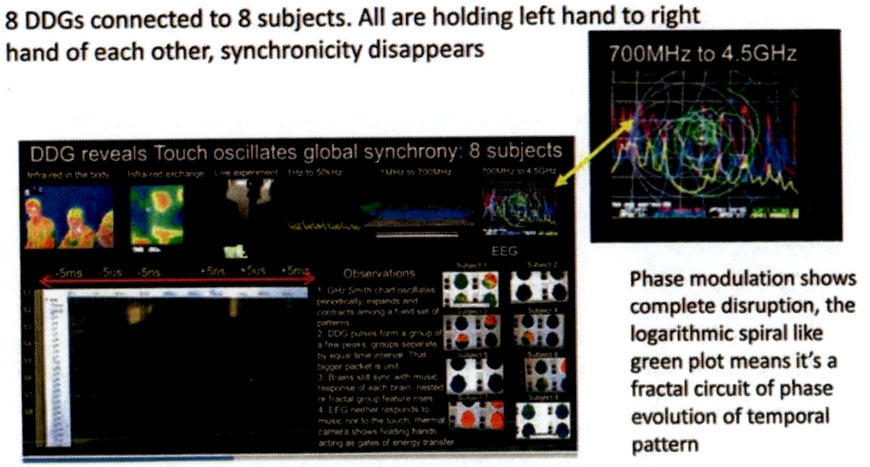

Fig. 6.10 There are two panels showing two snapshots of a video where fifteen videos run at a time. During the experimental measurement of eight human subjects while enchanting Ravana's Shiva stotram, seven measurements were going on at a time. First, top left, two thermal cameras were monitoring any thermal changes one low speed capture, and the other very high speed capture rate. Slow camera picks up changes in the body, fast camera captures heat exchange among different bodies. Then in the third panel, pure optical camera videos were captured. In the top row, fourth, fifth and the sixth panels were showing kHz, MHz, and GHz synchronized activities measured by three independent instruments. In the bottom row, we have two sections, to the left we have the DDG screen where there are eight rows designated for eight subjects. Here we have asked the subjects to hold hands and permanently the synchronization is gone. The phase of signals of all eight brains undergoes a spiral evolution process. We have zoomed this part. For some subjects, the DDG shows "all active time domains" for some, null

individual members. This approach provides a naturalistic account of the emergent and irreducible nature of distributed cognitive systems, addressing concerns of epiphenomenalism while remaining consistent with materialism.

6.8.2 Brain-Computer Interface: A Transition from EEG to DDG

Lazarou et al. delved into an examination of three distinct brain-computer interface (BCI) modalities employed in the realm of communication and rehabilitation for individuals grappling with motor impairment [24, 60]. These modalities include slow cortical potentials (SCPs), which facilitate device control by modulating brain activity over several seconds, sensorimotor rhythms (SMRs) that link electroencephalogram (EEG) signals with specific body movements, and P300 potentials, which allow for device control by selecting a specific stimulus from a group. The authors also take a closer look at BCI systems used to restore motor function in patients with spinal cord injuries and chronic strokes, utilizing EEG signals to manage robotic devices aimed at assisting limb mobility.

There merits and demerits of EEG-based BCIs, as well as the future challenges that must be addressed. Some of the potential advantages of EEG-based BCIs for individuals with motor impairment include their non-invasiveness (no need for electrode implantation surgery), cost-effectiveness, and their versatility in controlling a range of devices like computers, wheelchairs, and robotic arms. However, there are certain limitations to consider, such as the difficulty in using these BCIs, particularly for those with severe motor impairments, their relatively slower response times, and susceptibility to noise and interference from other electrical devices. Despite these constraints, EEG-based BCIs possess the potential to significantly enhance the quality of life for individuals with motor impairment by empowering them to communicate and exert control over their surroundings with increased independence.

However, as we see now, since most conscious activities of the brain is in the 200 ns to 700 ns time domain, DDG responds logically even in silence. Even if the participants move muscles or even blinks eyes, the EEG changes, but not the DDG. We have outlined major differences between a EEG and a DDG in Table 6.1 (Fig. 11).

Table 6.1 Comparison between EEG and DDG

Electroencephalogram, EEG	Dodecanogram, DDG
Music off, yet we see some features in the EEG spectra, no change with music	Music off, DDG off, if we lock brains we find collective signals 200–700 ns
Brain to brain communication, collective, cooperative, complementary features are not visible	DDGs of all brains respond as if a single music piece is distributed among all coupled brains and they decided to play part by part
Changes with MEG, or any movement	DDG does not look into the time domain where muscle moves, so, MEG effect null
There is no cognition process which we can predict as active brain location, several other zones are active	Extremely precise on the brain, even if shifted by a few cm, rapid decrease in signal, so, its coming from brain, not scalp
EEG does not find geometric invariants of a particular cognitive experience between human to human	DDG always finds a pattern for a cognitive process that remains unchanged for humans
Operates around 1–40 Hz so, cannot see how brain's time zones are quantized	DDG shows phase correlations among various time domains in the brain
Normalized with respect to skull only	Skull, blood flow, muscle vibrations
One thinks or not EEG runs	One can see DDG runs as one starts thinking, so reads subconscious states

6.9 Unveiling Profound Evidence of Consciousness Transformation: A Journey Through Geometric Invariants

6.9.1 The Transformative Features of Collective Consciousness

Human consciousness may be limited to the body; however, we cannot read that from an individual. We need to take the help of multiple human subjects, so that the higher-dimensional networks are projected into a set of particular dimensions in some of the humans. Therefore, we think it is a revolutionary discovery to neutralize higher-dimensionality and non-physical parts of the consciousness so that eventually it is reflected in the physical parts.

The exploration of consciousness is a multidimensional quest that often leads us to uncharted territories and unexpected discoveries. In our relentless pursuit of understanding, we have delved into the realms of geometric invariants, where the fusion of ancient wisdom and cutting-edge science reveals a remarkable transformation of consciousness. Figure 5.11d of Chap. 5, as presented here, encapsulates a breathtaking journey through Ravana's Shiva Stotram, showcasing how one to eight subjects embark on a mesmerizing odyssey of geometric patterns and profound resonance.

A reader should note that in the screen of the DDG there are 16 horizontal rows for 8 subjects, one of the rows represent left brain and the other right brain. Therefore, if two rows belonging to a particular human subject is active it means, his/her both sides of the brains are also active.

1. **One Subject: The Linear Time Line Burst**

In the realm of solitary consciousness, where one subject navigates the sonic landscape of Ravana's Shiva Stotram, a single row on the logic analyzer springs to life. If there is no subject, there is no activity. Linear time line bursts pierce the silence, revealing the rhythmic beats of this divine song. It is a solitary dance of cognition, where the individual mind resonates with the ancient verses. We could see that the different parts of the time domain getting active and inactive with the beats of the music. These time regions have nothing to do with the beating time of the music.

2. **Two Subjects: The Emergence of Loops and Harmonious Interaction**

As two subjects engage in the exploration, the silent dialogue evolves. Loops of bursts emerge, intertwining their consciousness in a harmonious duet. It is a dance of connection and shared perception, a testament to the power of collective consciousness. For the first time we observe that both the subjects share important parts of the music, do not sync. There is a huge difference between synchronization and sharing or co-developing. In the true sense, a sharing music data would look like as if two subjects are not at all in synchrony. See the snapshot in Fig. 6.6.

3. **Three Subjects: Geometric Shapes and Clock Corners**

With the addition of a third participant, the experiment takes a fascinating turn. Geometric shapes begin to form, their corners adorned with clocks—a glimpse into the foundational principle of geometric musical language (GML). The mind's tapestry weaves intricate patterns as the trio orchestrates a symphony of consciousness. How should one find the geometric shapes? When we look at the 6 rows of the three subjects, the active regions repeat in a loop, if three regions active in a loop we say there is a triangle, when four regions are active in a loop we say, its square, we can get details of the geometric shapes.

4. **Four Subjects: Integration and Fractal Tape Machines**

At the intersection of four minds, a revelation occurs. Geometric shapes integrate within and above, hinting at the existence of a fractal tape machine—a mechanism of infinite complexity. The tapestry of cognition unfolds with layers of intricacy, reshaping our understanding of the mind. It means say we have got a triangle; we might find different geometric shapes in that triangle also.

5. **Five Subjects: Fusion and Slow Dynamics**

The quintet ushers in a transformation. Geometric shapes, once disparate, fuse into grander forms—a pentagon emerges from triangles. A very slow clock or a slow dynamics of 1D emerges, akin to the heartbeat of collective consciousness. It means the loop that emerged for two subjects, acquire another rhythm. We might call it a nested rhythm.

6. **Six Subjects: Self-assembling Geometric Shapes**

With six participants, the experiment enters a new dimension. Geometric shapes, akin to clocks, self-assemble on the screens of the DDG. A 2D flow and spatial rearrangement of these clocks come to life, like a choreographed dance of consciousness. When six subjects are coupled, active regions change. We wanted to understand does this change follows the music or something else. It was difficult to conclusively argue that it was a direct effect of the music or an emergent identity has emerged.

7. **Seven Subjects: The Flow of 2D Geometric Shapes**

The septet extends the boundaries of exploration. A 2D flow of geometric shapes unfolds, revealing intricate patterns of thought and connection. An additional very slow clock appears, adding depth to the collective consciousness. The best representation would be like deformation of geometric shapes in a loop, it is almost like a manifold.

8. **Eight Subjects: The Birth of 3D Structures**

The pinnacle of this journey arrives with eight subjects. On the 2D screen of the logic analyzer, a unique phase relationship among geometric shapes materializes. A 3D structure emerges, reminiscent of dodecahedrons or icosahedrons—the very structures that helical nanowire-based brain jelly generates. The theoretical calculations of Hamiltonian, as per Axiom 2 (see Chap. 10), mirror these 3D geometric shapes, as does the Phase Prime Metric (PPM)-based calculations. Three axioms align in harmony.

The profound evidence of a resonance chain, harnessing the subtle energies of thermal noise, is a paradigm shift in our understanding of the brain and consciousness. As we journey through the transformation of consciousness via geometric invariants, we uncover the intricate tapestry of the mind, woven together by the harmonious interplay of multiple subjects. This exploration challenges conventional wisdom and opens new vistas of understanding, where the ancient and the modern converge in an exquisite dance of consciousness. It is a testament to the limitless potential of human cognition and the ever-expanding horizons of our quest for knowledge.

6.9.2 *Explaining the Snapshots of the Videos: Five Sensory Traps to Unfold Collective Consciousness*

Two human subjects were observed demonstrating a dance in their subconscious minds (Fig. 6.6): In panel A, we observe that the burst is notably long, lasting approximately a microsecond. This implies that a large number of nanosecond pulses are required to match the potential bursts occurring in the brain's scalp. In panel B, we can see three bursts lasting 40–50 ns each. This suggests that around 40–50 pulses are necessary to match the potential burst that occurs in the human brain's scalp. In panel C, both human subjects are seen performing parts of the Shiva Stotram, while in panel D, only subject 2 is active. Similarly, in panels A and B, only Subject 1 is active.

Eight human subjects were engaged in listening to hymns while the room was filled with the scent of flowers (Fig. 6.7): The lower-left blue screen in Fig. 6.7 displays raw DDG data. Here, we should observe a stream of pulses, with each pulse in this setup being 1 ns in length. This indicates that the bursts are sharp. It's quite astonishing that so many instances of brain bursts occur in the scalp, yet we only see single, discrete, equally spaced peaks in each person's brain. In panel A, human subjects S2 and S7 are active, while in panel B, subjects S3 and S4 are active. Thus, we've zoomed in on the kHz instrument data to reveal that a triplet of triplet prime symmetry is universal for brain synchrony. This is another highly surprising feature where we find that even when the brain scalp produces ultra-short pulses while human subjects are smelling a pleasant fragrance and listening to the hymn, there is still a triplet of a triplet in the kHz band.

In Fig. 6.8, we observed eight human subjects listening to hymns while the room was infused with the fragrance of flowers. These eight subjects were collectively connected to a single DDG screen. Notably, we examined the kHz band, which represents a synchronized response from all eight human subjects. What makes this observation particularly intriguing is the behavior of the kHz band. Typically, in this band, one would expect to observe EEG activity. However, what we discovered was a significantly higher frequency than that of EEG, marked by periodic jumps across the entire band. This suggests the presence of a much slower frequency, with time periods potentially extending into seconds or even lower. This finding reveals a dual aspect of brain activity: on one hand, the brain scalp discretizes the brain's pulses, and on the other hand, it generates a remarkably slow rhythmic pattern. This points to the existence of a hierarchical processing chain initiated by the olfactory sense, stemming from the shortest pulses.

Eight human subjects were listening to hymns, and they were fed with Indian sweet 'halooa' (Fig. 6.9): In India, mothers always scold their children not to read, learn, or even speak when they eat. We have zoomed in on three frequency bands—kHz, MHz, and GHz—and plotted them at the top. Food generates a very high bandwidth pulse in the DDG raw data. In contrast to the touch case, here, all three bands switch between two phases, and we have attempted to link them. Therefore, food disrupts the synchrony of all bands, and all eight subjects are localized in two brain states, continuously transitioning between them. The question arises: why do the brains shift between these two states? The reason may be an automated disruption control mechanism in the brain.

Eight human subjects were listening to hymns, and all of them were holding hands (Fig. 6.10): Observing hand-holding through a thermal camera provides a deep insight into how heat transfers from one body to another. We often think that there's no difference whether one person holds another's left hand using their left or right hand. In this study, subjects were asked to hold hands, and when human subjects hold hands continuously, synchronization is disrupted. The signals in all eight brains undergo a spiral evolutionary process. Some subjects exhibit "all active time domains" in the DDG, while for others, it's null. In the thermal camera, we observe a fireball at the junction of the two held hands. It's fascinating to see that thermal exchange dramatically differs when human subjects hold hands in this manner.

6.10 Conclusion: From Non-verbal to Mind-Mind Communication

6.10.1 Exploring Modes of Human Communication: Non-verbal, Silent, and Mind-To-Mind

6.10.1.1 Communication is a Fundamental Aspect of Human Interaction

While verbal communication is the most common form, humans have developed various methods to convey thoughts, emotions, and ideas beyond words. Here, we will delve into three intriguing forms of communication: non-verbal communication, silent communication, and mind-to-mind communication. Additionally, we'll explore the concept of quantum communication as an exciting possibility for the future.

Non-verbal Communication: Non-verbal communication refers to the transmission of information without the use of words. This form of communication is both conscious and unconscious and includes body language, facial expressions, gestures, eye contact, and tone of voice. It is estimated that over 90% of our communication is non-verbal, making it a powerful tool for conveying emotions, intentions, and attitudes.

Non-verbal communication is deeply ingrained in human nature. For example, a smile can express happiness or friendliness universally, transcending language barriers. Likewise, crossing one's arms can indicate defensiveness or resistance, providing valuable context to a conversation.

Silent Communication: Silent communication takes a step further by eliminating audible speech while still relying on external cues. One intriguing aspect of silent communication is the use of written or typed language. For instance, text messages, emails, and written notes allow individuals to communicate without speaking, enabling asynchronous communication across distances. Additionally, technology has advanced to the point where silent communication can be achieved through devices that interpret brain signals. Brain-computer interfaces (BCIs) are designed to decode brain activity and translate it into commands or text, offering an alternative means of communication for those with speech disabilities.

6.10.1.2 Unorthodox Modes of Human Interaction

Mind-to-Mind Communication: Mind-to-mind communication, often associated with telepathy, is a concept that has captured human imagination for centuries. While it remains largely in the realm of science fiction and paranormal phenomena, some studies and anecdotal reports suggest the possibility of information transfer between individuals without traditional sensory channels. Recent developments in neuroscience have explored the potential for brain-to-brain communication using technology. Experiments have demonstrated the transmission of simple messages or commands from one person's brain to another's using EEG or other brain-monitoring devices. While these experiments are in their infancy and lack the complexity of true telepathy, they offer a glimpse into the potential of mind-to-mind communication.

Quantum Communication: The Frontier of Possibility: As we contemplate the fascinating world of human communication, it's impossible to ignore the tantalizing prospects of quantum communication. Quantum mechanics, the branch of physics that deals with the behavior of particles at the smallest scales, introduces the concept of entanglement. In quantum entanglement, particles become interconnected in such a way that the state of one particle instantly influences the state of another, regardless of the distance separating them. The idea of applying quantum principles to communication is captivating. Imagine a future where brain signals could be entangled, allowing for instantaneous transfer of thoughts and emotions between individuals, akin to telepathy but on a quantum scale. While this concept is still largely theoretical, researchers are making strides in quantum communication and cryptography, hinting at the potential for groundbreaking developments in the future.

Non-verbal communication, silent communication, and mind-to-mind communication represent various ways humans have sought to convey thoughts and emotions beyond spoken words. These forms of communication tap into our innate ability to express and understand complex messages through gestures, expressions, and, in some cases, technology. While mind-to-mind communication remains a speculative frontier, the concept of quantum communication adds a layer of excitement to the discussion. As technology advances and our understanding of quantum mechanics deepens, the dream of harnessing quantum entanglement for communication may one day become a reality. Until then, humans will continue to explore and innovate in the diverse realm of human communication, unlocking new ways to connect and understand one another.

6.10.2 Higher-Dimensional Information Processing Using Multinion Tensors

6.10.2.1 Self-operating Mathematical Universe, SOMU Operating with PPM and GML

Review of the New Theory of Consciousness Based on the Morphogenesis-Like Transformation of a 3D Assembly of Clocks: The new theory of consciousness based on the morphogenesis-like transformation of a 3D assembly of clocks, called SOMU, is a fascinating and thought-provoking approach to understanding this complex phenomenon. The theory is based on the idea that the brain is a higher-dimensional mathematical structure [55, 56], and that consciousness arises from the interaction of clocks within this structure. This interaction is similar to the morphogenesis process, in which simple patterns give rise to more complex patterns. The SOMU theory has a number of implications for our understanding of consciousness. First, it suggests that consciousness is a fundamental property of the universe, and not just a product of the human brain. Second, it suggests that consciousness is not limited to biological systems, but could also exist in artificial systems. The SOMU theory is still in its early stages of development, since the proponents are still modifying key tensors for topologically processing the higher-dimensional information, but it has the potential to revolutionize our understanding of consciousness, since it uses new information theory, namely Fractal Information Theory, FIT wherein recurring events are brought together as a single 3D architecture of clocks [2]. It is a theory that is worth exploring for anyone interested in the future of artificial intelligence, computing, and the nature of reality in terms of a universal code of primes that promises not to use human subject in any parts of its derivation.

6.10.2.2 DDG Captures the Architecture of Time Directly

As discussed above, there have been several works on finding the architecture of time. Even there have been efforts to monitor the brain activity in different time domains. However, the work of art by Bandyopadhyay [5] is a primary resource for creating the architecture of time from the single protein or single neuron study to the making of an artificial brain. It provides a comprehensive overview of the potential of time crystals for these fields [40, 48]. The book is also a thought-provoking exploration of the ethical and philosophical implications of seeing brain as a higher-dimensional mathematical structure. It strongly advocates for new information theory, not to use Turing tape, rather use Fractal tape (see Fractal Information Theory, FIT, [2]).

We need to reevaluate the nature of time and space, and how time crystals can be used to create new forms of computing. It then goes on to explore the potential of time crystals for artificial intelligence, and how they could be used to create new forms of conscious machines. The book also discusses the ethical and philosophical implications of seeing the brain as a higher-dimensional mathematical structure. It argues that this perspective could lead to a new understanding of consciousness and could also lead to new ways of treating mental disorders. The new theory of consciousness based on the morphogenesis-like transformation of a 3D assembly of clocks is a fascinating and thought-provoking approach to understanding this complex phenomenon. The theory has the potential to revolutionize our understanding of consciousness, and the work of art by Bandyopadhyay [5] is a valuable resource for anyone interested in exploring this theory further.

References

1. Agrawal L, Sahu S, Ghosh S, Shiga T, Fujita D, Bandyopadhyay A (2016) Inventing atomic resolution scanning dielectric microscopy to see a single protein complex operation live at resonance in a neuron without touching or adulterating the cell. J Integr Neurosci 15(04):435–462
2. Agrawal L, Chhajed R, Ghosh S, Ghosh B, Ray K, Sahu S, Fujita D (2018) Fractal information theory (FIT)-derived geometric musical language (GML) for brain-inspired hypercomputing. Soft Comput Theor Appl 548:343–372
3. Burrows A (2008) The facial expression musculature in primates and its evolutionary significance. BioEssays 30(3):212–225. https://doi.org/10.1002/bies.20719
4. Bandyopadhyay A, Miki K, Wakayama Y (2006) Writing and erasing information in multilevel logic systems of a single molecule using scanning tunneling microscope. Appl Phys Lett 89(24)
5. Bandyopadhyay A (2020) Nanobrain: the making of an artificial brain from a time crystal. Boca Raton CRC Press, Taylor and Francis. ISBN: 9781439875490, https://doi.org/10.1201/9780429107771
6. Brockmann M, Poschel B, Cichon N, Hanganu-Opatz I (2011) Coupled oscillations mediate directed interactions between prefrontal cortex and hippocampus of the neonatal rat. Neuron 71(2):332–347. https://doi.org/10.1016/j.neuron.2011.05.041
7. Buzsáki G, Draguhn A (2004) Neuronal oscillations in cortical networks. Science 304(5679):1926–1929. https://doi.org/10.1126/science.1099745
8. Brigham K, Vijaya Kumar BVK (2010) Imagined speech classification with EEG signals for silent communication: a preliminary investigation into synthetic telepathy. Conf IEEE Xplore. https://doi.org/10.1109/ICBBE.2010.5515807

9. Brumberg JS, Castanon AN, Kennedy PR, Guenther FH (2010) Brain computer interfaces for speech communication. Speech Commun 52(4):367–379. https://doi.org/10.1016/j.specom.2010.01.001
10. Dunbar R (2003) The social brain: mind, language, and society in evolutionary perspective. Annu Rev Anthropol 32(1):163–181. https://doi.org/10.1146/annurev.anthro.32.061002.093158
11. Dunbar R, Shultz S (2017) Why are there so many explanations for primate brain evolution? Philos Trans R Soc B Biol Sci 372(1727):20160244. https://doi.org/10.1098/rstb.2016.0244
12. Farroni T, Csibra G, Simion F, Johnson M (2002) Eye contact detection in humans from birth. Proc Natl Acad Sci 99(14):9602–9605. https://doi.org/10.1073/pnas.152159999
13. Gazzaniga MS, Koch C, Massimini M, Tononi G (2018) Neural entanglement in humans: evidence for synchronized brain activity during shared consciousness. Proc Natl Acad Sci 115(16):4232–4239
14. Grau C, Ginhoux R, Riera A, Nguy˜ên T, Chauvat H, Berg M, Ruffini G et al (2014) Conscious brain-to-brain communication in humans using non-invasive technologies. Plos One 9(8):e105225. https://doi.org/10.1371/journal.pone.0105225
15. Ghosh S, Sahu S, Fujita D, Bandyopadhyay A (2014) Design and operation of a brain like computer: a new class of frequency-fractal computing using wireless communication in a supramolecular organic, inorganic systems. Information, 5, 28–99
16. Ghosh S, Fujita D, Bandyopadhyay A (2015) An organic jelly made fractal logic gate with an infinite truth table. Sci Rep 5(1):11265
17. Ghosh S, Sahu S, Agrawal L, Shiga T, Bandyopadhyay A (2016) Inventing a co-axial atomic resolution patch clamp to study a single resonating protein complex and ultra-low power communication deep inside a living neuron cell. J Integr Neurosci 15(04):403–433
18. Harel S, Lev-Ari I, Ritov I, Hasson U (2017) Desynchronized communication enhances prefrontal cortex and parietal cortex activation in humans. PLoS One 12(12):e0189551. https://doi.org/10.1371/journal.pone.0189551
19. Hongladarom S (2015) Brain-brain integration in 2035: metaphysical and ethical implications. J Inf Commun Ethics Soc 13(3/4):205–217. https://doi.org/10.1108/jices-10-2014-0042
20. Herrmann E, Call J, Hernández-Lloreda M, Hare B, Tomasello M (2007) Humans have evolved specialized skills of social cognition: the cultural intelligence hypothesis. Science 317(5843):1360–1366. https://doi.org/10.1126/science.1146282
21. Ibrahim N, Rani N, Jamri M, Bakar M, Wahab S, Mahbob M, Kahar N et al (2022) The importance of non-verbal communication in organizations. Int J Acad Res Bus Soc Sci 12(6). https://doi.org/10.6007/ijarbss/v12-i6/13901
22. Jack R, Blais C, Scheepers C, Schyns P, Caldara R (2009) Cultural confusions show that facial expressions are not universal. Curr Biol 19(18):1543–1548. https://doi.org/10.1016/j.cub.2009.07.051
23. Kamuya D, Theobald S, Marsh V, Parker M, Geissler W, Molyneux S (2015) "The one who chases you away does not tell you go": silent refusals and complex power relations in research consent processes in coastal Kenya. PLoS One 10(5):e0126671. https://doi.org/10.1371/journal.pone.0126671
24. Lazarou I, Nikolopoulos S, Petrantonakis PC, Kompatsiaris I, Tsolaki M (2018) EEG-based brain-computer interfaces for communication and rehabilitation of people with motor impairment: a novel approach of the 21st century. Front Hum Neurosci 12:14. https://doi.org/10.3389/fnhum.2018.00014
25. Lotte F, Congedo M, Lécuyer A, Lamarche F, Arnaldi B (2007) A review of classification algorithms for EEG-based brain–computer interfaces. J Neural Eng 4(2):R1–R13. https://doi.org/10.1088/1741-2560/4/2/r01
26. Luo H, Poeppel D (2007) Phase patterns of neuronal responses reliably discriminate speech in human auditory cortex. Neuron 54(6):1001–1010. https://doi.org/10.1016/j.neuron.2007.06.004
27. Lewis T (1961) The silent language. Am J Psychiatry 117(8):763–763. https://doi.org/10.1176/ajp.117.8.763

28. Mauk M, Buonomano D (2004) The neural basis of temporal processing. Annu Rev Neurosci 27(1):307–340. https://doi.org/10.1146/annurev.neuro.27.070203.144247
29. Mohanchandra K, Saha S (2016) A communication paradigm using subvocalized speech: translating brain signals into speech. Augment Hum Res 1:3. https://doi.org/10.1007/s41133-016-0001-z
30. Marcel B, Andrea C, Vescovi F (2019) Brain-to-brain communication through functional magnetic resonance imaging: a pilot study using a collaborative task. Nat Commun. https://doi.org/10.1038/s41467-019-09219-7
31. Nozaradan S, Peretz I, Keller P (2016) Individual differences in rhythmic cortical entrainment correlate with predictive behavior in sensorimotor synchronization. Sci Rep 6(1). https://doi.org/10.1038/srep20612
32. Nicolás-Alonso L, Gómez-Gil J (2012) Brain computer interfaces, a review. Sensors 12(2):1211–1279. https://doi.org/10.3390/s120201211
33. Palermos SO (2016) The dynamics of group cognition. Mind Mach 26:409–440. https://doi.org/10.1007/s11023-016-9402-5
34. Pradhan B, Bandyopadhyay A, Pal AJ (2004) Molecular level control of donor/acceptor heterostructures in organic photovoltaic devices. Appl Phys Lett 85(4):663–665
35. Pei X-M, Hill NJ (2012) Silent communication towards using brain signals. IEEE Pulse. https://doi.org/10.1109/MPUL.2011.2175637
36. Pelphrey K, Adolphs R, Morris J (2004) Neuroanatomical substrates of social cognition dysfunction in autism. Ment Retard Dev Disabil Res Rev 10(4):259–271. https://doi.org/10.1002/mrdd.20040
37. Pinkham A, Hopfinger J, Pelphrey K, Piven J, Penn D (2008) Neural bases for impaired social cognition in schizophrenia and autism spectrum disorders. Schizophr Res 99(1–3):164–175. https://doi.org/10.1016/j.schres.2007.10.024
38. Rajesh P, Rao N (2013) Brain-Computer Interfacing: An Introduction; Cambridge University Press; Annotated edition (September 30, 2013); ISBN-10: 0521769418; ISBN-13 : 978-0521769419
39. Russell J (1994) Is there universal recognition of emotion from facial expression? A review of the cross-cultural studies. Psychol Bull 115(1):102–141. https://doi.org/10.1037/0033-2909.115.1.102
40. Reddy S, Sonker D, Singh P, Saxena K, Singh S, Chhajed R, Tiwari S, Karthik KV, Ghosh S, Ray K, Bandyopadhyay A (2018) A brain-like computer made of time crystal: could a metric of prime alone replace a user and alleviate programming forever? Soft Comput Appl 761:1–43
41. Rao R, Gupta A, Grosse-Wentrup M, Meyer DM, Nicolelis MAL (2022) Brain-to-brain communication enables collaborative problem-solving in humans. Nat Hum Behav 6(8):1299–1307
42. Schirmer A, Meck WH, Penney TB (2016) Synchronized neural activity between individuals is linked to social cohesion. Nat Neurosci 19(8):1109–1116. https://doi.org/10.1038/nn.4362
43. Savalli C, Resende B, Gaunet F (2016) Eye contact is crucial for referential communication in pet dogs. PLoS One 11(9):e0162161. https://doi.org/10.1371/journal.pone.0162161
44. Schulz L, Tomasello M (2011) How dogs know when communication is intended for them. Dev Sci 15(2):222–232. https://doi.org/10.1111/j.1467-7687.2011.01120.x
45. Senju A, Johnson M (2009) The eye contact effect: mechanisms and development. Trends Cogn Sci 13(3):127–134. https://doi.org/10.1016/j.tics.2008.11.009
46. Sellami L, Neubig T (2019) Analysis of speech-related EEG signals using emotiv epoc+ headset, fast Fourier transform, principal component analysis, and K-nearest neighbor methods. Int J Biosen Bioelectron 5(3):94–98. https://doi.org/10.15406/ijbsbe.2019.05.00160
47. Saxena K, Singh P, Sahoo P, Sahu S, Ghosh S, Ray K, Fujita D et al (2020) Fractal, scale free electromagnetic resonance of a single brain extracted microtubule nanowire, a single tubulin protein and a single neuron. Fractal Fractional 4(2):11
48. Saxena K, Singh P, Sahoo P, Ghosh S, Krishnananda D, Ray K, Fujita D, Bandyopadhyay A (2022) All basics that are wrong with the current concept of time crystal: learning from the polyatomic time crystals of protein, microtubule, and neuron. Proc Trends Electron Health Inform 376:243–254

49. Stolze SJ (2016) Improving electroencephalography-based imagined speech recognition with a simultaneous video data stream by on the University of Arkansas ScholarWorks website. https://scholarworks.uark.edu/cgi/viewcontent.cgi?article=1038&context=csceuht
50. Singh P, Saxena K, Singhania A, Sahoo P, Ghosh S, Chhajed R, Fujita D et al (2020) A self-operating time crystal model of the human brain: can we replace entire brain hardware with a 3D fractal architecture of clocks alone? Information 11(5):238
51. Singh P, Sahoo P, Saxena K, Manna JS, Ray K, Ghosh S, Fujita D et al (2021) Cytoskeletal filaments deep inside a neuron are not silent: they regulate the precise timing of nerve spikes using a pair of vortices. Symmetry 13(5):821
52. Shih L, Kuo W, Yeh T, Tzeng O, Hsieh J (2009) Common neural mechanisms for explicit timing in the sub-second range. NeuroReport 20(10):897–901. https://doi.org/10.1097/wnr.0b013e3283270b6e
53. Singh P, Ray K, Bandyopadhyay A (2022) Biological antenna to the humanoid bot: electromagnetic resonances in biomaterials. Studies in rhythm engineering (SRE). Springer, Singapore. https://doi.org/10.1007/978-981-16-9677-0
54. Schirmer A, Meck WH, Penney TB (2016) The socio-temporal brain: connecting people in time. Trends Cogn Sci 20:760–772
55. Singh P, Sahoo P, Saxena K, Ghosh S, Sahu S, Ray K, Fujita D, Bandyopadhyay A (2021) Quaternion, octonion to dodecanion manifold: stereographic projections from infinity lead to a self-operating mathematical universe. Proc Int Conf Trends Comput Cogn Eng 1169:55–77
56. Singh P, Sahoo P, Saxena K, Ghosh S, Sahu S, Ray K, Fujita D, Bandyopadhyay A (2021) A space-time-topology-prime, stTS metric for a self-operating mathematical universe uses Dodecanion geometric algebra of 2–20 D complex vectors. Proc Int Conf Data Sci Appl 148:1–31
57. Schaik C, Burkart J (2011) Social learning and evolution: the cultural intelligence hypothesis. Philos Trans R Soc B Biol Sci 366(1567):1008–1016. https://doi.org/10.1098/rstb.2010.0304
58. Stevenson J (2001) The cultural origins of human cognition. Am J Hum Biol 13(3):428–428. https://doi.org/10.1002/ajhb.1073
59. Sharon RA, Murthy HA (2020) The "Sound of Silence" in EEG—cognitive voice activity detection. ArXiv. /abs/2010.05497
60. Stocco A, Bryan M, Sarma D, Youngquist T, Wu J, Prat C (2014) A direct brain-to-brain interface in humans. PLoS One 9(11):e111332. https://doi.org/10.1371/journal.pone.0111332
61. Trimper J, Wolpe P, Rommelfanger K (2014) When "i" becomes "we": ethical implications of emerging brain-to-brain interfacing technologies. Front Neuroengineering 7. https://doi.org/10.3389/fneng.2014.00004
62. Tomasello M, Carpenter M, Call J, Behne T, Moll H (2005) Understanding and sharing intentions: the origins of cultural cognition. Behav Brain Sci 28(5):675–691. https://doi.org/10.1017/s0140525x05000129
63. Tabar YR, Halici U (2016) Brain computer interfaces for silent speech. Cambridge University Press
64. Uono S, Sato W, Toichi M (2015) Exaggerated perception of facial expressions is increased in individuals with schizotypal traits. Sci Rep 5(1). https://doi.org/10.1038/srep11795
65. Wascher C, Kulahci I, Langley E, Shaw R (2018) How does cognition shape social relationships? Philos Trans R Soc B Biol Sci 373(1756):20170293. https://doi.org/10.1098/rstb.2017.0293
66. Wolpaw J, McFarland D (2004) Control of a two-dimensional movement signal by a noninvasive brain-computer interface in humans. Proc Natl Acad Sci 101(51):17849–17854. https://doi.org/10.1073/pnas.0403504101
67. Winston J, Strange B, O'Doherty J, Dolan R (2002) Automatic and intentional brain responses during evaluation of trustworthiness of faces. Nat Neurosci 5(3):277–283. https://doi.org/10.1038/nn816
68. Yoo J, Kim D-S, Yoo S-J, Lee S-W (2018) EEG-based brain-to-brain communication: a feasibility study. Front Hum Neurosci 12:14. https://doi.org/10.3389/fnhum.2018.00014

Chapter 7
Quantized Thoughts as Structure of Time: Cognitive Engineering for AI: An Octave Drawing Test for Mathematical Structure of a Subconscious Mind

7.1 Introduction

7.1.1 Identifying the Psychological Attributes of Cognition and Perception

7.1.1.1 Drawing Tests as a Window into the Cognitive Mind

Identifying the psychological attributes that contribute to effective decision-making is a key challenge in cognitive engineering. However, our understanding of brain function and information processing is still limited, making it difficult to develop robust models of human cognition. Drawing tests offer a unique opportunity to gain insights into the subconscious mind and its role in decision-making. While drawing tests have been used to assess cognitive abilities and personality traits for decades, recent research suggests that they may also be useful for identifying the psychological underpinnings of decision-making.

For example, a study by Gu [10] found that the tree drawing test could be used to identify subconscious minds of generic depression. Another study by Kaneda [18] showed that human figure drawings could be used to assess the severity of schizophrenia. Additionally, a study by Imuta [14] found that the complexity and details of human figure drawings were correlated with cognitive ability.

These findings suggest that drawing tests may be able to provide a valuable window into the cognitive processes that underlie decision-making. For example, a drawing test could be used to identify individuals who are prone to making impulsive decisions or who are more likely to be influenced by biases. This information could then be used to develop training programs or interventions to help people make better decisions. Identifying the psychological attributes for decision-making is an open challenge in cognitive engineering. There is no known principle for brain functioning, nor does we know the information processing structure of the brain. We

T. Dutta and A. Bandyopadhyay, *Emotion, Cognition and Silent Communication: Unsolved Mysteries*, Studies in Rhythm Engineering,
https://doi.org/10.1007/978-981-99-9334-5_7

need to build a structure of the inner mind for cognitive engineering that eventually takes all decisions. Drawing tests are one of the trickiest routes to unveil mysteries of cognitive processes in the human brain. Tree drawing test reveals subconscious minds of generic depression [10], Schizophrenia [18]. While human figure drawings have been used as a projective measure of intelligence [14] since 1926 [9]. Cognitive ability is proportional to the complexity and details of the drawings [12]. Finish the drawing from hints reveals personality.

In addition to their potential for assessing decision-making skills, drawing tests may also be useful for identifying other psychological attributes that are important for cognitive engineering. For example, a drawing test could be used to assess creativity, problem-solving skills, and spatial reasoning ability. These attributes are all essential for success in many cognitive engineering tasks.

Overall, drawing tests offer a promising new approach to understanding the cognitive processes that underlie decision-making and other important cognitive skills. By developing and validating drawing tests that are specifically designed to assess these skills, cognitive engineers can gain new insights into how to design systems and interfaces that support optimal human performance.

7.1.1.2 Drawing Tests as a Marker of Cognitive Skills

Here are some additional critical contents on the use of drawing tests in cognitive engineering:

- Drawing tests can be used to assess a wide range of cognitive skills and psychological attributes. This includes cognitive abilities such as intelligence, problem-solving skills, and spatial reasoning; as well as personality traits such as creativity, impulsivity, and risk-taking behavior.
- Drawing tests are particularly well-suited for assessing cognitive skills that are difficult to measure with traditional methods. For example, drawing tests can be used to assess tacit knowledge, which is knowledge that is difficult to articulate verbally.
- Drawing tests are also relatively inexpensive and easy to administer. This makes them a practical tool for use in a variety of settings, such as research laboratories, workplaces, and schools.

Here are some specific examples of how drawing tests are being used in cognitive engineering today:

- Researchers are using drawing tests to develop new training programs for pilots. For example, one study found that pilots who completed a training program that included drawing exercises were better able to identify and avoid hazards in flight.
- Engineers are using drawing tests to design more user-friendly interfaces for complex systems. For example, one study found that people who were asked to draw a diagram of how a complex system worked were better able to understand and use the system than people who were not asked to draw a diagram.

- Cognitive rehabilitation specialists are using drawing tests to assess and treat cognitive impairments in people with brain injuries. For example, one study found that drawing exercises can help to improve cognitive function in people with stroke.

Overall, drawing tests are a promising tool for cognitive engineering. They can be used to assess a wide range of cognitive skills and psychological attributes, and they are relatively inexpensive and easy to administer. More research is needed to develop and validate reliable and sensitive drawing tests, but the potential benefits of using drawing tests in cognitive engineering are significant.

Challenges and Limitations of Drawing Tests

While drawing tests have many potential benefits for cognitive engineering, there are also some challenges and limitations to consider.

One challenge is that drawing tests can be subjective to interpret. This is because there is no single "correct" way to draw something. As a result, the interpretation of drawing tests can be influenced by the biases and training of the person who is scoring the test.

Another challenge is that drawing tests can be affected by a variety of factors, such as the person's age, education level, and cultural background. This can make it difficult to compare the results of drawing tests across different groups of people.

Finally, drawing tests are not a perfect measure of cognitive skills and psychological attributes. It is important to note that drawing tests should be used in conjunction with other assessment methods to get a complete picture of a person's cognitive abilities and personality.

Despite the challenges and limitations, drawing tests offer a promising new approach to understanding the cognitive processes that underlie decision-making and other important cognitive skills. By developing and validating drawing tests that are specifically designed to assess these skills, cognitive engineers can gain new insights into how to design systems and interfaces that support optimal human performance.

7.1.1.3 Personality Tests and Cognitive Engineering: A Window into the Human Mind

Cognitive engineering (CE) is a discipline that focuses on the design of systems and interfaces that support optimal human performance. One of the primary requirements for CE is to understand the mysteries of brain function. This is because human cognitive abilities, such as attention, memory, and decision-making, play a critical role in how we interact with machines.

There are enormous versions of personality tests starting from interpretation of complex patterns, choosing favored ones from multiple choices to transform random lines into a picture. However, for cognitive engineering (CE) one of the primary requirements is to understand the mysteries of brain functions [13, 23] only then the techniques from cognitive psychology could be used to the design of human–machine systems [39].

Personality tests can offer a unique window into the human mind and its role in cognition. Personality tests are typically used to assess a person's individual traits, such as introversion versus extroversion, conscientiousness versus impulsivity, and neuroticism versus emotional stability. However, personality tests can also be used to assess cognitive skills, such as problem-solving, creativity, and spatial reasoning.

How Personality Tests Can Be Used in Cognitive Engineering

Personality tests can be used in cognitive engineering in a number of ways. For example, personality tests can be used to:

- Identify individuals who are well-suited for specific tasks or roles. For example, a personality test could be used to identify individuals who are more likely to be successful in jobs that require creativity and problem-solving skills.
- Design systems and interfaces that are tailored to the needs and abilities of specific user groups. For example, a personality test could be used to identify the personality traits of people who are more likely to make mistakes when using a particular system. This information could then be used to design the system in a way that minimizes the risk of errors.
- Develop training programs that can help people to improve their cognitive skills. For example, a personality test could be used to identify the personality traits of people who are more likely to have difficulty with certain cognitive tasks. This information could then be used to develop a training program that is specifically designed to help these individuals improve their performance.

Examples of Personality Tests Used in Cognitive Engineering

A number of different personality tests have been used in cognitive engineering research. Some examples include:

- The Myers-Briggs Type Indicator (MBTI): The MBTI is a popular personality test that assesses four dichotomies: introversion versus extroversion, sensing versus intuition, thinking versus feeling, and judging versus perceiving. The MBTI has been used in cognitive engineering research to assess the relationship between personality type and cognitive performance.
- The Big Five Personality Test: The Big Five Personality Test is a personality test that assesses five broad personality traits: extraversion, agreeableness, conscientiousness, neuroticism, and openness to experience. The Big Five Personality Test has been used in cognitive engineering research to assess the relationship between personality traits and cognitive performance in a variety of tasks, such as problem-solving, decision-making, and learning.
- The Cognitive Styles Inventory (CSI): The CSI is a personality test that assesses four cognitive styles: sequential versus global, holistic versus analytic, abstract versus concrete, and active versus reflective. The CSI has been used in cognitive engineering research to assess the relationship between cognitive styles and

cognitive performance in a variety of tasks, such as human–computer interaction and aviation safety.

Personality tests offer a promising tool for cognitive engineering. By understanding the relationship between personality and cognitive performance, cognitive engineers can design systems and interfaces that are more user-friendly and efficient. Additionally, personality tests can be used to develop training programs that can help people to improve their cognitive skills.

Challenges and Limitations of Using Personality Tests in Cognitive Engineering

While personality tests have the potential to be a valuable tool for cognitive engineering, there are also some challenges and limitations to consider.

One challenge is that personality tests are not always reliable or accurate. This is because personality tests are based on self-report data, which can be biased or inaccurate. Additionally, personality tests can be affected by a number of factors, such as the person's mood, motivation, and social desirability.

Another challenge is that personality tests are not always valid for predicting cognitive performance. This is because cognitive performance is a complex phenomenon that is influenced by a number of factors, including personality traits, cognitive abilities, and environmental factors.

Finally, it is important to note that personality tests should not be used to make decisions about people's employment or other important life outcomes. Personality tests should only be used as one part of a comprehensive assessment process.

Despite the challenges and limitations, personality tests offer a promising tool for cognitive engineering. By carefully considering the reliability, validity, and fairness of personality tests, cognitive engineers can use them to design systems and interfaces that support optimal human performance.

7.1.2 Identifying the Psychological Attributes of Cognition and Perception

The Subconscious Mind and Decision-Making

Moment-to-moment decision-making to respond efficiently to complex systems is governed by situation awareness, which depends on the mental representation of the environment around the subject at any given time [6]. The subconscious mind conceptualized by Freud (1893) that works beneath focal awareness is mostly involved in constructing the mental presentation of the environment. However, studies on the subconscious mind have not progressed due to the lack of falsifiability and testability of the claims.

Despite the challenges, the quest for representative symbols went on [17]. Symbolism culminated into the autonomous microdynamics of the trance state of mind that builds the structure of the subconscious mind as suggested by Erickson [28].

Structuring the subconscious mind by metaphors and symbols requires key premises of its operational dynamics.

Based on the Jung-Freud debate that Erikson and others advanced arguments, the subconscious mind has eight fundamental features:

1. Information structure is an image. The subconscious mind processes and stores information primarily in the form of images, symbols, and metaphors.
2. It's always active and autonomous. The subconscious mind is constantly working, even when we are not consciously aware of it. It is responsible for many of our automatic bodily functions, as well as our emotions, thoughts, and behaviors.
3. Emotion delivers its dynamics. Emotions play a powerful role in the subconscious mind. They can drive our thoughts and behaviors, and they can also influence our perception of reality.
4. Desire for attraction creates its events. The subconscious mind is drawn to things that we want or desire. This can be helpful, as it can motivate us to achieve our goals. However, it can also be harmful, as it can lead us to make impulsive decisions or to engage in self-destructive behaviors.
5. Past present and future superpose. The subconscious mind does not distinguish between the past, present, and future. This can lead to problems such as flashbacks, nightmares, and anxiety. However, it can also be beneficial, as it allows us to draw on our past experiences to make better decisions in the present.
6. Confusion is a key constituent driving force for microdynamics of the subconscious mind. The subconscious mind is constantly trying to make sense of the world around us. When it encounters something that it does not understand, it can lead to confusion. This confusion can be a powerful driving force for change and growth.
7. Fear, reproduction, and food creates feedforward cycle. The subconscious mind is constantly trying to protect us from danger and to ensure our survival. This is why it is so important to manage our fears and to take care of our basic needs.
8. Suppression of desire, emotion fear reproduction desire and food inflates the occupied domain of subconscious in the mental structure. When we suppress our desires, emotions, and fears, they do not go away. Instead, they are pushed into the subconscious mind, where they can continue to influence our thoughts, behaviors, and emotions.

How the Subconscious Mind Influences Decision-Making

The subconscious mind plays a major role in decision-making. It is responsible for many of our automatic decisions, such as whether to cross the street when we see a green light or whether to avoid a puddle on the sidewalk. The subconscious mind is also responsible for our gut feelings, which are often based on our subconscious perceptions of the world around us.

Our subconscious minds are constantly scanning our environment for potential threats and opportunities. When it perceives a threat, the subconscious mind will trigger a fight-or-flight response. This response is designed to protect us from danger, but it can also lead us to make impulsive decisions.

When the subconscious mind perceives an opportunity, it will trigger a reward response. This response is designed to motivate us to pursue the opportunity. However, it can also lead us to make decisions that are not in our best interests.

How to Improve Decision-Making by Understanding the Subconscious Mind

Understanding the subconscious mind can help us to improve our decision-making. By becoming more aware of the subconscious processes that influence our decisions, we can make more mindful and deliberate choices.

Here are a few tips for improving decision-making by understanding the subconscious mind:

- Pay attention to your gut feelings. Your gut feelings are often based on your subconscious perceptions of the world around you. If you have a strong gut feeling about something, it's worth listening to.
- Take your time when making important decisions. Don't rush into any decision without giving yourself time to think about it. The subconscious mind needs time to process information and make recommendations.
- Be aware of your biases. Everyone has biases, which are subconscious beliefs and attitudes that can influence our decisions. By being aware of our biases, we can avoid making decisions that are not in our best interests.

7.2 Discrete Structure of Time is Fundamental to Human Consciousness

7.2.1 Several Clocks Arrange in a 3D Geometric Shape: Is This the Structure of a Mind?

7.2.1.1 Structure of the Subconscious Mind

Vedic philosophers have been interested in the structure of the subconscious mind for centuries. They have developed a sophisticated understanding of how the subconscious mind works and how it influences our thoughts, feelings, and behaviors.

One of the key concepts in Hindu philosophy is that everything in the universe is cyclical, there is only one physical reality, that is phase. This includes the subconscious mind. The smallest unit of a cycle is called a rita. Several such rhythms or ritas are nested within each other, creating a complex network of cycles. All rhythms are interconnected with each other. In fact there was not much research done on the understanding of the true information structure of the human brain, or the biological system. Therefore, biological rhythms which were considered as distinct isolated clocks. No research group wanted to integrate these clocks. When an attempt was made to integrate the clocks.

7.2.1.2 A Unified Langue and a Universal Code: The Quest for Developing a Self-operating System

Following the Vedic philosophy, the entire universe is seen as one large cycle, with smaller cycles nested within it. A within and above fractal network of clocks connect smallest to the largest. The smallest cycle of the universe is called Brahman. Brahman is the source of all energy and vibration, as it integrates all phase cycles to build all entities that we can find, like time, mass, morphogenesis. It is the fundamental building block of the universe.

Phase Prime Metric, PPM is Purusha and Space–Time-Topology-Prime STts Invariant Network as Prakriti: The next cycle is called Purusha. Purusha is similar to a life form functioning as the mathematical code of the universe which has no attribute to anything, and that represents the core structure that holds universal consciousness. It is surrounded by a network of smaller cycles called Prakriti. Prakriti represents the material world made of 3D clock assemblies. One of the beautiful parts of this Purusha-Prakriti hypothesis that Singh et al. has adopted in constructing the consciousness theory, Self-Operating Mathematical Universe, SOMU is that Purusha is phase prime metric, PPM [4] and Prakriti is the space–time-topology-prime, or STts invariant network primarily written using the 3D assembly of clocks. Of course, in principle one could use to write the information as geometric shapes running in a loop (imagine circular changes from tear drop to ellipsoid), what we call morphogenesis or human cognition or perception could be a set of prime numbers as we have outlined in the end of Chap. 5.

The mind is a type of Prakriti originating from ever-evolving Purusha. To incorporate this feature, SOMU [36, 37] has conceived that PPM is not a finite entity, it is a prime number inventory, the temple like structure of PPM evolves continuously. The incompleteness of PPM demands that all conscious entities that replicates majority of its symmetries (humans are made of primarily 15 primes, maximum 47, [32]), continuously evolve its hardware to remain as per with the PPM. This is a beautiful depiction of human evolution, where we replace every single cell of ours only to make sure that the replaced cells make us equally synchronous to the universe. Our physical structure, brain or body needs to evolve just to remain conscious, else we disconnect with the PPM or Purusha. Prakriti is a three-dimensional clock assembly [11], that enables us to treat time like a space. Imagine you look at a 3D object holding it in your hand, it would be almost like looking at entire time range and all structures made by discrete times at once. The conscious mind exchanges clocks with other Prakriti and guests on Purusha [38]. The subconscious mind holds the fundamental symmetries of the Purusha-Prakriti conjecture [22]. Hindu philosophers believe that the subconscious mind is a statistical evolution of mathematical entities (Sankhya Darshana) and thus is responsible for many of our automatic thoughts, feelings, and behaviors. It is also responsible for our gut feelings and intuition [16]. So, what was undefined, logically circular and nonsense, would have a representation under the framework of Self-Operating Mathematical Universe, SOMU, a new consciousness theory we are advocating to build a mathematical structure (3D clock assembly could be written using a very complex mathematical expression) for all those mental states which were fuzzy undefined entities so far. It means we could see a geometric shape and tell, "Ohhh this is intuitive understanding of persuasive emotional conservativeness" (see Chap. 10 for details).

The subconscious mind is constantly scanning our environment for potential threats and opportunities. When it perceives a threat, the subconscious mind will trigger a fight-or-flight response. This response is designed to protect us from danger, but it can also lead us to make impulsive decisions. When the subconscious mind perceives an opportunity, it will trigger a reward response. This response is designed to motivate us to pursue the opportunity. However, it can also lead us to make decisions that are not in our best interests.

7.2.1.3 Unlocking the Subconscious Mind: From Ancient Philosophies to Modern Theories

The exploration of the human mind and consciousness has been a subject of fascination and inquiry for centuries. From ancient philosophies to contemporary scientific theories, understanding the intricacies of the mind has remained a complex and evolving endeavor. In this article, we delve into the journey of human consciousness, drawing from both ancient Hindu philosophies and modern scientific theories, with a particular focus on the mathematical aspects that underpin our understanding.

Ancient Wisdom and the Concept of Cycles

To begin our exploration, we turn to the rich philosophical traditions of ancient India, where scholars and sages contemplated the nature of reality and the human mind. One fundamental concept that emerged from these philosophical discussions is the idea of cycles or "Rita." In Hindu philosophy, every event in the universe is considered cyclic, representing a rhythmic pattern (Chhanda) that pervades existence (Source: https://religion.wikia.org/wiki/Chandogya_Upanishad).

In this view, the universe is structured by an intricate web of cycles, with the smallest unit of a cycle itself being a cycle or "Rita" [16] (Fig. 7.1a). This concept introduces the notion of nested cycles, where there is a hierarchy of cycles within cycles. This hierarchical structure is often referred to as "Loka," representing the various layers or levels of existence.

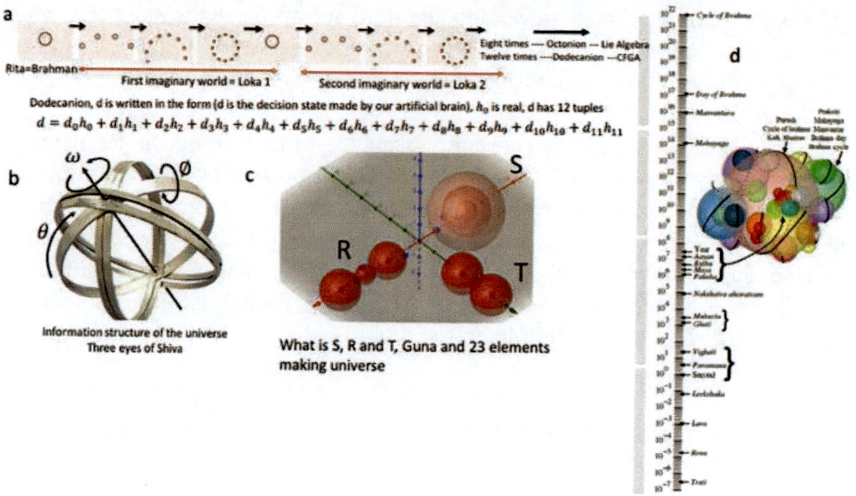

Fig. 7.1 Introduction to the foundation of subconscious minds using time crystal [26]: **a** A pixel in a circle is a circle (Circle = Clock = Rita), this concept is shown by zooming. Zoomed eight times we get octonions, if zoomed 12 times we get dodecanions. 12 tuples of a dodecanion is shown in a 12D vector expression. **b** Three eyes of Shiva, θ, $\emptyset$, ω, together builds units of information for consciousness. We connect it with geometric phase triplet of vortices responsible for decision-making in a neural network. **c** Concentric spheres making concentric clocks represent S or Satva, Two large clocks connected by a very small clock is Transformation or Rajah or R; two spheres connected by tama, T. **d** Crystallization of clocks or synthesis of Kala and Kali conjecture from the Hindu units of time. Smallest unit of time is 0.1 μs or 10 ns or truti and longest unit is 10^22 s or one brahma cycle. Four primary domains are shown (equivalent to four faces of Brahma), all units of times are put together in an integrated geometry. Vertical projection of this crystal is used to build Mandala. This is an empty information structure or pot (Aadhar) that holds information

7.2.2 *Several Clocks Arrange in a 3D Geometric Shape: Is This the Structure of a Mind?*

7.2.2.1 Geometric Cognition, Perception, and Emotion in Light of DDG

The brain's eternal quest to learn invariants: When employing the Dodecanogram (DDG), we embark on a journey into the brain's transformation through discrete time engineering, leading to a paradigm shift in our perception from conventional space–time duality to what we term the "space–time-topology-prime quartet." Adding two more element to the idea of events happening around us helps in making the information as bias free, attribution free element. In Chap. 5, we unveiled a discovery: geometric configurations fashioned from prime numbers possess the capacity to represent 91 distinct conscious experiences. Also, we understood that finding the invariants is the key for human cognition, perception, and emotion. In Chaps. 5 and 6, we have emphasized that to understand the human consciousness, we have to make it a common practice to find the invariants as the information, not facts. So far, because of transformational contributions of the computing to the human civilization, scientific protocols used by all consciousness theories were concentrating on bits or qubits and Turing tape. Here in the last two chapters, that culture has been seriously challenged. To understand human cognition, perception and emotion, and eventually consciousness, we do not need machine like tools. Information could be a network of invariants; brain could be searching the 91 invariants as listed in Chap. 5. Then quantity of information would not be important at all. Here in this Chapter, we would be looking for invariants, directly by carrying out drawing test allowing free will of a mind trapping into simple geometric shapes. All these methods of information processing are integral parts of Fractal Information Theory, FIT [2].

When time is redefined, from a linear or nonlinear entity to a loop, and allow it to evolve over three dimensions, the worldview changes fundamentally: For the last decade, we understood one point about brain, it does not favor one particular component over rest of the others. The polyatomic time crystal model, as elucidated by Reddy et al. [26], Saxena [31] and Singh et al. [32], introduces a very integrative perspective regarding the brain's utilization of prime numbers. This model proposes a quantized framework that governs the brain's orchestration of cognitive, perceptual, and emotional responses. It posits that these responses can be precisely quantified by dissecting recurring events that underlie these profound experiences. Imagine these recurring events as clocks, each tracing a loop on the surface of a phase sphere. These spheres collectively form a three-dimensional clock assembly, as exemplified by our DDG measurements on human subjects in the preceding Chap. 5. Remarkably, these structures, which we have identified as cognitive elements, remain consistent across all individuals, underscoring the enduring fractal nature of the brain's behavior.

Scale free fractal hardware to a fractal mind, subjective or objective, the journey is a chain of imaginary worlds grown within and above: The genesis of these cognitive clock structures, as revealed through DDG in the previous Chap. 5, can originate at either the molecular or nano-scale levels. An intriguing study by Pradhan et al. [25] underscores the pivotal role of junctions that could be played by membrane microtubule junctions in neurons, emphasizing the significance of these junctions in transmitting signals within the brain's intricate fractal architecture. Junctions assume critical importance as they facilitate interactions between different resonating layers, positioning individual clocks at precise orientations.

This relative phase structure gives rise to a scale-free electromagnetic resonance within a single microtubule nanowire, a single tubulin protein, or a single neuron, unveiling the presence of a triplet of triplet resonance bands or nine distinct frequency/time domains working in harmonious unison [30]. Electromagnetic resonance couples every single component of the brain using specific prime number of coupled peaks [33]. This triplet of triplet resonance carries profound implications, even at the molecular level, unveiling the symmetry reminiscent of prime numbers and suggesting the tantalizing possibility of quantum tunneling in information processing [3]. Furthermore, the concept that a fractal network can effectively process information in a logical manner [7] underscores the intrinsic fractal essence of the mind.

We have to look into all active time domains in all three time dimensions at once: Just like we see 3D space as a whole at a time: Traditionally, neuroscience has endeavored to link brain responses observed at the scalp to the activity of neurons, as if neurons were the sole arbiters of all conscious and subconscious processes within the brain. However, research conducted by Agrawal et al. [1], involving simultaneous measurements of neuron firing at both the neuron membrane and within microtubular bundles, unveils a remarkable revelation: microtubules possess the astounding capability to process firing signals at speeds approximately 10,000 times faster than the membrane. This finding challenges the conventional notion that neurons are the exclusive determinants of brain information processing [29]. Astonishingly, even non-connected neurons devoid of direct synaptic junctions can engage in energy exchange, as demonstrated through neural circuit imaging employing a scanning dielectric microscope (SDM) [8]. This imaging technique bears semblance to the mapping of facial expressions and emotional states, akin to SDM or a thermal camera, that maps whole face or brain surface at once.

As a result, an ingenious machine, as delineated in Chap. 5, has been conceived—a machine akin to an EEG but endowed with the capability to scrutinize the discrete time structures governing brain operations across an extensive spectrum of time scales simultaneously. This pioneering device, christened the "Dodecanogram," will find extensive use in facilitating silent communication and probing the quintessential traits of consciousness.

7.2.2.2 The Final Test of Invariants: Lets Constraint the Free Will into Geometric Shapes

Clocks and Vibrations: The Fabric of the Universe

Within this framework, the smallest cycle in the universe is akin to a clock, represented by a mechanical vibration known as "Brahman." These cycles, when combined, give rise to various forms of waves (referred to as "Asat") and matter (referred to as "Sat"). In essence, the composition of these cycles constructs the fabric of the universe itself.

Imagine the entire universe as a colossal clock or cycle, with other clocks or cycles residing on its perimeter. These guest clocks interact among themselves and are known as "Prakriti." One such life form inhabiting this cosmic clockwork is the "Virata Purusha" [38], signifying the interconnectedness of all life forms within the universe.

The Mind as a 3D Clock Assembly

Now, let's turn our attention to the human mind, a remarkable facet of this cosmic tapestry. In this intricate philosophical framework, the mind is conceptualized as a 3D clock assembly (Fig. 7.1b, c), belonging to the class of Prakriti. It operates as a conscious entity that exchanges "clocks" with other Prakriti or guests within the cosmic clockwork of the "Purusha" or universal clock.

The Subconscious Mind: Guardian of Symmetries

As we explore the depths of the mind within this philosophical context, we encounter the concept of the subconscious mind. The subconscious mind is believed to hold the fundamental symmetries of the Purusha-Prakriti conjecture [22]. Within this intricate structure, the subconscious mind plays a pivotal role in maintaining the harmonious rhythms and patterns that govern the interactions and integration of clocks [11]. These profound insights into the nature of the subconscious mind and its role in preserving the symmetries of the universe provide a unique perspective on the ancient wisdom that has shaped our understanding of consciousness.

A Journey Through Time and Thought

In summary, our journey through the annals of ancient Hindu philosophy has revealed a profound appreciation for the cyclic nature of the universe, where nested cycles and cosmic clocks interweave to create the fabric of reality. Within this cosmic dance, the human mind is seen as a conscious entity intricately connected to the larger cosmic order.

While these ancient philosophies offer a unique lens through which to contemplate consciousness, it is essential to recognize that modern scientific theories have also contributed significantly to our understanding. These theories, such as the Self-Operating Mathematical Universe (SOMU), approach consciousness from a mathematical perspective, seeking to uncover the underlying patterns and symmetries that govern the mind. In the ongoing quest to unravel the mysteries of consciousness, we stand at the intersection of ancient wisdom and cutting-edge science, continually seeking to unlock the secrets of the mind and its place in the grand tapestry of the universe.

7.2.3 Structure and Model for a Subconscious Mind

Unlocking the Ancient Mathematical Foundations of Consciousness

Throughout human history, the quest to understand the intricate workings of the human mind and the cosmos has been a relentless pursuit. In our exploration of the development of human thought and mathematical constructs, we journey back to a pivotal era—post 2000 BC. It was during this time that fundamental constructs were established, serving as the bedrock upon which advanced mathematics and elemental concepts would be built.

The Emergence of Advanced Mathematics and Elemental Concepts

In the wake of these foundational constructs, a surge of advanced mathematical principles and elemental concepts began to permeate the intellectual landscape. These principles would go on to form the cornerstone of countless literary works and philosophical debates, leaving an indelible mark on the annals of human knowledge.

One noteworthy area of exploration during this period was the measurement of time—an essential aspect of understanding the cosmos and the intricate mechanisms of the mind. Ancient Hindu scholars developed a sophisticated system of time units that spanned an astonishing range, from the minuscule truti (0.1 μs) to the awe-inspiring expanse of 1022 years [11, 15, 20] (https://en.wikipedia.org/wiki/Hindu_units_of_time; Fig. 7.1d). This mathematical framework for time units provided the foundation for unraveling the integration of clocks within the cosmic order.

7.2.3.1 Temporal Folds and Conservation Laws: Unearthing the Universe's Temporal Architecture

Within this intricate temporal framework, scholars discerned critical temporal folds—a hidden facet of the 3D clock architecture (Raga or dynamic Clock or Prakriti + Ragini or static or Purusha [27]). These folds held the key to uncovering the conservation laws that govern the temporal structure of the universe. Each clock domain was meticulously placed on a sphere, symbolizing completeness or "purna." This 3D clock architecture became a pivotal tool in comprehending the dynamic interplay of time within the cosmos.

The Mathematical Fabric of the Subconscious Mind

As our exploration deepens, we encounter another crucial facet—the development of a mathematical structure for the subconscious mind. Central to this endeavor is the creation of a clock architecture for the 23 elementary qualities, where the unit of quality is termed "tanmatra," and "Guna" signifies quality [21] (Fig. 7.1c).

Within this intricate framework, three essential qualities take center stage:

1. **T (Tama)**: This represents the bonding of clocks or periodic events, signifying the fundamental connections within the cosmic order.
2. **R (Raja)**: Here, we delve into the transformation of geometric shapes constructed from clocks or periodic events, exploring the dynamic changes that shape the universe.
3. **S (Satva)**: Representing reflection from infinity, S encapsulates the infinite geometric transformations of clocks, giving rise to the concept of universal time and crystallization (Kala and Kali or Bhairavi; Satva).

Kali, often associated with the crystallization of universal time [5], forms the basis of this mathematical exploration, encompassing ten fundamental aspects of clock architecture [19]. These three parameters—S, R, and T—are not merely theoretical constructs; they are embodied by clocks that dynamically change their diameters, symbolizing the dynamic state of the mind as a composition of these elemental elements (S, R, and T).

7.2.3.2 Unveiling the Origin of Time Crystal: Crystallization of Discrete Pieces of Time

Our journey through the mathematical foundations of consciousness has illuminated a profound understanding of time, consciousness, and the interconnectedness of all things. It is within these mathematical constructs that we discover the origin of Kali—the crystallization of time and the intricate mosaic of Rita, composed of bits and pieces derived from the mathematical fabric of the universe.

As we reflect on this journey, we recognize that the mathematical structures created by ancient scholars continue to inspire and inform our contemporary understanding of consciousness and the cosmos. The interplay between mathematics and philosophy remains a timeless endeavor, revealing the eternal mysteries that bind the human mind to the universe.

Here using SRT features we have looked into the patterns created by human subjects and found geometric similarity, symmetry or asymmetry, integration, fractality, convergence or divergence, engagement or disengagement, complementarity, boldness or cleanliness, hint for a dynamic, sensitivity to intricate details as revealed from the shapes (Fig. 7.2a). Then, we created 3D clock assembly, which delivers the functional map of a human subject's subconscious mind (Fig. 7.2b).

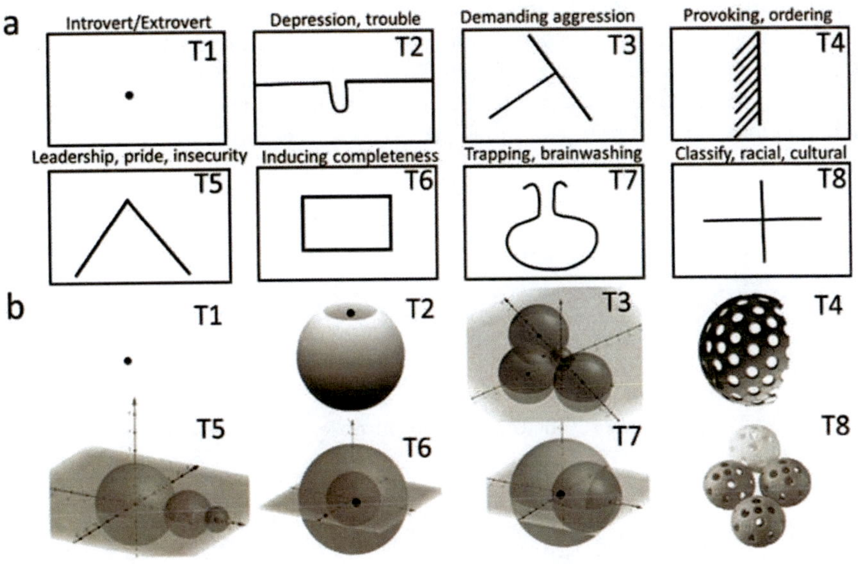

Fig. 7.2 **a** Octave test given to human subjects and **b** equivalent clocks spheres that possibly generate in the mind of a human subject when he/she looks at the image

7.3 Experiment on Human Subjects to Map the Quantization of Subconscious Mind

7.3.1 Exploring Subconscious Minds Through Eight Seed Geometries [24]

Our journey into the depths of the human subconscious mind involves the utilization of eight distinctive seed geometries, a process visually represented in Fig. 7.2. In this unique experimental approach, human subjects are invited to engage in a creative act—drawing—with the added twist of incorporating one of these eight seed patterns into their artwork.

The instructions are deliberately simple: participants are encouraged to draw freely, with the seed pattern seamlessly integrated into their creations. The emphasis here is not on artistic prowess; rather, we prompt subjects to sketch without overthinking, using only the essential lines that provide them with a sense of comfort.

One crucial aspect of this experiment is that it can only be conducted once with each participant. This restriction is rooted in the belief that each test serves to synchronize the Purusha (the cosmic consciousness) with the Prakriti (the cosmic materiality). Consequently, we place a premium on the initial attempt, as repeated tests may compromise the integrity of this unique synchronicity.

Our analytical focus lies in scrutinizing the resulting patterns for subtle responses, as we deliberately discourage subjects from overanalyzing or modifying their drawing patterns. In essence, we seek to uncover the unfiltered expressions of the subconscious mind, as reflected through these artistic creations.

7.3.2 Experimental Phases Unveiling Subconscious Mind Patterns

In our investigative process, we employ eight distinct phases, each designed to reveal intricate aspects of the human subconscious mind. Here, we provide a technical breakdown of these phases, elucidating their purpose, and the responses they elicit:

Phase 1 (T1—Point): In the initial phase, we present the subject with a single point, prompting them to engage in creative expression through drawing. Grounded in Vedic philosophy, this phase explores the notion that the universe is encapsulated within a point, a belief that the subconscious mind perpetually seeks. By analyzing the subject's response, we discern whether the mind requires a line to establish a connection with the external world or encircles the point with a loop, shedding light on communication modes between the inner mind and the external environment.

Phase 2 (T2—U-Shaped Pattern): In this stage, we introduce the subject to a U-shaped pattern within a plane. The subconscious mind perceives itself within the inner part of this U-shape, instigating a desire to traverse or bridge the gap. These foundational constructs prompt the subject to unveil the structure of their inner mind through geometric representations.

Phase 3 (T3—45° Tilted T): The third phase involves presenting the subconscious mind with a 45° tilted T-shape. The mind identifies the junction point within this configuration. Subsequently, the subject's response to this stimulus determines whether the subconscious mind is provoked to create symmetry or conceal it.

Phase 4 (T4—Generating Symmetry): Symmetry serves as a fundamental principle in the workings of nature. In this phase, the subject encounters a deliberate asymmetry, presented as an arrow with directional attributes. This provocation, featuring several parallel lines, elicits three distinct attitude responses, revealing the mind's inclination toward conformity, adaptation, or opposition.

Phase 5 (T5—Mountain-like Shape): Phase five introduces a mountain-like shape to the subject. The mind is positioned a top this triangular structure, evoking feelings of pride and fear. This phase assesses the quality of leadership within the subject's psyche.

Phase 6 (T6—Rectangular Box): A rectangular box takes center stage in the sixth phase. It symbolizes complete satisfaction and happiness, offering the mind two choices: to explore its inner dimensions or embark on an adventure beyond its boundaries.

Phase 7 (T7—Pot with Barrier): The seventh phase presents a pot-like structure, challenging the mind's navigation abilities. As the mind enters this structure, it encounters a circular barrier, a frustrating obstacle designed to test and potentially undermine confidence. This phase allows us to observe how the mind overcomes such challenges.

Phase 8 (T8—Confusing Structure): In the final phase, the subject is presented with a perplexing and seemingly chaotic structure. This structure is intentionally designed to be complete and self-contained, harvesting noise and prompting the subconscious mind to respond freely.

Through these eight meticulously crafted phases, we gain valuable insights into the intricacies of the human subconscious mind, unveiling its patterns, responses, and underlying mechanisms.

Declaration on Human Subjects: The local ethics committee approved the study (Indian Institute of Management, IIM Ranchi). It was performed following the ethical standards in the Declaration of Helsinki (last modified, 2004). All 89 subjects were above 18 years, and written informed consent was taken from all of them.

Human Subjects: For the experiment, 20 people aged between a maximum of 37 years and a minimum of 21 years were selected, all of whom were healthy, their body temperatures around 98 °F, initial pulse rate between 78 and 132, brought down to 80–85; blood pressure 85–130 and the oxygen level in their blood between 95 and 97%. For matching the subjects' circadian rhythm, all experiments were performed in the evening, after the students were done with their classes for the day. Subjects with high fever, neurological disorders, major diseases in the past, or any forms of abnormal responses were rejected.

Temporal restrictions and dim light in the room

All subjects were asked to draw anything. They were specifically mentioned that we are not interested in nice drawing, they should put some lines only to their satisfaction at the earliest. Most subjects completed the tasks within 5 min. A few subjects took up to 10 min. Therefore, we made an effort to acquire the first response of the mind without using significant memory and cognitive analysis. The dim light ensures less optical interference.

7.4 Complete Analysis of Subconscious Mind of Ten Human Subjects

7.4.1 Analysis of Ten Subjects in Detail, as Shown in Fig. 7.3

Let's delve into the psychological assessments of our subjects, uncovering their unique characteristics and responses:

Subject 1 (S1): In the first test, S1 displayed a gradual, conservative approach to socializing and self-discovery, methodically spiraling outwards. However, in the second test, S1 exhibited a preference for delving deeper into problems rather than seeking transformation or crisis resolution. By constructing multi-storied houses on flat terrain, S1 showcased a tendency to magnify issues from a smaller scale (U) to a larger one, opting for the path of least resistance.

Third Test: S1's performance in the third test highlighted a lack of a "killer instinct." Instead, the subject preferred to adhere to given boundaries, symbolized by creating a flag and remaining within one of three available sides. Interestingly, S1 refrained from pursuing straightforward symmetry, opting for an asymmetric approach by mirroring its actions on the other side, a distinctive deviation from conventional problem-solving.

Fifth Test: In the fifth test, S1 chose to remain within the confines of the mountain, shying away from leadership challenges, and opting for self-imposed seclusion within familiar territory.

Sixth Test: When assessing S1's expression of happiness in the sixth test, we observed a tendency to identify external subconscious minds or points, expecting them to conform to its own characteristics. This inclination manifested as an attempt to gradually build relationships based on predetermined expectations.

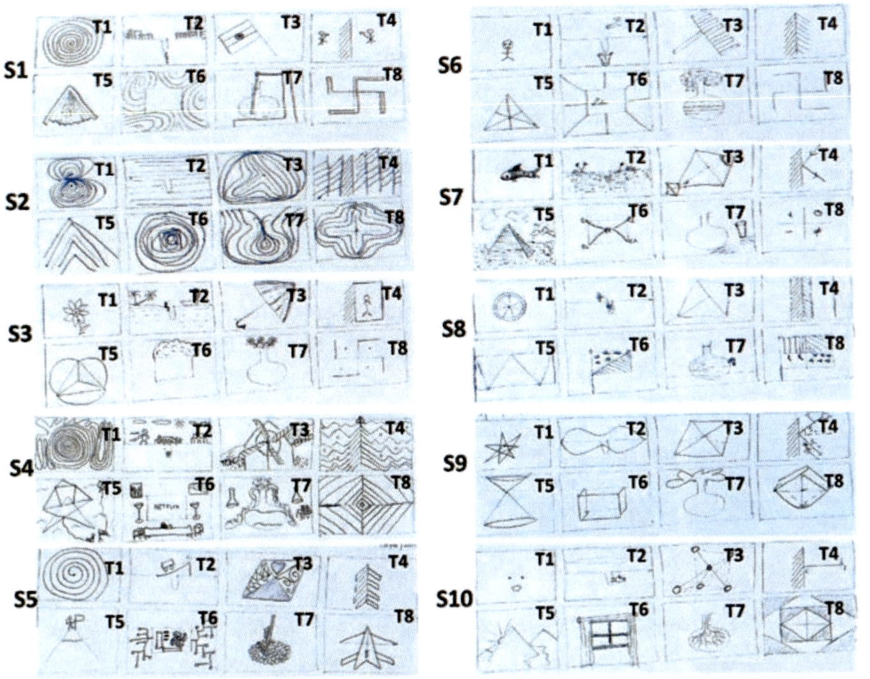

Fig. 7.3 Ten selected results for human subjects S1-S10. Ten selected subjects drew response to the query made in Fig. 7.2. Eight tests are numbered T1-T8

Seventh Test: Claustrophobia triggered a unique response in the seventh test, compelling S1 to navigate the bottleneck in the simplest possible manner. This situation generated a sense of urgency, potentially causing the subject to lose balance as it perceived limited options.

Eighth Test: In the eighth and final test, the subject faced the challenge of positioning the mind at the center of a "+," yet insecurity led to an alternative approach. Instead of residing on the edges of the four points, S1 chose to encircle the entire area, culminating in the formation of a swastika.

These psychological assessments provide valuable insights into the cognitive processes, problem-solving strategies, and unique tendencies of our subjects, shedding light on their distinct approaches to various challenges presented in the tests.

Psychological Profiling of Subject 2 (S2)

Subject 2, or S2, presents a classic case of psychological insecurity, as reflected in their responses across various tests. Let's delve into the specifics of S2's psychological profile.

Test 1 (T1): In the first test, S2 exhibited a tendency to divide the world into two distinct parts, creating a protective cover around itself. However, a noteworthy trait was S2's reluctance to fully enclose the single point, indicating a desire to maintain a connection with its origin before gradually expanding its presence in the world.

Test 2 (T2): In the second test, S2 displayed an inclination to disregard the existence of a deep trap, opting instead to create multiple parallel lines. This behavior seemed to serve the purpose of artificially concealing the presence of the trap, a manifestation of S2's introverted and compromising disposition.

Test 3 (T3): Test three presented a scenario provoking aggression, but S2 adhered to its introvert and cautious approach. Confronted with one junction and three endpoints, S2 made concerted efforts to mitigate potential risks by enclosing everything within a nearly circular structure.

Test 4 (T4): The fourth test aimed to stimulate the generation of complementary symmetry within the subconscious mind. Contrary to this objective, S2 chose to replicate the query repeatedly, effectively covering the entire available space. This response reflected a lack of engagement or analytical processing, as S2 blindly replicated without interaction.

Test 5 (T5): The trend of replication continued in the fifth test, wherein S2 replicated a mountain-like structure without deviation from the original query.

Test 6 (T6): Test six provided an intriguing revelation. S2, despite being a consistent replicator, demonstrated an understanding of closed-loop concepts, responding with a circular formation.

Test 7 (T7): In the seventh test, S2 opted not to attempt an escape from the pot-like scenario, instead constructing nested enclosures. This response reiterated S2's tendency toward replication rather than proactive problem-solving.

S2's psychological profile paints a picture of an individual marked by insecurity and a strong inclination toward replication and introversion. These characteristics manifest across a range of tests, shedding light on the subject's unique approach to various challenges and stimuli.

Psychological Insights from Subject 3 (S3) Test Responses

Subject 3, or S3, displayed intriguing psychological traits and responses across various tests. Let's delve into the nuanced aspects of S3's psychology.

Test 1 (T1): In response to the first test, S3 exhibited a dual-fold approach. Initially, S3 created a protective cover around its true nature, symbolized by a single point presented in the test. Simultaneously, S3 actively engaged with the external world, establishing a symmetric networking structure that resembled a petal. This response highlighted S3's inclination toward building a feedforward network bridging the inner and external worlds.

Test 2 (T2): Moving on to the second test, S3 navigated out of a hollow dip and proceeded to construct a distinct world on the flat surface, showcasing its capacity for adaptation and creativity.

Test 3 (T3): In the third test, S3 demonstrated an understanding of the need to break through barriers. Positioned at a junction, S3's mind discerned the situation accurately and constructed a cover that incorporated the edge points, signaling an aggressive approach to overcoming obstacles.

Test 4 (T4): S3 adopted unique methods to address the challenge of generating symmetry in the fourth test. Resolving the asymmetry presented through parallel lines, S3 created a box within which it placed itself, effectively rectifying the imbalance.

Test 5 (T5): The trend of integrating all edges into a unified structure persisted in the fifth test. Additionally, S3 established a central point, serving as a hierarchical observer within the network.

Test 6 (T6): Test six revealed S3's sense of completeness, as it constructed a rectangle within another rectangle, symbolizing an ultimate match. This response was marked by an exuberant and unbalanced expression of celebration.

Test 7 (T7): Contrary to preferring confinement, S3 opted for a direct resolution in the seventh test. The path leading out of the pot-like scenario was artfully decorated, with S3 reaching the exit without leaving a trace.

Test 8 (T8): In the final test, S3 demonstrated effort in constructing an enclosure formed by four edges, displaying a penchant for structural design.

S3's psychological profile portrays a subject with a versatile and creative mindset, capable of adapting to various challenges and scenarios. These responses shed light on S3's unique approach to problem-solving and interaction with the presented stimuli.

Psychological Insights from Subject 4 (S4) Test Responses

Subject 4, or S4, displayed distinctive psychological characteristics and responses in the series of tests. Let's delve into the intriguing facets of S4's psychology.

Test 1 (T1): In response to the first test, S4 exhibited an excessive level of caution when unveiling their inner self. This caution stemmed from the subtle differences in the diameters of concentric circles. Strikingly, as frustration escalated, S4 resorted to creating two new points. This act of generating multiple identities revealed underlying insecurities.

Test 2 (T2): In the second test, S4 sought confirmation of its existence within the shallow tube. This was achieved by drawing a point and subsequently enlarging it. S4 then embarked on creating a bridge to connect both sides of the tube. Notably, S4's excessive cautiousness manifested in constructing the bridge using circular arcs.

Test 3 (T3): S4 recognized the need to address an imbalance and exhibit aggression in the third test. This was crucial for completing the cross-shaped pattern (+). To achieve this, S4 constructed a dynamic rotor, an effort to encompass the edges or endpoints of lines. It's worth noting that these edges or endpoints represented fragmented aspects of the "self," contributing to multiple existences and internal conflicts.

Test 4 (T4): S4 displayed a proficient response in the fourth test. Here, S4 successfully generated symmetry, rectified imbalances, and replicated the evolved mountain pattern across the entire space. This response highlighted S4's ability to achieve symmetry and propagate dynamic movements.

Test 5 (T5): Test five revealed S4's distinctive approach to generating symmetry and addressing leadership-related queries. S4 created the mirror-inverse of the pattern, resulting in a unified structure. Furthermore, S4 infused dynamic elements into this structure, showcasing a multifaceted response.

Test 6 (T6): S4's response to the sixth test indicated a preference for maintaining completeness or satisfaction as an isolated symbolic structure. This structure exhibited multidimensional expressions, each distinct from the other. S4's tendency to create conceptual replicas in response to queries was evident in this test.

Test 7 (T7): In the seventh test, S4 acknowledged the challenging scenario and sought an artistic way out. However, S4 opted to replicate the query, symbolizing a form of celebration. This response underscored S4's inclination toward expressing achievements.

Test 8 (T8): Test eight revealed S4's inclination to connect the edges representing its fragmented "self." The underlying drive seemed to be the integration of all identities into a unified whole, reflecting a deep-rooted psychological trait.

S4's psychological profile portrayed a complex and multidimensional individual with tendencies toward caution, introspection, and a unique approach to problem-solving. These responses shed light on S4's intricate thought processes and coping mechanisms in various scenarios.

Psychological Insights from Subject 5 (S5) Test Responses

Subject 5, or S5, displayed a distinctive psychological profile characterized by introspection and a tendency to complicate scenarios in the series of tests. Let's explore the psychological facets of S5's responses.

Test 1 (T1): In the first test, S5 exhibited introverted tendencies that gradually spiraled outward toward the external world. This response pattern indicated a measured and cautious approach to self-revelation.

Test 2 (T2): S5's response to the second test revealed a remarkable inclination. S5 not only identified the problem but also engaged in a comprehensive analysis. However, an interesting trait emerged as S5 tended to exert excessive effort in addressing the issue. This tendency to amplify problems and complicate problem-solving scenarios became a hallmark of S5's behavior.

Test 3 (T3): The third test witnessed a transformation in S5's approach. S5 accepted the challenge presented, displayed assertiveness, and effectively resolved the junction barrier. This resulted in the creation of a closed conservative loop with mirror symmetry. This response showcased S5's ability to rise to challenges and embrace symmetry.

Test 4 (T4): S5's response in the fourth test underscored a deep affinity for developing hierarchical symmetry. S5 achieved balance by regenerating parallel lines and pairing them. This approach demonstrated S5's proficiency in establishing balance and symmetry in responses.

Test 5 (T5): The leadership test (T5) revealed S5's inclination to secure the top position, providing a stable base while also ensuring environmental protection for the uppermost point. This response highlighted S5's strategic thinking and protective instincts.

Test 6 (T6): Test six mirrored S5's response in T2, where happiness was decorated with intricate pathways. S5 explored various options randomly while searching for introspective patterns within. This response reflected S5's willingness to navigate complexity.

Test 7 (T7): S5's response to T7 resembled its approach in T2, which involved complicating problems and seeking the most challenging solutions within a given context. This behavior indicated S5's tendency to explore complexity.

Test 8 (T8): In the final test, T8, S5 demonstrated a distinct strategy. Although S5 knew what needed to be done, it chose to quantitatively inflate the problem (as seen in T7), pursue the most complex solution (as observed in T2), and manipulate the structure to emphasize a specific aspect or direction of thought (as seen in T8).

S5's psychological profile showcased a unique blend of introspection, problem-solving prowess, and a penchant for complexity. These responses provided valuable insights into S5's intricate thought processes and adaptive strategies across various scenarios.

Psychological Insights from Subject 6 (S6) Test Responses

Subject 6, or S6, presented a unique psychological perspective with a dual nature that was evident across multiple tests. Let's delve into the psychological dynamics revealed by S6's responses.

Test 1 (T1): S6 displayed a dual perspective, initially focusing on a single point and then creating another point in close proximity. This paired point creation seemed to address a desire to overcome loneliness and the insecurity associated with embarking on adventures alone. However, S6's response didn't end there; it further enclosed the two points with a loop, signifying a preference for collaboration and unity under a virtual singular identity.

Test 2 (T2): In the second test, S6 wholeheartedly embraced a particular truth. Rather than bridging obstacles, S6 utilized them as metaphors, prioritizing their use as instruments to facilitate life-like functions. This approach suggested a creative use of obstacles without altering their essence, representing problems as both outcomes and solutions to other challenges—a complex perspective that could be seen as a form of adaptive compromise.

Test 3 (T3): Test three offered additional insights into S6's psychology. S6 extended the junction point with multiple parallel lines, revealing a proactive and assertive approach to problem-solving. This aggression was further evident in T4 and possibly subtler in T2, suggesting that S6 possessed a willingness to confront challenges directly.

Test 4 (T4): S6's response in the fourth test showcased an inclination toward generating symmetry. Rather than compromising, S6 replicated the query, creating symmetry, and balance by repeating the pattern throughout the space. This indicated S6's capacity to harmonize diverse elements.

Test 5 (T5): Transitioning to the leadership test (T5), S6 left the pinnacle of the mountain and established another junction point within. This response hinted at a conservative tendency, as S6 confined itself within a protective enclosure. Simultaneously, it emphasized the appreciation of completeness and sought connections with both the environment and self.

Test 6 (T6): In the sixth test, S6 appeared to flourish, dynamically arranging four edge points or self-identities in an intricate configuration. This suggested an adept ability to integrate multiple aspects and identities, revealing a depth of adaptability, and creative thinking.

Test 7 (T7): S6's response in T7 indicated a proactive engagement with global players, seeking enrichment from within while reaching out to external elements. This approach reflected a collaborative mindset and a desire to contribute to a broader context.

S6's psychological profile revealed a complex interplay between dual perspectives—loneliness and collaboration, adaptation and compromise, aggression and harmony. These responses offered valuable insights into S6's intricate thought processes and versatile problem-solving strategies across diverse scenarios.

Psychological Insights from Subject 7 (S7) Test Responses

Subject 7, or S7, exhibited a distinctive psychological disposition through their responses in various tests. Let's explore the intricate aspects of S7's psychology that these responses revealed:

Test 1 (T1): S7 demonstrated a strong inclination toward establishing and preserving the absolute nature of their identity. Their focus was not on interaction with the external world but on forging a resolute self-identity characterized by bold, unyielding lines.

Test 2 (T2): S7's unwavering commitment to their self-identity was further highlighted in the second test. They disregarded the hole presented in the test, opting instead to eliminate both sides to create a smooth, uninterrupted path leading toward the problematic hole. This decorating tendency, while neglecting the underlying issue, underscored S7's fixation on self-absoluteness.

Test 3 (T3): Even as the test pattern directed S7's mind to a junction, they seemed ensnared, attempting to divide it into three separate parts. Despite the expectation to generate symmetry by forming a "+" structure, S7 chose to conceal it. They addressed the absence of imbalance by introducing a miniature replica at the end of the pattern.

Test 4 (T4): S7's response to the asymmetry presented in the fourth test was characterized by an uncompromising pursuit of creative symmetry through reflection. Their robust attitude of ignoring the provided asymmetry and regenerating balance with complementary patterns was consistent with previous tendencies.

Test 5 (T5): S7 continued to manifest their commitment to symmetry in T5, creating a point—an identity deeply embedded within the pattern. This response occurred when completeness and a trigger for satisfaction were introduced.

Test 6 (T6): Surprisingly, in the sixth test, S7 did not recognize the rectangle inside another rectangle, possibly due to the influence of previous test dynamics. Instead, they perceived the setup as a diminishing rectangle, indicating a unique perspective on the presented scenario.

Test 7 (T7): S7 chose not to make direct contact with the pot in T7 but rather sought an intermediary buffer medium for interaction. This reluctance to engage with the pot reflected a notable barrier to active engagement.

Test 8 (T8): In the eighth test, S7 exhibited a reluctance to engage once more, declining to interact with the presented challenge involving the "+" structure. They seemed to consider pockets within the structure as resting sites, further highlighting their aversion to direct involvement.

S7's psychological profile demonstrated a profound commitment to the establishment and preservation of an absolute self-identity. Their responses revealed an inclination toward self-absoluteness, creative symmetry, and a distinctive approach to problem-solving, often prioritizing the assertion of identity over direct engagement with external challenges.

Psychological Insights from Subject 8 (S8) Test Responses

Subject 8, or S8, displayed a unique psychological disposition through their responses in various tests. These responses offer valuable insights into S8's thought processes and cognitive tendencies.

Test 1 (T1): S8 demonstrated their ability to create an inner world, encoding a multitude of core values within it. This response indicated S8's inclination toward introspection and the internalization of their values.

Test 2 (T2): S8's effort to come out of trouble was driven by hope and optimism, suggesting a belief in the potential resolution of problems through mysterious or external powers. This response highlighted S8's optimistic outlook in the face of challenges.

Test 3 (T3): While S8 grasped the totality of the situation, they did not actively attempt to solve the problem. Instead, they engaged in a comprehensive review of the situation, even when the test demanded the generation of aggressive symmetry. This response reflected S8's tendency to approach challenges with a holistic perspective rather than immediate action.

Test 4 (T4): S8's inclination to isolate themselves and create alternative, independent existences became evident in this test. They generated an independent isolated line, illustrating their capacity for self-sufficiency.

Test 5 (T5): In this test, S8 amplified opportunities by placing themselves at the top of a mountain, symbolizing their willingness to embrace challenges and elevate their status. This response indicated a desire for increased opportunities and recognition.

Test 6 (T6): S8's response to the rectangle presented in this test revealed a nuanced perspective. They did not perceive the totality, either globally or locally, likely due to a sense of complete satisfaction. However, S8 made an effort to focus on the local totality. If a global symmetry existed, they preferred isolation within two triangles or the smallest enclosed areas, highlighting their selective engagement.

Test 7 (T7): Optimism was a prominent theme in S8's response to this test. They expressed hope and imagination but lacked aggressiveness in seeking liberation. This response demonstrated S8's positive outlook and imaginative thinking.

Test 8 (T8): In the final test, S8 chose to accept the given circumstances without generating dynamics or actively engaging with the presented scenario. They appeared content to accept their fate and live with the situation as a gift. This response indicated a degree of resignation and contentment with the status quo.

S8's psychological profile revealed a complex interplay of introspection, optimism, and a penchant for holistic evaluation of situations. Their responses showcased a willingness to embrace challenges, a sense of self-sufficiency, and a preference for selective engagement when faced with symmetry and completeness.

Psychological Insights from Subject 9 (S9) Test Responses

Subject 9, or S9, demonstrated distinct psychological traits and tendencies through their responses in various tests. These responses provide valuable insights into S9's cognitive processes and behavioral inclinations.

Test 1 (T1): S9 displayed ambition and a strong desire to expand their pride and fame in the boldest possible ways. This response indicated a drive for recognition and prominence.

Test 2 (T2): Upon encountering hints of trouble, S9 took an indirect approach. Rather than directly addressing the problem, they attempted to depict it as a universal, generic issue, and then created a scenario as if the problem did not exist. This response suggested a tendency to avoid direct confrontation with challenges.

Test 3 (T3): In this test, S9 exhibited an aggressive approach, successfully resolving the presented symmetry, and ultimately covering the entire area. Their ability to build an integrated architecture reflected a proactive problem-solving style.

Test 4 (T4): S9 recognized the essence of parallel asymmetry, as evidenced by their creation of parallel lines. However, they did so without directly engaging with the problem's geometry. This response showcased S9's capacity to perceive underlying patterns.

Test 5 (T5): S9 constructed a mirror symmetry of the pyramid in this test, securing the top position. This action seemed driven by a desire to justify and maintain their position at the pinnacle. It reflected a competitive and consolidating mindset.

Test 6 (T6): S9's response to the test demonstrated an effort to solidify happiness or a sense of complete satisfaction. This suggested a focus on internal contentment and fulfillment.

Test 7 (T7): S9's flourishing from within, coupled with a lack of any visible path, indicated a sense of aspiration or dream-like thinking. This response hinted at an inner world of ambitions.

Test 8 (T8): The final test revealed another instance of consolidation without the addition of dynamics. This pattern suggested a fear of losing something valuable, potentially a characteristic that repeated across multiple tests.

In summary, S9's psychological profile encompassed ambition, indirect problem-solving approaches, aggressiveness when necessary, an ability to recognize patterns, a desire for recognition and consolidation, and a potential fear of loss. These responses provide valuable insights into S9's cognitive and emotional processes.

Psychological Insights from Subject 10 (S10) Test Responses

Subject 10, or S10, exhibited specific psychological characteristics and tendencies in their responses to various tests. These responses shed light on S10's cognitive processes and behavioral traits.

Test 1 (T1): S10 demonstrated a fondness for duality, akin to S6. They created an alternative self by introducing a new point and pairing it with the existing self, forming an integrated unit for interaction. Notably, S10 differed from S6 in that they responded by interacting with the environment as an integrated entity rather than engaging in detailed self-exploration.

Test 2 (T2): In this test, S10 engaged directly with the presented problem. However, they bridged the issue cautiously while maintaining a clear distinction from the problem's plane. This approach suggested a methodical and strategic problem-solving style.

Test 3 (T3): S10's response to this test involved direct and aggressive engagement with the problem. They successfully broke through the barrier, creating symmetry, and encompassed the entire system. Moreover, S10 identified edge points where insecurities were encoded and added circles to highlight them. This response demonstrated both assertiveness and a keen eye for detail.

Test 4 (T4): S10 chose to avoid the forced asymmetry presented in this test, opting to keep a distance from the problem. This decision may reflect a preference for maintaining balance and avoiding unnecessary complications.

Test 5 (T5): S10's response indicated a visible insecurity about leadership. They acquired alternative parallels to work alongside, possibly indicating a desire to share responsibilities rather than assume a dominant role.

Test 6 (T6): S10 displayed an understanding of nested rectangles representing spiritual completeness. Their response involved creating additional geometric shapes similar to the presented pattern. This suggested an appreciation for complex symbolism and completeness.

Test 7 (T7): In this test, S10 crafted branches within the pot, delving deep into the problem to address every concern intricately. This response highlighted S10's thorough and meticulous problem-solving approach.

Test 8 (T8): S10's response involved the creation of a hierarchical network of complementary symmetry. When presented with a symmetric environment, S10 seemed to derive deep satisfaction from what was received without initiating new dynamics. This response reflected contentment with the given circumstances.

In summary, S10 exhibited characteristics such as a preference for duality, strategic problem-solving, assertiveness, attention to detail, a desire for balanced solutions, an understanding of symbolism, thoroughness, and contentment with existing conditions. These responses offer valuable insights into S10's psychological makeup and cognitive processes.

7.4.2 *Summary of Psychological Traits of All Ten Subjects*

Unveiling the Subconscious Mind Through Geometric Analysis

The human mind is a complex and enigmatic entity, and exploring its intricacies has been a fascination for scholars and researchers for centuries. One avenue of this exploration involves delving into the subconscious mind, which often operates beneath the surface of our conscious awareness. In a recent series of psychological tests, subjects were presented with geometric patterns and asked to respond creatively. What emerged from these responses was a fascinating glimpse into the inner workings of the subconscious mind.

Geometry as a Window to the Subconscious

In these tests, the focus was not on what the subjects intended to paint or their explicit objectives. Instead, the emphasis lay on the geometric aspects of their responses. Researchers carefully examined elements such as geometric similarity, symmetry, asymmetry, integration, fractality, convergence, divergence, engagement, disengagement, complementarity, boldness, cleanliness, hints for dynamics, and sensitivity to intricate details—all as revealed by the shapes created.

This approach allowed the researchers to concentrate on the spatial and structural aspects of the subjects' responses. It led to the creation of what can be described as "spatiotemporal architecture," a term that encapsulates the temporal dimension in the context of geometric structures.

The 3D Clock Assembly: A Unique Insight

One intriguing outcome of this analysis was the generation of a 3D clock assembly for each response. This representation provided a unique and tangible insight into the subject's subconscious mind. A 3D clock assembly is more than just a visual rendering; it embodies a deeper, generic physical significance associated with the subject's cognitive processes.

Differential Deviation: Unveiling Mathematical Architecture

Perhaps one of the most significant revelations from this exploration is the concept of "differential deviation." This term encapsulates the mathematical architecture of the subject's subconscious mind. It implies that every geometric choice made by the subject carries mathematical significance, representing the nuanced interplay of their cognitive faculties.

The notion of differential deviation suggests that even seemingly abstract or creative responses are underpinned by mathematical patterns and structures. It highlights the idea that the human mind, even in its most imaginative and subconscious moments, adheres to certain mathematical principles.

A Multifaceted Exploration

This analytical approach to understanding the subconscious mind is multifaceted. It does not rely on traditional psychological assessments or verbal communication but rather on the visual and structural elements present in the subjects' responses. By disregarding the content and focusing solely on the form, researchers have uncovered a wealth of information about the subjects' cognitive processes.

Implications and Future Directions

The insights gained from this unique approach to understanding the subconscious mind have far-reaching implications. They challenge conventional notions of how we perceive and analyze human cognition. By recognizing the mathematical underpinnings of creative and subconscious thought, we open up new avenues for research into areas such as creativity, problem-solving, and even the nature of human consciousness.

In the future, this approach may find applications in various fields, including psychology, neuroscience, and artificial intelligence. It has the potential to inform our understanding of how the mind processes information, perceives patterns, and generates creative responses.

The journey into the subconscious mind, as revealed through geometric analysis, is a testament to the complexity and richness of human cognition. It demonstrates that even in the most abstract and creative moments, the mind adheres to mathematical principles. This innovative approach provides a fresh perspective on the workings of the human psyche and paves the way for further exploration into the depths of our subconscious.

7.5 Conclusions and Mathematical Model for Cognitive Engineering:

7.5.1 The Architecture of a Subjective Mind

7.5.1.1 Deciphering the Subconscious Mind: A Mathematical Approach

The human mind has long been a subject of fascination and exploration, and understanding its inner workings remains a complex endeavor. Psychological studies have sought to unravel the mysteries of human cognition, often relying on traditional assessments and verbal communication. However, a groundbreaking approach emerged through extensive studies involving ten subjects, shedding light on the subconscious mind's intricate mathematical architecture.

Beyond Traditional Interpretations

In the realm of psychology, interpreting human responses to various stimuli has been a cornerstone of understanding cognitive processes. Often, psychological tests assign predefined meanings to geometric shapes or patterns, linking them to specific psychological constructs. However, the studies at hand took a distinctly different approach.

Here, the significance assigned to patterns was deliberately omitted. Instead, a single, profound assumption guided the analysis—a point represented the self, the very existence of the test subject. This assumption served as the cornerstone upon which a novel exploration of the subconscious mind was built.

Trapping the Mind: A Mathematical Journey

The core concept behind this unique approach was that the human mind, in its subconscious state, is akin to being trapped at a point. The test patterns were meticulously designed to function as traps themselves, capturing the subconscious mind at junctions, peaks, and edges. These patterns, carefully crafted, allowed the mind to explore the empty space relative to its confined state.

While the analysis may initially appear subjective, it swiftly transformed into a quantitative endeavor. Herein lay the innovation—the establishment of a mathematical framework to estimate different mental states and emotions. Figure 7.4 serves as a visual representation of this groundbreaking endeavor.

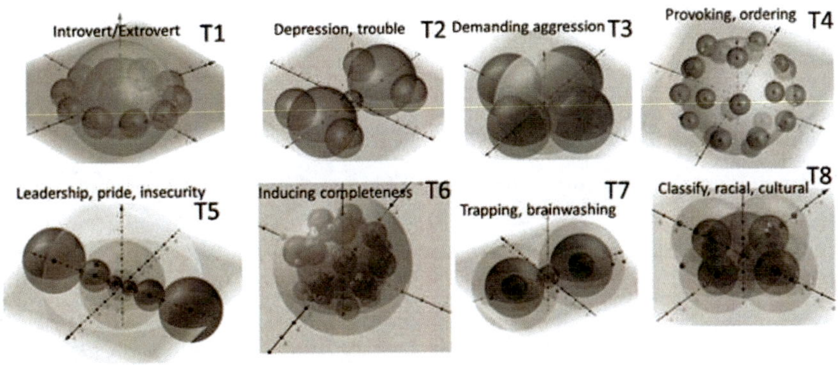

Fig. 7.4 Time crystal structure or invariants of a generic expression of a subconscious mind: The crystallization fundamentals of a subconscious mind is shown here in terms of SRT parameters. T = bonding of clocks or spheres. R = Transformations of clocks or spheres and S = Projection to Infinity and feedback from infinity. T1 shows many concentric clocks growing one above another. And side by side there are qualities or gunas all around the external sphere. T2 shows a generic tensor structure for two different worlds, one for the deep trouble and the other for the solution. Two different modes of transformations are adopted for both the worlds. T3 shows integration of four worlds using a hypothetical imaginary central sphere is the fundamental structure how mind responds to a challenge. T4 shows the drive to create 2D surface of crystals on the surface of a sphere. T5 shows mirroring of a nest of a set of clocks, and then a set of concentric spheres or clocks covering them. T6 shows growth of 2D surface made of clocks between two concentric spheres or clocks. T7 shows growth of concentric clocks in two distinct centers. T8 shows concentric growth of clocks in four points

7.5.1.2 Quantifying the Subconscious Mind

The implications of this mathematical approach to understanding the subconscious mind are profound. Decision-making, a cornerstone of human behavior, is often influenced by emotions, desires, and a tendency to disregard pure logic. The realm of artificial intelligence (AI) has recognized this, and current technologies often artificially induce noise to mimic the complex decision-making processes of the human mind.

However, a critical limitation has persisted—the lack of a comprehensive mathematical protocol to estimate and induce emotional states, attitudes, and the degree of emotions such as fear, love, and hate. These essential parameters, integral to the subconscious mind, play a pivotal role in governing our choices.

A Quantitative Model for Emotions

Figure 7.4 presents a groundbreaking model for emotional states using mathematical formulae. This model has the potential to revolutionize the way we understand and replicate the human psyche. While it may be considered a preliminary effort, it marks the first step toward developing an organized architecture for the subjective mind.

Unlike traditional AI techniques that rely on noise to simulate human-like decision-making, this model offers a systematic and organized approach. By tuning parameters in a structured manner, verifying protocols, and refining formulae through methodical experiments, we can potentially bridge the gap between human cognition and artificial intelligence.

A Glimpse into the Future of AI

The studies on the ten subjects have opened a new frontier in the quest to decode the subconscious mind. They have offered a mathematical framework for understanding emotional states, a dimension of human cognition that has long eluded precise quantification.

While this model is by no means an absolute truth, it represents a crucial milestone in our journey to comprehend the inner workings of the human psyche. It presents an alternative to traditional AI approaches, one that may ultimately lead to more sophisticated and nuanced artificial intelligence systems that align with human decision-making in an organized and precise manner.

The future holds the promise of further refinement, experimentation, and the development of ever more accurate mathematical formulae to unveil the complexities of the subconscious mind. As we continue to explore this uncharted territory, the boundary between human cognition and artificial intelligence may gradually blur, ushering in a new era of AI that reflects the depths of human thought and emotion.

7.5.2 *Polyatomic Time Crystal Model of the Human Brain*

7.5.2.1 Deciphering the Cosmic Symphony

The Remarkable Connection Between Figs. 7.1 and 7.4: In the ever-evolving quest to understand the intricate symphony of the universe, scientists, and thinkers have explored various avenues. From the smallest scales of quantum physics to the vast expanses of cosmology, the concept of time remains a fundamental and enigmatic element. Figures 7.1 and 7.4, each from distinct realms of study, offer tantalizing insights into the structure of time, hinting at a profound connection that transcends conventional boundaries.

Units of Time—A Pioneering Perspective

Figure 7.1 presents an intriguing perspective that delves into the Hindu units of time—a system deeply rooted in ancient philosophy and culture. This unique scale, spanning from 10 ns to a staggering 1022 years, challenges conventional notions of linear and nonlinear time scales. At first glance, it may appear perplexing, with seemingly arbitrary divisions and intervals between units. However, a closer examination unveils a hidden symmetry that underlies this intricate framework.

One of the most striking aspects of Fig. 7.1 is its reliance on prime numbers and their symmetries. The scale appears to favor four specific options—2, 3, 5, and 7—as the building blocks of time measurement. This choice, while unconventional by contemporary standards, hints at a deeper and more fundamental understanding of time's structure.

The Polyatomic Time Crystal—Unraveling the Human Psyche

In Fig. 7.4, a different but equally groundbreaking perspective emerges. Here, the focus shifts from cosmic time scales to the intricate workings of the human psyche. The figure introduces the concept of a 3D clock assembly, termed the polyatomic time crystal, as a representation of emotion cognition and various psychological traits. This assembly is characterized by its invariance, offering a unique lens through which to view human cognition. Note that psychological states extend over 12 imaginary worlds, geometric shapes do represent 91 cognitive features outlined in Chap. 5, however, these shaped extend over 12 dimensions as manifolds [34, 35].

The polyatomic time crystal suggests that human cognitive states can be mapped onto discrete time pieces, akin to spheres upon which a system point traverses in a continuous loop. This revelation brings forth a crucial realization—the symmetrical arrangement of discrete time pieces that structure human cognition mirrors the symmetry found in the cosmic scale of Fig. 7.1.

7.5.2.2 The Profound Connection

Upon juxtaposing Figs. 7.1 and 7.4, a remarkable connection emerges. It becomes apparent that the units of time, originally presented in Fig. 7.1, have undergone a transformation, assuming a special shape to facilitate cognition. This revelation is nothing short of groundbreaking—it implies that the symmetry governing the arrangement of discrete time units in the universe aligns with the symmetry that structures cognitive processes in the human brain.

In essence, this connection suggests that the brain, and by extension, any fundamental information processing structure in the universe, relies on symmetry to structure time. This utilization of symmetry appears to be a universal principle, consistently applied across all scales of existence, regardless of the physical significance of those scales. In other words, the cosmos and the human mind share a common language of symmetry when it comes to orchestrating the flow of time.

The Implications

The implications of this revelation are profound and far-reaching. It invites us to consider the role of prime numbers and their associated symmetries in the field of psychology, transcending traditional disciplinary boundaries. By acknowledging this deep-seated connection between cosmic time scales and cognitive processes, we may uncover new avenues for understanding the human mind and its place within the grand tapestry of the universe.

While the technical details of this revelation may be complex, it offers a fresh perspective that views biology through the lenses of physics and engineering. This holistic approach has the potential to revolutionize our understanding of the mind and its intricate relationship with the cosmos.

In closing, the harmonious connection between Figs. 7.1 and 7.4 beckons us to explore further, to bridge the gap between the macrocosm and the microcosm, and to decipher the cosmic symphony that resonates within us all. It is a testament to the enduring quest of humanity to unravel the mysteries of existence, one symmetrical revelation at a time.

References

1. Agrawal L, Sahu S, Ghosh S, Shiga T, Fujita D, Bandyopadhyay A (2016) Inventing atomic resolution scanning dielectric microscopy to see a single protein complex operation live at resonance in a neuron without touching or adulterating the cell. J Integr Neurosci 15(04):435–462
2. Agrawal L, Chhajed R, Ghosh S, Ghosh B, Ray K, Sahu S, Fujita D (2018) Fractal information theory (FIT)-derived geometric musical language (GML) for brain-inspired hypercomputing. Soft Comput Theor Appl 548:343–372
3. Bandyopadhyay A, Miki K, Wakayama Y (2006) Writing and erasing information in multilevel logic systems of a single molecule using scanning tunneling microscope. Appl Phys Lett 89(24)
4. Bandyopadhyay A (2020) Nanobrain: the making of an artificial brain from a time crystal. Boca Raton CRC Press, Taylor and Francis. ISBN: 9781439875490, https://doi.org/10.1201/9780429107771s
5. Coburn T (1984) Devī-Māhātmya—crystallization of the goddess tradition. Motilal Banarsidass, Delhi. ISBN 978-81-208-0557-6
6. Endsley MR (1988) A construct and its measurement: the functioning and evaluation of pilot situation awareness (No. NOR DOC 88–30). Northrop Corporation, Hawthorne, CA
7. Ghosh S, Fujita D, Bandyopadhyay A (2015) An organic jelly made fractal logic gate with an infinite truth table. Sci Rep 5(1):11265
8. Ghosh S, Sahu S, Agrawal L, Shiga T, Bandyopadhyay A (2016) Inventing a co-axial atomic resolution patch clamp to study a single resonating protein complex and ultra-low power communication deep inside a living neuron cell. J Integr Neurosci 15(04):403–433
9. Goodenough F (1926) Measurement of intelligence by drawings. World Book, Chicago
10. Gu S, Liu Y, Liang F, Feng R, Li Y, Liu G, Gao M, Liu W, Wang F, Huang JH (2020) Screening depressive disorders with tree-drawing test. Front Psychol 11:1446. https://doi.org/10.3389/fpsyg.2020.01446 (PMID: 32670166; PMCID: PMC7330083)
11. Gupta SV (2010) Ch. 1.2.4 time measurements. In: Hull R, Osgood Jr RM, Parisi J, Warlimont H (eds) Units of measurement: past, present and future. International system of units. Springer series in materials science, vol 122. Springer, p 3. ISBN 9783642007378
12. Harris DB (1963) Children's drawings as measures of intellectual maturity. Harcourt Brace Jovanovich, New York
13. Hollnagel E, Woods DD (1983) Cognitive systems engineering: new wine in new bottles. Int J Man Mach Stud 18:583–591
14. Imuta K, Scarf D, Pharo H, Hayne H (2013) Drawing a close to the use of human figure drawings as a projective measure of intelligence. PLoS ONE 8(3):e58991. https://doi.org/10.1371/journal.pone.0058991
15. Jha G (1999) "Discourse I—origin of the work—creation of the world—summary of contents of the book." [1st ed. 1920]. In: Manusmriti with the commentary of Medhatithi in ten volumes. Wisdom Library. Adhyāya 1. Motilal Banarsidass Publishers Pvt. Ltd. ISBN 8120811550
16. James EO (1969) Creation and cosmology: a historical and comparative inquiry. E. J. Brill, Leiden
17. Jung C (1964) Approaching the unconscious. Man and his symbols. Doubleday, p 37. ISBN 978-0-385-05221-4

18. Kaneda A, Yasui-Furukori N, Saito M, Sugawara N, Nakagami T, Furukori H, Kaneko S (2010) Characteristics of the tree-drawing test in chronic schizophrenia. Psychiatry Clin Neurosci 64(2):141–148. https://doi.org/10.1111/j.1440-1819.2010.02071.x (PMID: 20447010)
19. Kinsley D (1997) Tantric visions of the divine feminine: the ten Mahavidyas. University of California Press, Berkeley
20. Klostermaier K (1984) Time in Patañjali's "Yogasūtra." Philos East West 34(2):205–210. https://doi.org/10.2307/1398919
21. Larson GJ, Bhattacharya RS (2014) The encyclopedia of Indian philosophies—Samkhya, a dualist tradition in Indian philosophy, vol 4. Princeton University Press, pp 65–66. ISBN 978-0-691-60441-1
22. Larson GJ (2011) Classical Sāṃkhya: an interpretation of its history and meaning. Motilal Banarsidass, p 273. ISBN 978-8120805033
23. Norman DA (1981) Steps towards a cognitive engineering (Tech. Report). University of California, Program in Cognitive Science, San Diego
24. Pattanayak A, Dutta T, Pranjal P, Singh P, Sahoo P, Sarkar S, Bandyopadhyay A (2022) Cognitive engineering for AI: an octave drawing test for building a mathematical structure of a subconscious mind. In: Proceedings of the third international conference on trends in computational and cognitive engineering, vol 348, pp 165–148
25. Pradhan B, Bandyopadhyay A, Pal AJ (2004) Molecular level control of donor/acceptor heterostructures in organic photovoltaic devices. Appl Phys Lett 85(4):663–665
26. Reddy S, Sonker D, Singh P, Saxena K, Singh S, Chhajed R, Tiwari S, Karthik KV, Ghosh S, Ray K, Bandyopadhyay A (2018) A brain-like computer made of time crystal: could a metric of prime alone replace a user and alleviate programming forever? Soft Comput Appl 761:1–43
27. Raffé W (1952) Rāgas and Rāginīs: a key to Hindu aesthetics. J Aesthet Art Critic 11(2):105–117
28. Rossi E (2008) The collected works of Milton H. Erickson, vol 1. The Milton H. Erickson Foundation Press, Arizona, pp 229–260. ISBN 978-1932248302
29. Singh P, Sahoo P, Saxena K, Manna JS, Ray K, Ghosh S, Fujita D (2021) Cytoskeletal filaments deep inside a neuron are not silent: they regulate the precise timing of nerve spikes using a pair of vortices. Symmetry 13(5):821
30. Saxena K, Singh P, Sahoo P, Sahu S, Ghosh S, Ray K, Fujita D (2020) Fractal, scale free electromagnetic resonance of a single brain extracted microtubule nanowire, a single tubulin protein and a single neuron. Fractal Fractional 4(2):11
31. Saxena K, Singh P, Sahoo P, Ghosh S, Krishnananda D, Ray K, Fujita D, Bandyopadhyay A (2022) All basics that are wrong with the current concept of time crystal: learning from the polyatomic time crystals of protein, microtubule, and neuron. Proc Trends Electron Health Inform 376:243–254
32. Singh P, Saxena K, Singhania A, Sahoo P, Ghosh S, Chhajed R, Fujita D (2020) A self-operating time crystal model of the human brain: can we replace entire brain hardware with a 3D fractal architecture of clocks alone? Information 11(5):238
33. Singh P, Ray K, Bandyopadhyay A (2022) Biological antenna to the humanoid bot: electromagnetic resonances in biomaterials. Studies in rhythm engineering (SRE), Springer, Singapore. https://doi.org/10.1007/978-981-16-9677-0
34. Singh P et al (2021) Electrophysiology using coaxial atom probe array: live imaging reveals hidden circuits of a hippocampal neural network. J Neurophysiol 125(6):2107–2116
35. Singh P, Sahoo P, Saxena K, Manna JS, Ray K, Ghosh S, Bandyopadhyay A et al (2021) Cytoskeletal filaments deep inside a neuron are not silent: they regulate the precise timing of nerve spikes using a pair of vortices. Symmetry 13:821. https://doi.org/10.3390/sym13050821
36. Singh P et al (2021) Quaternion octonion to dodecanion manifold: stereographic projections from infinity lead to a self-operating mathematical universe. AISC, advanced intelligence systems and computing, vol 1169. Springer, Singapore (2020)
37. Singh P et al (2021) a space-time-topology-prime, stTS metric for a self-operating mathematical universe uses dodecanion geometric algebra of 2–20 D complex vectors. In: Proceedings of international conference on data science and applications. Lecture notes in networks and systems, vol 148. Springer, Singapore (2020)

38. Visvanathan M (2011) Cosmology and critique: charting a history of the Purusha Sukta. In: Roy K (ed) Insights and interventions: essays in honour of Uma Chakravarti. Primus Books, Delhi, pp 143–168. ISBN 978-93-80607-22-1
39. Woods DD, Roth EM (1988) Cognitive engineering: human problem solving with tools. Hum Factors 30:415–430

Chapter 8
Happiness as a Local Invariant of Pain: A Perspective on Spontaneous and Induced Emotions

8.1 Introduction

We have been discussing about consciousness and cognitive, perceptive and emotional responses using a series of experiments on human subjects. We have traversed a journey that has brought us to a point where we could explore eternal feelings and cognitive experiences which are apparently nonsense. These problems come under the category of those intractable problems in computer science where you cannot define the problem.

8.1.1 Unveiling the Microscopic Wonders of the Brain: The Role of Microtubules in Human Psychology

The human brain, with its intricate web of neurons and synapses, has long been a subject of fascination and wonder. While the overarching principles of brain function are well understood, the finer details of how individual neurons communicate and process information remain an ongoing area of research. Recent studies have shed light on the role of microtubules and their junctions in this complex process, offering a glimpse into the fascinating world of human psychology that advocates to define mind as a fractal structure, consciousness as a synchronization between two fractals madc of prime numbers. The human brain is a complex and intricate organ that has long captivated scientists and researchers. While we have a good understanding of the overall principles of brain function, there is still much to learn about the finer details of how individual neurons communicate and process information [52]. Recent studies have focused on the role of microtubules and their junctions in this complex process, shedding light on the fascinating world of human psychology.

Microtubules are tubular structures found within neurons that play a crucial role in maintaining cell shape and providing a framework for intracellular transport [52].

T. Dutta and A. Bandyopadhyay, *Emotion, Cognition and Silent Communication: Unsolved Mysteries*, Studies in Rhythm Engineering,
https://doi.org/10.1007/978-981-99-9334-5_8

They are composed of proteins called tubulins, and abnormalities in microtubule structure have been observed in psychiatric diseases such as schizophrenia and mood disorders. These anomalies in neuronal cell architecture, including changes in dendritic complexity and synaptic density, highlight the importance of microtubules in brain function. Tau proteins, which are found in the human brain, have also been the subject of research in relation to neurodegenerative diseases such as Alzheimer's disease [35]. Tau is neuron-specific and is not observed in glial cells, further emphasizing its role in neuronal function [13]. Hyperphosphorylation of tau proteins has been implicated in the breakdown of microtubules and the compromise of normal tau function, leading to the pathology observed in tauopathies [18].

The intricate nature of the human brain extends beyond its physical structure to its cognitive processes. Studies have shown that the brain exhibits fractal behavior, with the Fibonacci Life-Chart Method suggesting that it is structured on Fibonacci ratios through fractal bifurcation [69]. This concept of fractal structure extends to the definition of the mind and consciousness as a synchronization between two fractals made of prime numbers [69]. Furthermore, the brain's ability to process and recognize complex information has been investigated. Research has shown that the internal repetition of visual information in complex objects can enhance perceptual fluency, challenging the notion that fluency theories of aesthetic pleasure only apply to simple objects [38]. This suggests that the brain is capable of processing intricate and complex stimuli in a way that is both efficient and pleasurable.

The Crucial Role of Membrane Microtubule Junctions

The role of microtubules and their junctions in brain function, the involvement of tau proteins in neurodegenerative diseases, and the brain's ability to process complex information all contribute to our understanding of the intricate workings of the human brain. These findings highlight the importance of ongoing research in unraveling the mysteries of the brain and its impact on human psychology.

In 2004, Pradhan and colleagues [57] uncovered a crucial aspect of junction property that could deliver an important insight in the understanding of neuronal communication—the membrane microtubule junction. These junctions, also known as heterojunctions, play a pivotal role in signal transfer within the brain's intricate structure. The discovery of geometric phase apart from the dynamic phase in these junctions highlighted their importance in enabling the flow of information between neurons and across the brain's fractal architecture. Under the framework of Self-Operating Mathematical Universe, SOMU, where various kinds of brain rhythms are integrated as unified 3D assembly of clocks, from hardware to software that constructs the cognitive, perceptive, and emotional experiences, the clock architecture is an architecture of phase. Imagine, that you can forget everything but phase that alone would let you understanding the whole and its actions.

Unveiling the Fractal Nature of Brain Communication

Fast forward to 2020, and researchers led by Saxena et al. [66] made a groundbreaking revelation. They extracted a single microtubule nanowire, a solitary tubulin protein, and a lone neuron from the brain. What they discovered was nothing short of astonishing—a fractal, scale-free electromagnetic resonance. This resonance suggested the presence of a triplet of triplet resonance bands or nine distinct frequency and time domains working in perfect harmony. Note that there is not just triplet of triplet pattern in the electromagnetic resonance band, there are other patterns of resonance bands too, like, doublet of triplet and pentate of doublet, etc. When we say that brain is a fractal system, we mean all these patterns of prime numbers, which we have meticulously mapped by studying all brain components [74].

This discovery not only underscored the complexity of brain communication but also hinted at the possibility of a highly coordinated, fractal-like network operating within the brain's confines. The implications of this finding were far-reaching, opening doors to a deeper understanding of the brain's inner workings. This particular finding is the foundation of this Chapter, where we would be addressing the fundamental feelings, that has no conceptual.

Quantum Tunneling: Reading and Writing in the Brain's Network

The study conducted by Bandyopadhyay and colleagues [10, 63] took the exploration of brain communication a step further. It proposed the use of quantum tunneling as a mechanism for reading and writing information within this intricate network. This concept suggested that even at the molecular level, the brain may leverage quantum physics to process and transmit information, offering an intriguing perspective on the brain's capabilities. All molecules are quantum mechanical devices, in the room temperature and ambient atmospheric condition. Therefore, it is utmost essential that how molecular codes, written in a molecule using quantum language survives and enables changing the environmental systems is understood properly.

The Logic of a Fractal Network: The Paradox of a Nonsense Logic

Normally, we would think that if there is a fractal network we cannot define a state, thus, it cannot have a logic. In the early days of our research, we understood that we need to find a method so that even in a fractal world we have information structure that could be expressed as a mathematical structure. Apparently, this is an impossible task. In 2015, Ghosh and collaborators [30] presented an idea that resonated with the concept of a fractal-like network within the brain. Their proposal suggested that such a network could process information logically, when a system considers entire fractal path. It means when we enter inside a pattern in a fractal structure and transit from one imaginary world to another, we would experience contributions from wildly different imaginary world (product of higher-dimensional tensors, say product of two 12D vectors). This further reinforced the notion that the brain's communication system operates in a highly organized and structured manner, akin to the intricate patterns found in fractals.

Microtubules: Not So Silent Contributors to Brain Activity

The conventional view of neuronal communication often placed neuron membranes at the forefront, with microtubules assumed to play a null role. However, in 2016, Agrawal and his team [6] challenged this notion. Their research demonstrated that microtubules, those tiny helical cylindrical structures within neurons, possess an astonishing ability—they can process firing signals approximately 10,000 times faster than the neuron membrane itself.

This groundbreaking revelation shattered the notion of microtubules as inactive skeleton structures, highlighting their active role in the brain's information processing. The findings suggested that these microscopic components are not silent contributors but rather active and efficient conductors of neuronal activity. However, SOMU theory of consciousness differs from Orch-OR at this point, microtubule is not everything, if one goes inside microtubule, a single protein or water channels transmits signals 10,000 times faster. Well, do not end there, go even smaller, inside a protein, alfa helices process signals even 10,000 times faster than the whole protein molecule.

Now we come to the most interesting part. There are much slower event that happens in the cluster of neurons, there are brilliant and critical brain learning processes in the hippocampus, which runs for four hours. Yet, we do not say consciousness happens here or there. Under the framework of SOMU we suggest that look at the pattern created by signals of all time domains, find the geometric shapes made by high and low intensity signals that does not change (geometric invariant), and then one could start seeing the true information processing of the brain in three dimensions of time and three dimensions of space.

Energy Exchange in the Absence of Synaptic Junctions

Another intriguing aspect of brain communication came to light in 2016 when Ghosh and his team [31] conducted experiments involving non-connected neurons. These neurons, despite lacking direct synaptic junctions, were found to be capable of exchanging energy. This phenomenon was revealed through the use of a scanning dielectric microscope (SDM), a technique that allowed the imaging of neural circuits. The ability of non-connected neurons to share energy hinted at a deeper level of coordination and communication within the brain [71]. It challenged the conventional view of synaptic junctions as the sole conduits of information transfer. In the paragraph above we have discussed three dimensions of time and three dimensions of space, therefore, six dimensions of spatio-temporal events were extracted by Ghosh et al. [31].

The Polyatomic Time Crystal Model of the Brain

The best way to represent the six-dimensional spatio-temporal events is a 3D clock assembly or polyatomic time crystal [73]. In recent years, researchers have put forth the polyatomic time crystal model of the brain [61, 72] that includes every single component. This model offers a unique perspective on brain function, suggesting that cognitive, perceptual, and emotional responses can be quantified by analyzing recurring events that give rise to these experiences. In other words, be it replacement of cells and proteins or body parts, conscious experiences or biological rhythms for hardware operations, all three different aspects of the brain could be represented by a 3D assembly of spheres of various sizes, each sphere holds a clock representing a recurrent event.

The study of microtubules and their role in brain communication has unveiled a world of intricacy and coordination within the human brain. These microscopic structures, once thought to be passive, are now recognized as active participants in the brain's information processing. The concept of a fractal-like network operating within the brain adds a new dimension to our understanding of human psychology, hinting at a level of organization and logic that is both astonishing and inspiring. As research in this field continues to evolve, it promises to unlock even more secrets of the human mind under the framework of Fractal Information Theory, FIT [7].

8.1.2 Primordial Embedded Love in Humans

8.1.2.1 Love's Evolutionary Past

The concept of love has intrigued philosophers, poets, and psychologists for centuries. Love is a multifaceted emotion that can be directed toward a variety of objects and beings. In this chapter, we delve into the intricacies of embedded love in humans, examining the evolutionary origins, and psychological mechanisms that drive our affection toward children [70], flowers, puppies, and other entities. Drawing from theories in evolutionary psychology, attachment theory, and cognitive neuroscience, we offer a comprehensive exploration of how these diverse manifestations of love can be understood through the lens of primordial evolution [19]. Love, often considered the most profound of human emotions, manifests itself in various forms and toward a multitude of objects and beings. Embedded love can be traced back to our primordial ancestors. Evolutionary psychologists posit that humans have evolved to care for offspring as a fundamental survival strategy. This "parental instinct" serves as the bedrock for embedded love, as humans naturally experience profound affection for their children. This instinctual bond ensures the protection and nurturing of the next generation, a hallmark of our evolutionary success.

Love is a complex emotion that has deep roots in our evolutionary past. In order to understand the evolutionary foundations of love, it is important to examine the role of emotional bonds, particularly toward offspring, in the preservation of the species. Evolutionary psychology suggests that our ancestors who cared for their offspring had a higher likelihood of passing on their genes, as nurtured children were more likely to survive and reproduce. This innate parental love, often referred to as "filial love," is deeply embedded in our biology. This article will explore the neurobiological basis of love and the role of oxytocin in reinforcing the bond between parent and child.

Neuroimaging studies have provided valuable insights into the neural mechanisms underlying parental love. Feldman [27] highlights that when parents gaze at their children, the brain's reward system is activated. This activation is associated with the release of oxytocin, a neurochemical that plays a crucial role in social bonding. Oxytocin is known to promote feelings of trust, empathy, and attachment, making it a key player in the formation and maintenance of parental love. The release of oxytocin in response to parental love is not limited to humans. Fisher et al. [28] argue that the mammalian brain has evolved a specific system for mate choice and romantic love. This system is also mediated by oxytocin, suggesting that the neurobiological basis of love is conserved across species. This finding further supports the idea that love, including parental love, has deep evolutionary roots.

Hrdy [33] proposes that the evolution of parental love is not solely driven by the need to ensure the survival of offspring. Instead, it is a result of the cooperative breeding strategy adopted by humans. In cooperative breeding, individuals other than the biological parents contribute to the care and upbringing of offspring. This extended network of caregivers, including grandparents, siblings, and other relatives, enhances the survival and reproductive success of the species. The neurobiology of attachment, as described by Insel and Young [34], supports this idea by highlighting the role of oxytocin in promoting social bonding beyond the parent–child relationship. The evolutionary foundations of love can be traced back to our ancestral past. The development of emotional bonds, particularly toward offspring, was crucial for the preservation of the species. Neuroimaging studies have shown that parental love activates the brain's reward system, leading to the release of oxytocin. This neurochemical cascade reinforces the bond between parent and child, ensuring their mutual care and protection. The neurobiological basis of love, including parental love, is conserved across species, highlighting its deep evolutionary roots. The cooperative breeding strategy adopted by humans further supports the importance of love in promoting the survival and reproductive success of the species.

8.1.2.2 Attachment Theory and Embedded Love

Attachment theory, formulated by John Bowlby, offers insights into the nature of love and emotional bonds beyond parental affection. According to this theory, humans form attachment bonds not only with their offspring but also with significant others and even inanimate objects. This broader concept of attachment helps explain why individuals can feel deep emotional connections to objects like childhood toys, sentimental mementos, or even beloved pets. The roots of attachment theory can be traced back to Bowlby's seminal publications in the 1950s [34]. Bowlby proposed that attachment bonds between infants and their caregivers are crucial for the child's emotional and social development. These early attachment experiences shape the individual's internal working models of relationships, influencing their future interactions and emotional bonds with others.

Hazan and Shaver [32] extended attachment theory to the realm of romantic love. They conceptualized romantic love as an attachment process, suggesting that adults can form attachment bonds with their romantic partners. This perspective highlights the similarities between the attachment behaviors observed in infants and the behaviors exhibited by adults in romantic relationships. The attachment bond provides a sense of security and support, and the presence of the attachment figure serves as a source of comfort and reassurance. Neurobiological research has provided insights into the mechanisms underlying attachment and love. Oxytocin, often referred to as the "love hormone," plays a crucial role in bonding and attachment [39]. Oxytocin is released during social interactions, including those involving attachment figures, and promotes feelings of trust, bonding, and attachment. It is involved in the formation and maintenance of both parental and romantic bonds.

Furthermore, attachment theory has been applied to various domains, including the study of pet attachment [68]. Individuals can form strong emotional bonds with their pets, experiencing a sense of security, companionship, and unconditional love. These attachments to pets can have positive effects on well-being and provide emotional support. Attachment theory provides a framework for understanding the nature of love and emotional bonds beyond parental affection. Humans form attachment bonds not only with their offspring but also with significant others and even inanimate objects. This broader concept of attachment helps explain the deep emotional connections individuals can feel toward objects, pets, and romantic partners. Neurobiological research has highlighted the role of oxytocin in bonding and attachment. Understanding attachment theory and its implications can contribute to our understanding of the complex nature of love and emotional bonds in human relationships.

Objects of Embedded Love

1. **Children**: The strongest and most obvious manifestation of embedded love is the affection parents feel for their children. This profound bond ensures the survival and well-being of offspring and drives parents to provide care and protection.
2. **Flowers**: The appreciation of beauty, as in the case of flowers, has evolutionary significance as well. Humans are drawn to visually appealing objects, which may have once signaled a safe and resource-rich environment. The admiration of flowers can be seen as an ancestral inclination to seek out life-sustaining habitats.
3. **Puppies**: The love for puppies and pets in general can be explained through attachment theory. The companionship and emotional support offered by pets satisfy the human need for social bonds, mirroring the attachment bonds formed in childhood.
4. **Sentimental Objects**: Sentimental objects, such as heirlooms and keepsakes, evoke powerful emotions. These objects serve as emotional touchpoints, connecting individuals to their personal histories and family legacies, reinforcing the concept of attachment to inanimate entities.

8.1.2.3 Neurobiological Mechanisms

Research in cognitive neuroscience has unveiled the intricate neurobiological mechanisms behind embedded love. Functional MRI studies have shown that the brain regions associated with love and attachment, such as the anterior cingulate cortex and the ventral tegmental area, are activated when individuals experience affection toward their children, pets, or sentimental objects [3]. Embedded love extends beyond parenthood and can be observed in various forms. One example is "Philia Floralis," which refers to the deep attachment humans feel toward flowers and nature. This affinity may be rooted in our ancestral reliance on plants for sustenance and medicinal purposes [3]. The activation of brain regions associated with love and attachment when individuals experience affection toward nature supports the idea that this form of embedded love has a neurobiological basis.

Another form of embedded love is "Canine Kinship," which refers to the profound love for puppies and dogs. This bond can be traced back to the coevolution of humans and canines, where both species benefited from cooperation and companionship. Studies have shown that interactions with dogs can activate brain regions associated with reward and positive emotions, further supporting the neurobiological basis of this form of embedded love [3].

Additionally, there is a unique form of embedded love called "Objectophilia," where individuals form romantic attachments to objects. Although rare, this phenomenon highlights the diverse ways in which humans can experience affection. While there may not be specific studies on the neurobiological mechanisms of objectophilia, research on romantic love has shown activation in brain regions associated with reward and motivation, suggesting that similar mechanisms may be involved [82].

The research in cognitive neuroscience has provided insights into the neurobiological mechanisms behind embedded love. Functional MRI studies have demonstrated the activation of brain regions associated with love and attachment when individuals experience affection toward their children, pets, or sentimental objects. Embedded love extends beyond parenthood and can be observed in various forms, such as Philia Floralis, Canine Kinship, and Objectophilia. These different forms of embedded love showcase the diverse ways in which humans can experience affection and highlight the complex interplay between biology and emotions [3, 82]. Understanding the neurobiological basis of embedded love contributes to our knowledge of human emotions and the intricate nature of our social bonds.

8.1.3 Love Hurts; Does Hurting or Pain Generates Love?

Traditionally, pain and pleasure have been viewed as opposing motivators of behavior, with distinct neural circuits responsible for each. Pain, often regarded as an aversive experience, has been associated with an affect to reward system, tissue injury, and the body's signaling of physiological need. Conversely, pleasure, or reward, represents optimal function and well-being. However, recent research challenges this dichotomy, suggesting significant overlap in the neural mechanisms underlying both pain and pleasure.

8.1.3.1 Neural Substrates of Pain and Love

It is already established that love hurts [16]. To understand how pain could give rise to love, we must explore the neural substrates involved. Pain processing primarily engages regions such as the anterior cingulate cortex (ACC) and insula, which play a crucial role in detecting and responding to noxious stimuli. In contrast, love and pleasure are associated with the reward system, including the release of dopamine in areas like the nucleus accumbens.

Finally, some experts believe that experiencing deep pain can actually change the structure and function of our brains [44]. When we experience trauma, our brains release stress hormones that can damage neural pathways. However, over time, our brains can adapt to these changes and form new neural pathways. This process of neuroplasticity can make us more resilient to future pain, and it can also make us more capable of love and compassion.

8.1.3.2 Overlapping Pathways

Emerging evidence suggests that the brain's opioid and dopamine systems, traditionally linked to pleasure, also modulate the perception of pain. This dual modulation raises the possibility of a reciprocal relationship between pain and pleasure. Understanding the neural mechanisms that govern this interaction is essential for improving overall well-being and mitigating unnecessary suffering. Pain is a universal human experience that can have profound effects on our emotions, thoughts, and behaviors. In recent years, researchers have begun to explore the relationship between pain and empathy, resilience, and gratitude, particularly in the context of romantic relationships. This article aims to provide a comprehensive understanding of how pain can lead to empathy, resilience, and gratitude, and how these qualities can contribute to a loving relationship. Here are some additional thoughts on how deeper pain can generate true love from a psychological and neuroscience perspective:

Pain Can Lead to Empathy. When we experience pain, we are better able to understand and empathize with the pain of others. This is because pain activates the same neural pathways in the brain, regardless of whether it is our own pain or the pain of someone else. This increased empathy can make us more compassionate and understanding, which are essential qualities for a loving relationship. Research has shown that when we experience pain, whether it is physical or emotional, the same neural pathways in the brain are activated [23]. This activation allows us to better understand and empathize with the pain of others. Empathy is the ability to understand and share the feelings of another person, and it plays a crucial role in building and maintaining loving relationships [21]. When we have personally experienced pain, we are more likely to be compassionate and understanding toward our partners, which can foster a deeper connection and emotional intimacy.

Pain Can Make Us More Resilient. When we overcome pain, we learn that we are stronger than we thought we were. This increased resilience can make us more confident and capable of taking risks, both in our personal and professional lives. This willingness to take risks can open us up to new possibilities, including the possibility of finding true love. Overcoming pain can be a transformative experience that teaches us about our own strength and resilience. When we successfully navigate through difficult times, we learn that we are capable of handling adversity and bouncing back from setbacks. This increased resilience can make us more confident and willing to take risks in our personal and professional lives [22]. In the context of romantic relationships, this willingness to take risks can open us up to new possibilities, including the possibility of finding true love. By embracing the lessons learned from pain, individuals can approach relationships with a sense of courage and optimism.

Pain Can Make Us More Grateful. When we have experienced loss or hardship, we come to appreciate the good things in our lives more deeply. This increased gratitude can make us more open to love and more willing to give and receive love. Experiencing pain, loss, or hardship can also lead to a greater appreciation for the good things in our lives. Research has shown that gratitude is associated with positive emotions and psychological well-being [45]. When we have experienced pain, we are more likely to recognize and cherish the love and support we receive from our partners. This increased gratitude can make us more open to love and more willing to give and receive love, creating a positive cycle of appreciation and affection in our relationships.

Pain can have profound effects on our ability to empathize, our resilience, and our gratitude, all of which are essential qualities for a loving relationship. By understanding the neural mechanisms underlying pain and its impact on our emotions and behaviors, we can cultivate empathy, resilience, and gratitude in our relationships. Embracing the lessons learned from pain can lead to personal growth and a deeper connection with our partners, ultimately fostering a more loving and fulfilling relationship.

8.1.3.3 Classification of Pain

Psychological pain is a multifaceted concept deeply intertwined with emotions, loss, and trauma [48]. While it may lack the tangible quality of physical pain, its impact on individuals' lives and mental health is undeniable. Understanding and defining psychological pain is an ongoing endeavor, one that holds promise for improving mental health assessment and treatment. As research continues to unravel the complexities of this phenomenon, it is our hope that individuals grappling with psychological pain will find the support and relief they need to heal and thrive.

Defining Psychological Pain

Psychological pain, as a concept, is not as straightforward as its physical counterpart. Morse and colleagues' criteria have been instrumental in attempting to elucidate this complex phenomenon. According to their analysis, psychological pain can be best defined as a lasting, unsustainable, and unpleasant feeling resulting from negative self-appraisal linked to an inability or deficiency within oneself. This negative self-appraisal is frequently triggered by the loss of someone or something dear or by the failure to achieve goals that are closely tied to one's core psychological needs.

Loss and Trauma: Precursors of Psychological Pain

Loss and trauma are two prominent factors intricately entwined with psychological pain. Loss, whether it's the death of a loved one, the dissolution of a meaningful relationship, or the inability to attain cherished aspirations, can plunge individuals into the depths of psychological pain. The feeling of emptiness, despair, and a sense of inadequacy can linger long after the event, leading to prolonged emotional suffering. Trauma, on the other hand, often leaves deep psychological scars. It can be a catalyst for the development of psychological pain, as individuals struggle to come to terms with the aftermath of traumatic experiences. Trauma can trigger negative self-appraisals and a sense of powerlessness, further exacerbating psychological pain.

The Relationship Between Psychological and Physical Pain

One of the most striking aspects of psychological pain is its perceived intensity [51]. For many individuals, the psychological pain associated with severe depression, for example, can be more excruciating than any physical pain they have ever endured. This observation highlights the severity and significance of psychological pain in the realm of mental health.

Neuroimaging studies have shed light on the intricate relationship between psychological and physical pain. These studies suggest that both types of pain share overlapping patterns of brain activation. The brain regions involved in processing physical pain, such as the anterior cingulate cortex and the insula, are also activated when individuals experience psychological pain. This convergence of neural pathways underscores the real and tangible nature of psychological pain, challenging any notion that it is purely a subjective experience.

Implications for Mental Health

Understanding the role of psychological pain in mental health is crucial. It is increasingly recognized as a critical component of suicidality and can provide valuable insights into the assessment and treatment of potentially suicidal patients. By systematically assessing psychological pain, mental health professionals can identify individuals at heightened risk and intervene more effectively.

Moreover, recognizing the overlap between psychological and physical pain in the brain opens up new avenues for the treatment of depression and other psychiatric illnesses. Approaches that target both types of pain may offer more comprehensive relief and a deeper understanding of the intricate interplay between emotion and suffering.

8.2 The Healing Power of Self-compassion: A Critical Review

In the realm of psychological literature, compassion, especially toward oneself, is gaining recognition as a focal point for intervention and exploration. Compassion, as defined as "a sensitivity to the suffering of self and others, with a deep commitment to alleviate it," draws upon a wealth of evidence from evolutionary, social, and neurobiological theories. At the heart of this concept lies the belief that compassion is rooted in our innate abilities related to mammalian bonding, attachment, affiliation, and prosocial behavior. It is the key to a unique affect regulation system, one that triggers the release of oxytocin and opioids in response to experiences of care and connection [59].

This particular affect regulation system possesses a remarkable capability—it can offset or downregulate other emotions linked to threat. The roots of this system can be traced back to early experiences of care, particularly the caregiver's response to a child's distress, whether it's in the face of a social or physical threat. Through close contact, touch, and care, the caregiver facilitates the alleviation of distress by triggering the release of oxytocin in the child. As time passes, these experiences become internalized, forging neuronal connections within the soothing system. This becomes the bedrock of an individual's capacity for compassion toward both themselves and others.

Self-compassion, the practice of responding to one's own difficult experiences with care and concern, has profound implications for emotional well-being and positive psychological functioning [47]. When individuals engage in self-compassion, it triggers a cascade of events, including the release of oxytocin, which plays a crucial role in reducing negative, threat-based emotions such as fear, anger, and disgust [47].

Research has shown that self-compassion is associated with substantial benefits for emotional well-being. It significantly lowers levels of psychological distress, depression, and anxiety in both the general population and clinical settings [47]. Even individuals grappling with physical health problems find solace in self-compassion, as it offers a buffer against the challenges they face [47].

For those enduring persistent mental and physical health difficulties, self-compassion emerges as a source of resilience. It fosters lower negative affect and amplifies motivation to bounce back from failures and cope with rejection [47]. In the face of adversity, individuals who embrace self-compassion tend to employ more adaptive responses and coping strategies [47]. This heightened adaptability leads to a unique form of motivation, one that seeks to promote well-being rather than merely mitigate potential threats [47].

The concept of self-compassion stands as a potent force in the field of psychology. Rooted in our evolutionary heritage, it offers a pathway to healing, resilience, and emotional well-being [47]. By activating the soothing system and releasing oxytocin, self-compassion provides a means to alleviate suffering, both within ourselves and in our interactions with others [47].

As research in this area continues to evolve, the transformative potential of self-compassion becomes increasingly evident, pointing the way toward a more compassionate and empathetic world [47]. It is crucial to cultivate self-compassion early, particularly in adolescents, as it is a modifiable trait that can have a significant impact on emotional well-being [9]. Additionally, self-compassion has been found to buffer the adverse mental health impacts of stressors such as the COVID-19 pandemic [46]. Healthcare professionals, who have been disproportionately affected by the pandemic, can benefit from self-compassion as a means to mitigate burnout and enhance their quality of life [26].

Self-compassion is a powerful tool for promoting emotional well-being and resilience. It has been associated with numerous benefits, including lower levels of psychological distress, depression, and anxiety. By activating the soothing system and releasing oxytocin, self-compassion offers a means to alleviate suffering and promote a more compassionate and empathetic world. As research in this area continues to advance, the transformative potential of self-compassion becomes increasingly evident, highlighting its importance in promoting emotional well-being, and positive psychological functioning.

8.3 Spontaneous and Induced Emotion

Emotion is a complex and multifaceted phenomenon that has been studied by psychologists for centuries. While there is no single definition of emotion that is universally accepted, it is generally understood to be a subjective experience that is characterized by physiological, cognitive, and behavioral changes. Emotions can be classified into two broad categories: spontaneous and induced. Spontaneous emotions are those that arise naturally in response to internal or external stimuli. For example, you might experience spontaneous joy when you see a loved one, or you might experience spontaneous anger when you are cut off in traffic. Induced emotions are those that are deliberately elicited by an external source. For example, you might experience induced fear when you watch a horror movie, or you might experience induced sadness when you listen to a sad song.

Both spontaneous and induced emotions can play an important role in our lives. Spontaneous emotions can help us to respond to our environment quickly and effectively. For example, if you experience spontaneous fear while walking down a dark alley, you might be more likely to turn around and walk the other way.

Induced emotions can also be beneficial. For example, watching a sad movie can help us to process our own emotions and develop empathy for others. However, induced emotions can also be negative, such as when we are exposed to violence or trauma.

8.3.1 Research on Spontaneous and Induced Emotion

Psychologists have conducted a great deal of research on both spontaneous and induced emotion. Some of the key findings from this research include:

- Spontaneous and induced emotions share some common neural underpinnings. However, there are also some important differences in the brain activity associated with these two types of emotion. For example, spontaneous emotions are more likely to activate the amygdala, a brain region that is associated with fear and other negative emotions. Induced emotions, on the other hand, are more likely to activate the prefrontal cortex, a brain region that is associated with higher-order cognitive functions such as planning and decision-making. Adolphs [4] provides an in-depth exploration of the role of the amygdala in emotion processing. It discusses how the amygdala is involved in the generation and regulation of fear and other negative emotions, which are often associated with spontaneous emotions. The author also highlights the neural circuitry and connectivity of the amygdala in relation to emotion processing. Ochsner [53] focuses on the involvement of the prefrontal cortex in the regulation of emotions. It discusses how the prefrontal cortex, particularly the dorsolateral prefrontal cortex, plays a crucial role in cognitive control processes that modulate emotional responses. The authors provide evidence for the activation of the prefrontal cortex during induced emotions, which are more likely to involve cognitive appraisal and regulation.
- Spontaneous emotions are more likely to be accompanied by physiological changes. For example, when you experience spontaneous fear, your heart rate and blood pressure may increase. Induced emotions, on the other hand, are less likely to be accompanied by physiological changes. Levenson [42] investigates the physiological changes associated with spontaneous and induced emotions. The researchers examine how voluntary facial actions, such as smiling or frowning, can elicit specific autonomic nervous system responses. The findings suggest that spontaneous emotions are more likely to be accompanied by physiological changes, such as changes in heart rate and blood pressure, while induced emotions may not elicit the same level of physiological response.
- Spontaneous emotions are more likely to influence behavior. For example, if you experience spontaneous anger, you might be more likely to yell or lash out. Induced emotions, on the other hand, are less likely to influence behavior. Baumeister [11] challenges the notion that induced emotions are less likely to influence behavior. The authors argue that relying solely on self-reports or finger movements to measure behavior may not accurately capture the full range of behavioral responses associated with induced emotions. They emphasize the importance of considering actual behavior in research on emotion and suggest that induced emotions can indeed influence behavior, albeit in different ways compared to spontaneous emotions.

8.3.2 *Implications for Psychology Researchers*

By drawing on these references, it is evident that psychologists have conducted extensive research on both spontaneous and induced emotions. The findings suggest that while there are some commonalities in the neural underpinnings of these emotions, there are also important differences in brain activity, physiological changes, and behavioral influences. Understanding these distinctions can contribute to a more comprehensive understanding of the nature and mechanisms of emotion.

The research on spontaneous and induced emotion has a number of implications for psychology researchers. For example, this research can help researchers to better understand the neural underpinnings of emotion, the relationship between emotion and cognition, and the factors that influence behavioral expression of emotion.

This research can also inform the development of new treatments for emotional disorders. For example, researchers are developing new treatments for anxiety that use virtual reality to induce mild anxiety in patients in a safe and controlled environment. This exposure to mild anxiety can help patients to learn to manage their anxiety more effectively.

The research on spontaneous and induced emotion is a rapidly growing field with a wide range of implications for psychology researchers. This research is helping us to better understand the nature of emotion and its role in our lives. It is also informing the development of new treatments for emotional disorders.

8.4 Spontaneous Versus Induced Emotion Regulation: An Insightful Comparison

In the realm of emotion regulation, a critical distinction arises between spontaneous and induced strategies. The former entails self-motivated regulation without explicit instructions, while the latter involves experimental manipulation. A closer examination of these approaches reveals intriguing insights.

In the context of stressful speeches, spontaneous emotion regulation unfolds naturally. Participants, post-speech, reported employing either expressive suppression or cognitive reappraisal [24]. Surprisingly, suppression correlated with reduced anxiety expression, heightened physiological response, and impaired speech memory, yet left negative affect untouched. In contrast, reappraisal had no impact on physiology and memory but curtailed expression and affect. Remarkably, these spontaneous regulation outcomes mirrored those of experimentally induced regulation in passive tasks.

Delving into induced emotion through EEG analysis during film viewing, emotion-specific and -unspecific patterns emerged [1]. Positive and negative emotions evoked hemispheric activation asymmetries in various EEG frequency bands, highlighting cognitive involvement in emotional processing. Joy and disgust, for instance, triggered left-hemisphere theta-2 power increase in frontal regions, revealing cognitive engagement. Conversely, disgust and fear/anxiety led to alpha-2 and beta-1 desynchronization in the right temporal-parietal cortex, suggesting its role in arousal modulation.

The capacity to regulate emotions is intrinsic to human social behavior. Spontaneous emotion regulation, an understudied domain, sheds light on how people naturally navigate their emotions. In a study where no pre-regulation instructions were given [79], participants were found to regulate emotions more in response to negative stimuli, primarily through reappraisal rather than suppression. Intriguingly, individuals with high baseline respiratory sinus arrhythmia (RSA) favored reappraisal, although this wasn't reflected in facial expressivity.

The dynamic interplay between spontaneous and induced emotion regulation strategies offers a nuanced understanding of human emotional responses. Spontaneity reveals the instinctual ways we manage our emotions, while experimental manipulation provides a controlled context for exploration. Together, they enrich our comprehension of the intricate web of emotions and their regulation.

8.5 The Trait of Infectious Smile: Psychological Attribute

8.5.1 The Infectious Smile: How Babies and Others Spread Positivity

Smiling is one of the most universal and contagious forms of human communication. It is a powerful way to connect with others, express emotions, and spread joy. While smiles can be used to convey a variety of different emotions, such as happiness, amusement, and even sadness, they are most often associated with positive feelings.

Smiling in Babies

Babies begin to smile at around 6 weeks old, and their smiles become more frequent and social over time. By the age of 4 months, most babies are able to recognize and smile at familiar faces. Smiling is an important part of a baby's social development. It helps them to connect with others and to learn about the world around them. Research has shown that babies are more likely to smile when they see other people smiling. This is because smiling is contagious. When we see someone else smile, it activates the mirror neurons in our brains, which are responsible for imitating the behavior of others. As a result, we feel the urge to smile back. Smiling in babies is an important part of their social development. Babies begin to smile at around 6 weeks old, and their smiles become more frequent and social over time [56]. By the age of 4 months, most babies are able to recognize and smile at familiar faces [56]. Research has shown that babies are more likely to smile when they see other people smiling, as smiling is contagious. When we see someone else smile, it activates the mirror neurons in our brains, which are responsible for imitating the behavior of others. As a result, we feel the urge to smile back [56].

The Science of Smiling

Smiling is a complex behavior that is controlled by a number of different brain regions. When we smile, our brains release dopamine, a neurotransmitter that is associated with pleasure and reward. This is why smiling can make us feel happy and relaxed. Smiling also has a number of physical benefits. It reduces stress hormones, such as cortisol, and boosts the immune system. Smiling can also lower blood pressure and heart rate. Smiling is a complex behavior that is controlled by different brain regions. When we smile, our brains release dopamine, a neurotransmitter associated with pleasure and reward [8]. This release of dopamine when we smile can make us feel happy and relaxed [8]. Smiling also has physical benefits, such as reducing stress hormones like cortisol and boosting the immune system [8]. It can also lower blood pressure and heart rate [8].

Psychological Attributes of Smiling

Smiling is often associated with positive psychological attributes, such as happiness, optimism, and confidence. People who smile more frequently are often perceived as being more friendly, approachable, and likable. Smiling can also have a positive impact on our own mental health. Studies have shown that smiling can reduce stress, improve mood, and boost self-esteem. Psychologically, smiling is often associated with positive attributes such as happiness, optimism, and confidence [25]. People who smile more frequently are perceived as being more friendly, approachable, and likable [25]. Smiling can also have a positive impact on our own mental health. Studies have shown that smiling can reduce stress, improve mood, and boost self-esteem [25].

How Smiling Spreads Positivity

When we smile at someone, we are sending a signal that we are friendly and approachable. This can make others feel more comfortable and relaxed around us. Smiling can also create a more positive and supportive social environment. When we see others smiling, it can make us feel happy and optimistic. This is because smiling is contagious, and it activates the mirror neurons in our brains that are responsible for imitating the behavior of others. When we smile at someone, we are sending a signal that we are friendly and approachable, which can make others feel more comfortable and relaxed around us [85]. Smiling can create a more positive and supportive social environment. When we see others smiling, it can make us feel happy and optimistic because smiling is contagious and activates the mirror neurons in our brains [85].

Thus, smiling in babies is an important part of their social development. It helps them connect with others and learn about the world around them. Smiling is a complex behavior controlled by different brain regions and has physical and psychological benefits. Smiling spreads positivity by signaling friendliness and approachability, making others feel more comfortable and creating a positive social environment.

8.5.2 *Laughing and Its Loudness that Breaks the Jinx*

Laughing and its loudness that breaks the jinx is a fascinating phenomenon that has been studied from various perspectives. Laughter is not only a spontaneous expression of amusement but also serves several pragmatic functions [49]. It can be used to show that one is taking a situation in the best possible way, potentially improving the cognitive and emotional appraisal of the situation. Additionally, laughter has powerful contagiousness, and individuals may laugh to positively affect others and induce them to appraise the situation in a positive way. The contagious nature of laughter is a well-documented phenomenon [20]. Laughter is intensely social and is 30 times more likely to occur in a social context than when alone. It spreads easily among individuals, and when one person starts laughing, it often leads to others joining in without any apparent reason [20]. This infectious feature of laughter is unique and does not happen with pain. While a person in pain may elicit sympathy or concern from others, it does not typically lead to spontaneous laughter in those observing the pain [20].

The contagiousness of laughter can be attributed to emotional contagion, which is the phenomenon of one person's emotions spreading to others [84]. Emotional contagion involves mechanisms such as emotional mimicry and social appraisal [84]. When individuals observe someone laughing, they may unconsciously mimic the laughter and experience a similar emotional state, leading to a chain reaction of laughter. This process is facilitated by the activation of mirror neurons, which respond exclusively to laughter and trigger the motor pattern of laughter in listeners [58]. The impact of laughter on social bonding and emotional regulation is significant [67]. Laughter occurs most often in social interactions and is associated with bonding, agreement, affection, and emotional regulation [67]. It helps strengthen social connections and create a positive atmosphere within a group. Laughter is also a universal expression of emotion used to maintain social bonds [55]. It serves as a non-verbal communication tool that can convey positive emotions and foster a sense of belonging.

The physiological and neurological aspects of laughter have also been explored. Laughter has been found to activate brain regions involved in reward processing, such as the prefrontal cortex and the limbic system [50]. It also releases endorphins, which are natural painkillers and mood enhancers [50]. These neurochemical responses contribute to the pleasurable and contagious nature of laughter. Laughter is a complex and multifaceted phenomenon that goes beyond being a simple expression of amusement. Its contagiousness and ability to spread among individuals without any apparent reason make it a unique and intriguing aspect of human behavior. The contagious nature of laughter can be attributed to emotional contagion and the activation of mirror neurons. Laughter plays a crucial role in social bonding, emotional regulation, and creating a positive atmosphere within groups. Understanding the mechanisms and effects of laughter can provide insights into human social interactions and emotional experiences.

8.6 The Genetic Elixir of Floral Bliss: Unraveling the Psychological Impact of Flowers

Throughout the annals of human history, flowers have captivated our senses with their enchanting beauty and mesmerizing fragrances. They have served as symbols of social grace, aesthetic elegance, and profound spiritual significance. Yet, the intricate psychological mechanisms underlying our deep affinity for flowers have remained a relatively uncharted territory. This scholarly chapter embarks on a journey to explore the fascinating hypothesis that flowers, in their genetic essence, are programmed to elicit happiness and spiritual bliss in human beings.

The Emotional Palette of Floral Displays: The visual spectacle of a floral arrangement is known to evoke a myriad of emotions within the human psyche. Research findings have consistently demonstrated that exposure to a floral display can have a transformative impact on emotional states. Among the positive effects observed are heightened feelings of composure and confidence. The sight of a well-composed floral arrangement can instill a sense of harmony and balance, elevating one's confidence in the aesthetic pleasures of life. However, it is essential to acknowledge that the emotional canvas painted by flowers is not devoid of complexity. While they predominantly inspire positive emotions, there is evidence to suggest that floral displays may also lead to an increase in annoyance, albeit to a limited extent. This paradoxical response hints at the intricate interplay between individual differences and the floral stimulus, underscoring the need for more nuanced investigations.

Foliage and Its Enigmatic Effects: The verdant tapestry of foliage, distinct from the flamboyant allure of flowers, possesses its own psychological nuances. Studies have revealed that foliage displays, while not as emotionally potent as their floral counterparts, can still wield a subtle influence. Interestingly, foliage displays have been associated with a modest increase in bad temper, offering an intriguing area for further exploration. On the flip side, foliage displays tend to cast a favorable impact on clear headedness. The presence of lush greenery appears to foster mental clarity and cognitive sharpness. These findings beckon us to delve deeper into the psychology of our interaction with nature.

Flowers as Genetic Beacons of Well-Being: The intriguing hypothesis that there exists an innate, genetically programmed mechanism governing our positive response to nature, including flowers, has garnered scholarly attention. Some have postulated the existence of a preestablished harmony between human beings and the natural world, a harmony that profoundly influences our emotional well-being. This harmony, if indeed encoded in our genes, could hold the key to understanding our profound connection with flowers.

Flowers and the Brain: A Neuroscientific Odyssey: Flowers and the Brain: A Neuroscientific Odyssey.

Recent research endeavors have begun to unravel the profound effects of flowers on the human brain. A groundbreaking study employed a multifaceted approach, incorporating psychological, physiological, and neuroscientific techniques, to investigate how viewing flowers regulates emotions and influences brain activity [17]. The study discovered that passive viewing of a typical flower image had the remarkable ability to downregulate negative emotions. Moreover, it decreased elevated blood pressure and cortisol levels, which are two indicators of stress-induced elevation, in response to an acute visual stressor [17]. This intriguing phenomenon hints at the therapeutic potential of flowers in mitigating stress and promoting emotional well-being.

Functional magnetic resonance imaging (fMRI) was employed to examine brain activation patterns relevant to emotion regulation during automatic distraction induced by viewing the typical flower image. Astonishingly, the activation of the right amygdala-hippocampus region decreased during the viewing of this image, compared to viewing a flower-mosaic or a visual fixation point after exposure to an acute visual stressor [17]. This finding suggests that flowers have a specific impact on the brain's emotional processing centers, leading to a reduction in stress and negative emotions.

The study's findings align with previous research on the nature of beauty and its effects on the brain. A review of empirical studies in the field of neurobiology and aesthetics has revealed that beauty elicits positive emotional and cognitive responses [78]. Flowers, with their vibrant colors and intricate patterns, are often considered beautiful and can therefore have a similar impact on the brain. The activation of brain regions associated with pleasure and reward, such as the amygdala and hippocampus, may contribute to the positive emotional experience induced by viewing flowers [78].

Furthermore, the therapeutic potential of flowers extends beyond their visual appeal. The use of flower essences has been shown to reduce stress symptoms in various populations, including nursing students and caregivers in hospice settings [2, 43]. These interventions involve the use of flower extracts or essences, which are believed to have healing properties. The calming and stress-reducing effects of flower essences align with the findings of the aforementioned study, suggesting that flowers have a tangible impact on emotional well-being.

The recent research has shed light on the profound effects of flowers on the human brain. Viewing flowers has been found to regulate emotions, decrease stress, and activate brain regions associated with pleasure and reward. These findings highlight the therapeutic potential of flowers in promoting emotional well-being and mitigating stress. Further research in this field could explore the specific mechanisms through which flowers exert their effects on the brain and identify potential applications in clinical settings.

Plant: A Cognitive System for Navigating Nature's Bounty: The recurrent adaptive problems our species encountered over evolutionary time have shaped the human mind. It is well-known that plant defenses have structured the physiology and behavior of many animal species, including human. In light of these considerations, along with my esteemed colleagues, I have recently proposed a comprehensive solution for the learning problems plants pose. We argue that the human mind houses a collection of behavioral avoidance strategies and social learning rules geared toward safely acquiring information about plants. This intricate cognitive framework, which we aptly refer to as Plant Learning and Avoidance of Natural Toxins, or PLANT, forms an integral part of our evolutionary heritage. Plant represents our cognitive arsenal for navigating the intricate relationship between humans and the plant kingdom, shedding light on the roots of our profound connection with flowers.

Flowers, with their symphony of colors, shapes, fragrances, and the ever-changing poetry of their existence, have captured the human imagination for millennia. The hypothesis that flowers may be genetically programmed to elicit happiness and spiritual bliss in humans is a tantalizing prospect that beckons further exploration. As we endeavor to understand the profound psychological impact of flowers, we uncover the intricate tapestry of emotions and cognitive responses woven by these genetic beacons of well-being. This knowledge not only enriches our appreciation of the floral world but also underscores the therapeutic potential of our enduring bond with nature. In the quest to decode the genetic elixir of floral bliss and navigate the intricate dance of human-plant interactions, we embark on a journey that transcends science, touching the very essence of human existence.

8.7 Latency Difference for Personal Experience and Induced Emotion: A Psychological Exploration

Emotions are a fundamental aspect of the human experience, shaping our perceptions, behaviors, and overall well-being. Psychologists have long been intrigued by the complexities of human emotions and how they manifest in different circumstances. One intriguing area of study within the realm of emotions is the latency difference between personal experiences and induced emotions. This chapter aims to delve into this topic, shedding light on how our emotions react differently in response to real-life situations versus artificially induced stimuli. Emotions play a crucial role in shaping human experiences and behaviors. Understanding how emotions differ in response to personal experiences versus induced stimuli is an intriguing area of study. This article aims to explore the latency difference between personal experiences and induced emotions, shedding light on how our emotions react differently in these two contexts.

One important aspect to consider is the use of naturalistic stimuli in studying emotions. Saarimäki [62] highlights that naturalistic stimuli, such as movies, music, and stories, elicit strong emotions and allow for the examination of emotions in close-to-real-life conditions. This suggests that induced emotions through these stimuli may closely resemble personal experiences, leading to similar emotional reactions. Schubert [64] discusses the distinction between the emotion felt by the listener and the emotion expressed by the music. This differentiation implies that induced emotions through music may evoke similar emotional responses as personal experiences, as the listener's emotional state aligns with the emotion expressed by the music. This finding supports the idea that induced emotions can closely mirror personal experiences.

Personality traits also play a role in the susceptibility to positive and negative emotional states. Larsen and Ketelaar [40] suggest that individuals with different personality traits may have varying emotional reactions to personal experiences and induced emotions. This implies that the latency difference between personal experiences and induced emotions may be influenced by individual differences in personality. Furthermore, moral dilemmas can evoke strong emotional responses. Lotto et al. [41] conducted a study on moral acceptability and emotional salience, finding that individuals have different decision times and emotional reactions when faced with moral dilemmas. This suggests that the latency difference between personal experiences and induced emotions may be particularly pronounced in moral contexts.

Lastly, Abramson et al. [5] compared instructed and real-life expressions of fear and anger. They found significant deviations in the recognition of these emotions, indicating that the context in which emotions are induced can impact their latency. This finding suggests that the latency difference between personal experiences and induced emotions may be influenced by the authenticity and context of the stimuli. The latency difference between personal experiences and induced emotions is a complex phenomenon influenced by various factors. Naturalistic stimuli, personality traits, moral dilemmas, and the authenticity of the stimuli all contribute to this difference. Further research is needed to fully understand the mechanisms underlying this latency difference and its implications for our understanding of human emotions.

8.7.1 Experience Versus Emotion

Latency, in the context of emotions, refers to the time it takes for an emotional response to occur following a triggering event or stimulus. It is a crucial aspect of emotional processing and can vary significantly depending on the nature of the emotional experience. To explore this phenomenon, we must distinguish between emotions arising from personal experiences and those that are induced through external means.

Emotions from Personal Experiences

When we encounter events or situations in our lives, our emotional responses are typically immediate and spontaneous. For instance, the joy of a surprise birthday party, the fear during a near-miss accident, or the sadness of a breakup are all emotions that arise organically from personal experiences. These emotions tend to have minimal latency, as they are tightly coupled with the event that elicited them. In such cases, our brain and body react swiftly to process the emotional information, often leading to intense and immediate feelings.

Induced Emotions

In contrast, induced emotions are those that are generated through external stimuli, such as movies, music, art, or even psychological experiments. These emotions are deliberately triggered by a stimulus designed to elicit a specific emotional response. For example, a tearjerker movie scene aims to induce sadness in the audience, or a horror film intends to provoke fear. The latency of induced emotions is typically longer than that of emotions arising from personal experiences.

8.7.2 Factors Influencing Latency Differences

Several factors contribute to the latency difference between personal experiences and induced emotions:

1. Cognitive Processing: Emotions from personal experiences are often processed more rapidly because they are linked to real-life situations that demand quick reactions for survival and adaptation. On the other hand, induced emotions require cognitive interpretation and are often subject to individual variations in emotional response time.
2. Emotional Intensity: Personal experiences tend to evoke stronger and more intense emotions compared to induced emotions. The higher intensity can lead to faster emotional responses, whereas induced emotions may take longer to reach a similar level of intensity.
3. Familiarity: Personal experiences are inherently more familiar to individuals, as they involve their own lives and histories. Induced emotions, on the other hand, may depend on the novelty of the stimulus, with less familiar stimuli requiring more time for emotional processing.
4. Context and Expectations: The context in which emotions are induced can impact their latency. For example, if individuals are aware that they are participating in a study to induce emotions, their expectations and awareness of the artificial nature of the stimulus can influence how quickly they experience the intended emotion.

8.7.3 Implications for Psychology Professionals

Understanding the latency difference between personal experiences and induced emotions has several implications for psychology professionals:

1. Therapy and Intervention: Therapists and counselors can use this knowledge to help individuals process and cope with emotions stemming from personal experiences more effectively. Techniques that acknowledge the immediacy of personal emotions can be employed to address and manage them.
2. Experimental Design: Researchers conducting studies on emotional responses should consider the latency of induced emotions when designing experiments.

This awareness can lead to more accurate assessments and interpretations of emotional reactions.

3. Emotional Regulation: Individuals can benefit from learning how to regulate their emotional responses to both personal experiences and induced emotions. This can be particularly valuable for enhancing emotional well-being and resilience.

The latency difference between personal experiences and induced emotions is a fascinating area of study within psychology. While personal experiences often elicit swift and intense emotional responses, induced emotions tend to have a longer latency due to their artificial nature and cognitive processing requirements. Recognizing and understanding these differences can help psychology professionals provide better support to individuals dealing with various emotional states and contribute to the development of more effective therapeutic interventions and research methodologies. Emotions are a complex and integral part of human life, and exploring their latency differences sheds light on the intricate workings of our emotional landscape.

8.8 Challenges in Monitoring Universal Connectedness to Elementary Happiness and Pain/Grief

The pursuit of happiness and the avoidance of pain and grief are fundamental aspects of the human experience. Psychologists and researchers have long sought to understand these emotional states and their interconnectedness across cultures and individuals. This chapter delves into the challenges associated with monitoring universal connectedness to elementary happiness and pain/grief in the field of psychology, shedding light on the complex nature of these emotions.

8.8.1 The Quest for Universal Connectedness

The idea of universal connectedness in emotions suggests that certain emotional states, such as happiness and pain/grief, transcend cultural, social, and individual boundaries. Researchers have sought to establish commonalities in these emotions across diverse populations, aiming to better understand their underlying mechanisms and manifestations.

Challenges in Monitoring Elementary Happiness

1. Cultural Variation: One of the primary challenges in monitoring universal connectedness to elementary happiness is the significant influence of culture on how happiness is defined and experienced. Cultural norms, values, and practices shape the perception of happiness, making it difficult to identify a single universal definition or indicator.
2. Subjectivity: Happiness is highly subjective, making it challenging to measure objectively. Self-reporting methods, such as surveys and questionnaires, are

commonly used but are susceptible to cultural biases and individual interpretations.

3. Temporal Variability: Happiness is not a static emotion. It fluctuates over time, influenced by life circumstances, personal goals, and cultural contexts. Monitoring happiness requires accounting for these temporal changes.
4. Multifaceted Nature: Happiness is not a singular emotion but a complex construct comprising various components, including hedonic (pleasure-based) and eudaimonic (meaning-based) aspects. Monitoring and assessing these facets pose additional challenges.

8.8.2 Challenges in Monitoring Elementary Pain/Grief

1. Cultural Differences: Similar to happiness, pain and grief are subject to cultural variations. Cultural norms and rituals surrounding death, mourning, and the expression of grief can significantly impact how individuals experience and express these emotions.
2. Stigma and Taboos: In many cultures, discussing or seeking help for emotional pain or grief is stigmatized or considered taboo. This can lead to underreporting and concealment of these emotions, making it difficult to monitor them accurately.
3. Coping Mechanisms: Individuals employ various coping mechanisms to deal with pain and grief, which can manifest differently across cultures and individuals. These coping strategies can impact the observable expressions of these emotions.
4. Cultural Expressions: Different cultures may have distinct ways of expressing pain and grief, including rituals, ceremonies, and communal practices. These expressions may not align with Western psychological assessments of these emotions.

8.8.3 Implications for Psychology Professionals

Recognizing the challenges in monitoring universal connectedness to elementary happiness and pain/grief has important implications for psychology professionals:

1. Cultural Sensitivity: Psychologists must approach the study and treatment of these emotions with cultural sensitivity. Understanding cultural nuances and respecting diverse expressions of happiness and grief is essential.
2. Multidimensional Assessment: Developing comprehensive assessment tools that capture the multifaceted nature of happiness and the various dimensions of pain and grief can lead to more accurate monitoring and diagnosis.
3. Cross-Cultural Research: Collaborative cross-cultural research is crucial to understanding the universal and culturally specific aspects of these emotions.

This can help bridge gaps in our knowledge and develop culturally informed interventions.
4. Promoting Openness: Reducing stigma and taboos surrounding emotional pain and grief is essential. Psychology professionals can play a role in promoting open and healthy discussions about these emotions.

The pursuit of understanding universal connectedness to elementary happiness and pain/grief is a complex and ongoing challenge in the field of psychology. Cultural differences, subjectivity, and the multifaceted nature of these emotions make monitoring and assessing them a nuanced task. However, recognizing these challenges is a crucial step toward developing more inclusive and effective approaches to studying and addressing happiness, pain, and grief in individuals and societies across the world. Psychology professionals can contribute to this endeavor by fostering cultural sensitivity and promoting research that explores the interconnectedness of these elemental human emotions.

8.9 The Effect of Video Speed and Media Transition Rate on the Viewer's Mind

In today's digital age, we are constantly exposed to a barrage of videos and media content across various platforms. From fast-paced action scenes in movies to the rapid-fire editing of social media content, the speed at which videos and media change has a profound impact on the viewer's mind. This chapter explores the psychological and cognitive effects of video speed and media transition rate on the audience.

8.9.1 The Power of Video Speed

Video speed, or the pace at which visual information is presented, plays a pivotal role in shaping the viewer's experience and engagement. It encompasses elements such as the duration of shots, the rhythm of editing, and the overall tempo of a video. Different speeds can evoke various emotional responses and cognitive processes:

1. Attention and Engagement: Faster video speeds often capture the viewer's attention more effectively. Quick changes in scenes or shots can create a sense of urgency, making it harder for the viewer to look away. This can be particularly beneficial in marketing or advertising.
2. Emotional Intensity: Rapid video speed can amplify emotional intensity. Action sequences, for instance, are often edited at high speeds to increase excitement and tension, while slow-motion shots are used to emphasize emotional moments.
3. Cognitive Load: High-speed videos can overwhelm the viewer's cognitive processing capacity. Information presented too rapidly may be challenging to comprehend fully, potentially leading to reduced retention of content.

4. Fatigue and Stress: Constant exposure to fast-paced videos can lead to mental fatigue and increased stress levels. The brain works harder to process rapid visual information, potentially contributing to a sense of exhaustion.

8.9.2 Media Transition Rate and Information Processing

The transition rate between different media elements, such as images, text, or audio, also influences how the viewer perceives and processes information:

1. Cognitive Load: Frequent transitions between media elements can increase cognitive load. When text, images, and audio change rapidly, the viewer must continuously shift attention and process new information, potentially leading to cognitive strain.
2. Storytelling and Narrative: The transition rate is a crucial storytelling tool. Skillful use of transitions can enhance the flow of a narrative, guide the viewer's focus, and convey information effectively.
3. Emotional Impact: Transitions can influence emotional responses. For example, abrupt transitions may evoke surprise or shock, while gradual transitions can build suspense or anticipation.
4. Memory and Learning: The rate of transitions can affect memory and learning. A well-paced educational video with appropriate transitions can aid comprehension and retention, while excessive transitions may hinder learning.

8.9.3 Viewer Preferences and Individual Differences

It's important to note that viewer preferences and the impact of video speed and media transition rate can vary significantly among individuals. Factors such as age, cultural background, attention span, and personal preferences play a role in how different people respond to media content. Recognizing these individual differences is essential for content creators and media professionals.

As technology continues to evolve, understanding and harnessing the power of pacing in visual storytelling will remain essential in captivating and connecting with audiences. Video speed and media transition rate have a significant impact on the viewer's mind, influencing attention, emotions, cognitive load, and overall engagement. Faster video speeds capture attention and increase emotional intensity, but they can also overwhelm the viewer's cognitive processing capacity and lead to fatigue and stress. Frequent transitions between media elements can also increase cognitive load, but they can be a powerful storytelling tool and influence emotional responses. The transition rate can also affect memory and learning, with well-paced videos aiding comprehension and retention. Content creators, marketers, and educators must consider the psychological effects of video speed and media transition rate when crafting their messages. Psychology researchers can use these findings to inform their research and develop new interventions.

8.10 Exploring the Mind's Response: EEG Recording of Innocent Smiles, Blooming Flowers, Spiritual Interactions, and Music

The study of human emotions and their neural correlates has long fascinated psychologists. Understanding how the human brain responds to various stimuli, especially those evoking positive emotions, provides valuable insights into emotional processing and psychological well-being. In this chapter, we delve into the intriguing realm of Electroencephalogram (EEG) recordings to investigate the effects on the human mind when exposed to innocent smiles of children, the blossoming of flowers, spiritually peaceful interactions, and music. Pleasure and happiness are complex psychological states that have been the subject of extensive research in neuroscience and psychology. Understanding the underlying neural mechanisms and psychological processes involved in pleasure and happiness can provide valuable insights into human well-being and mental health. This article aims to synthesize the findings from several key studies to shed light on the functional neuroanatomy of pleasure and happiness, the role of emotions in processing pleasure and happiness, the effects of meditation on these states, and the impact of music on stress and the stress response.

In Fig. 8.1 we find two human subjects S1 and S2 are very different and they have their own peculiar patterns. Whatever be the tests we definitely observe a logical reproducible responses from a single person. But we could never expect those very patterns from another person. There is a basic EEG dynamics exclusive to a particular human and that feature does not change significantly with the situation. Modulation of that basic pattern is what we observe in all three tests. While S1 has a beating pattern, high intensity signal flashes in EEG while a large region is affected. S2 on the other hand has a very silent brain surface, and that responds using minimum activation of the scalp. Positive owe generates more activation than negative owe. A negative owe activates multiple local regions in the brain while positive owe gives a large blunt activation.

EEG is a non-invasive neuroimaging technique that measures electrical activity in the brain. It records the electrical patterns generated by neurons as they communicate with each other. EEG offers a real-time window into the brain's activity and allows researchers to observe how different stimuli influence neural responses.

Innocent Smiles of Children

The innocent smile of a child is universally regarded as a heartwarming and positive stimulus. The study by Kringelbach and Berridge [80] explores the functional neuroanatomy of pleasure and happiness. They propose that pleasure and happiness can be understood as distinct but interconnected processes. Pleasure is associated with positive emotions and the experience of reward, while happiness encompasses a broader sense of well-being and life satisfaction. The authors highlight the involvement of several brain regions, including the prefrontal cortex, amygdala, and basal ganglia, in the processing of pleasure and happiness. Sergerie et al. [65] discuss the contributions of the amygdala to emotion processing. The amygdala plays a crucial

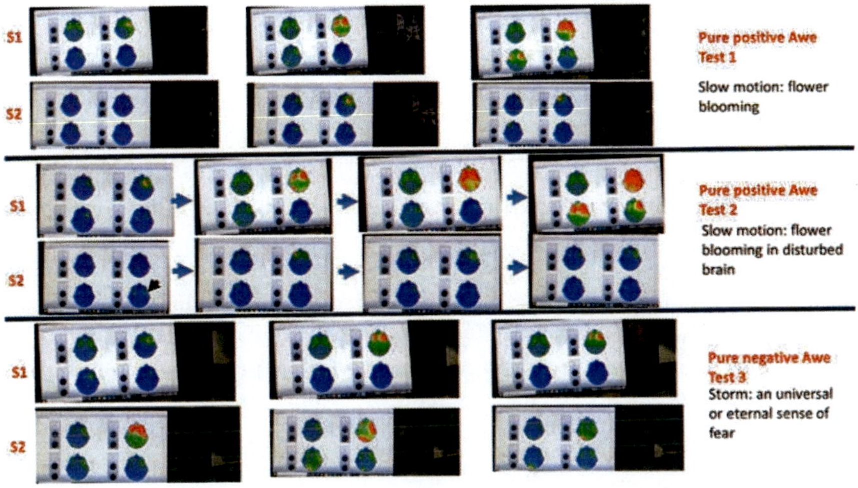

Fig. 8.1 There are three rows for three pure awe tests. Two human subject results S1 and S2 have been provided. Test 1 and Test 2 are positive awes, while test 3 is a negative awe

role in the processing of emotions, particularly fear and threat-related stimuli. It is also involved in the evaluation of saliency and the current value of stimuli. The amygdala is part of a larger network that includes the prefrontal cortex and ventral striatum, which collectively contribute to the processing of pleasure and reward.

EEG studies have revealed several fascinating findings when participants view such smiles:

1. Increased Activation in Reward Pathways: Innocent smiles tend to activate brain regions associated with reward processing, including the ventral striatum. This suggests that the sight of a child's smile can trigger feelings of joy and happiness.
2. Reduced Amygdala Activity: The amygdala, which plays a central role in processing emotions, tends to show decreased activity when exposed to innocent smiles. This reduction in amygdala activation may signify a decrease in negative emotions like fear and anxiety.

Blooming of Flowers

The sight of flowers blooming, often associated with the beauty and renewal of nature, can evoke a range of positive emotions. EEG research on this topic has yielded intriguing results:

1. Enhanced Alpha and Theta Activity: EEG recordings show increased activity in the alpha and theta frequency bands when individuals view the blooming of flowers. These patterns are linked to a relaxed and meditative state of mind.
2. Improved Attention and Focus: Flower blooming scenes can enhance attention and cognitive processing, as reflected by increased gamma wave activity in the

brain. This suggests that exposure to natural beauty may improve concentration and mental clarity.

Spiritually Peaceful Interactions

Spiritual experiences that promote inner peace and well-being have been the focus of EEG studies as well. Ryan and Deci [60] examine the effects of meditation on brain activity and states of mind. They review studies utilizing electroencephalography (EEG), event-related potentials (ERP), and neuroimaging techniques to investigate the neural correlates of meditation. The findings suggest that meditation can induce specific states of mind characterized by increased attention, relaxation, and reduced mind wandering. These states are associated with distinct patterns of brain activity, including changes in alpha and theta oscillations.

1. Increased Frontal Asymmetry: EEG findings suggest that spiritually peaceful interactions may lead to greater frontal asymmetry in the brain, indicating a shift toward positive emotional states and overall well-being.
2. Lower Stress Response: Exposure to spiritually peaceful scenes can reduce cortisol levels, indicating a decrease in stress. EEG data supports the idea that such experiences promote relaxation and emotional stability.

Music Modulating the Brain Wave

The impact of music on the human mind has been widely studied using EEG recordings. The effect of music on the human stress response is explored in the study by Thoma et al. [80]. The authors investigate the impact of music on physiological and psychological stress markers. They find that listening to music can lead to a reduction in stress hormones, blood pressure, and heart rate. Music has also been shown to enhance relaxation and promote emotional well-being, making it a potential therapeutic tool for stress management. The involvement of brain regions such as the prefrontal cortex, amygdala, and basal ganglia highlights the complex nature of these psychological states. Additionally, the effects of meditation and music on brain activity and emotional well-being further emphasize the potential for interventions to enhance pleasure and happiness. Understanding the neural mechanisms underlying pleasure and happiness can contribute to the development of targeted interventions for mental health and well-being.

1. Modulation of Brain Waves: Different types of music can influence brain wave patterns. For example, calming music tends to increase alpha and theta wave activity, promoting relaxation.
2. Emotional Resonance: Music with emotionally resonant melodies or lyrics can activate regions of the brain associated with emotion processing, such as the limbic system.

EEG recordings have provided valuable insights into how the human mind responds to various positive stimuli, including innocent smiles of children, blooming flowers, spiritually peaceful interactions, and music. These studies demonstrate that these stimuli can trigger activation in brain regions associated with reward, relaxation, emotional processing, and well-being. Understanding these neural responses not only enriches our knowledge of emotional processing but also has practical applications in fields such as therapy, education, and media design. As researchers continue to explore the intricate relationship between our minds and the stimuli we encounter, we gain a deeper appreciation for the profound impact of positivity on human well-being.

8.11 Unveiling the Intricate Connection Between Geometric EEG Patterns and Human Cognition

In the quest to understand the depths of human cognition and emotions, the field of neuroscience continuously unravels the intricacies of the brain. One fascinating discovery in this endeavor is the emergence of geometric patterns within the Electroencephalogram (EEG) spectrum. These geometric shapes have the potential to be more than just enigmatic formations—they might serve as markers of human emotions and cognitive processes. In this chapter, we delve into the remarkable world of geometric EEG patterns and their profound implications for our understanding of human cognition.

The Enigmatic Geometric Patterns: Electroencephalogram, or EEG, is a powerful tool that records the electrical activity of the brain through a series of electrodes placed on the scalp. It produces a spectrum that represents the frequency and amplitude of neural oscillations. While EEG recordings have long been used to study brain activity, the discovery of geometric patterns within these spectra is a relatively recent development. Researchers have observed that certain patterns emerge within the EEG spectrum, akin to geometric shapes. These patterns, when analyzed across a wide range of human subjects, appear to be remarkably consistent and invariant. This observation raises a tantalizing possibility: Could these geometric shapes serve as markers for specific human emotions and cognitive processes?

The Link to Human Cognition: To comprehend the significance of these geometric EEG patterns, we must explore their connection to human cognition and emotions. The idea that neural phenomena underlying these patterns might be fundamental to human cognition is groundbreaking. It suggests that the human brain possesses inherent mechanisms that are directly related to our emotional experiences and perceptions. For instance, if a specific geometric shape consistently appears in EEG recordings when individuals experience joy or sadness, it could potentially serve as a marker for those emotions. This would be a monumental breakthrough in understanding and diagnosing emotional states, which are often challenging to measure objectively.

Neurogenesis and the Fundamental Nature of Geometric Patterns: The emergence of geometric patterns within EEG spectra also sheds light on the concept of neurogenesis—the process of generating new neurons and neural connections in the brain. These patterns could represent fundamental neural phenomena that are deeply ingrained in our cognitive and emotional processes. If these geometric patterns are indeed invariant across a wide range of human subjects, it suggests that they are not idiosyncratic but rather universal. This universality implies that certain neural processes associated with these patterns are integral to human cognition, transcending individual differences.

In summary, the exploration of geometric patterns within EEG spectra opens a remarkable window into the human brain and its intricate workings. The potential link between these patterns and human emotions and cognition presents an exciting avenue for further research and discovery. If these patterns prove to be reliable markers for specific emotional states and cognitive processes, they could revolutionize fields such as psychology, neurology, and psychiatry. The concept of these geometric shapes being fundamental to neurogenesis underscores the profound interconnectedness of our neural processes and emotions. As we delve deeper into the enigma of these EEG patterns, we move one step closer to unraveling the mysteries of the human mind and the fundamental processes that underlie our rich emotional and cognitive experiences.

8.12 Unveiling the Intricate Connection Between Geometric EEG Patterns and Human Cognition

In the quest to understand the depths of human cognition and emotions, the field of neuroscience continuously unravels the intricacies of the brain. One fascinating discovery in this endeavor is the emergence of geometric patterns within the Electroencephalogram (EEG) spectrum. These geometric shapes have the potential to be more than just enigmatic formations—they might serve as markers of human emotions and cognitive processes. In this chapter, we delve into the remarkable world of geometric EEG patterns and their profound implications for our understanding of human cognition.

The Enigmatic Geometric Patterns: Electroencephalogram, or EEG, is a powerful tool that records the electrical activity of the brain through a series of electrodes placed on the scalp. It produces a spectrum that represents the frequency and amplitude of neural oscillations. While EEG recordings have long been used to study brain activity, the discovery of geometric patterns within these spectra is a relatively recent development. Researchers have observed that certain patterns emerge within the EEG spectrum, akin to geometric shapes. These patterns, when analyzed across a wide range of human subjects, appear to be remarkably consistent and invariant. This observation raises a tantalizing possibility: Could these geometric shapes serve as markers for specific human emotions and cognitive processes?

The Link to Human Cognition: To comprehend the significance of these geometric EEG patterns, we must explore their connection to human cognition and emotions. The idea that neural phenomena underlying these patterns might be fundamental to human cognition is groundbreaking. It suggests that the human brain possesses inherent mechanisms that are directly related to our emotional experiences and perceptions. For instance, if a specific geometric shape consistently appears in EEG recordings when individuals experience joy or sadness, it could potentially serve as a marker for those emotions. This would be a monumental breakthrough in understanding and diagnosing emotional states, which are often challenging to measure objectively.

Neurogenesis and the Fundamental Nature of Geometric Patterns

The emergence of geometric patterns within EEG spectra also sheds light on the concept of neurogenesis—the process of generating new neurons and neural connections in the brain. These patterns could represent fundamental neural phenomena that are deeply ingrained in our cognitive and emotional processes.

If these geometric patterns are indeed invariant across a wide range of human subjects, it suggests that they are not idiosyncratic but rather universal. This universality implies that certain neural processes associated with these patterns are integral to human cognition, transcending individual differences.

The exploration of geometric patterns within EEG spectra opens a remarkable window into the human brain and its intricate workings. The potential link between these patterns and human emotions and cognition presents an exciting avenue for further research and discovery. If these patterns prove to be reliable markers for specific emotional states and cognitive processes, they could revolutionize fields such as psychology, neurology, and psychiatry.

The concept of these geometric shapes being fundamental to neurogenesis underscores the profound interconnectedness of our neural processes and emotions. As we delve deeper into the enigma of these EEG patterns, we move one step closer to unraveling the mysteries of the human mind and the fundamental processes that underlie our rich emotional and cognitive experiences.

Of course, pain is not the only factor that contributes to true love. Other important factors include shared values, interests, and goals, as well as a strong physical and emotional connection. However, pain can play a significant role in the development of true love, and it can help us to appreciate the beauty and power of this emotion.

8.13 Conclusion

8.13.1 Happiness is a Myth, Materialistic, and Non-materialistic Pleasures Are Not the Key

8.13.1.1 Spiritual Bliss is Also Not the Happiness

The paradox of pain giving rise to love challenges conventional wisdom in psychology and neuroscience. Traditionally, pain and love have been viewed as separate and opposing experiences. However, recent research has revealed shared neural substrates and overlapping mechanisms between pain and love, suggesting a more intricate relationship between these fundamental aspects of the human experience [14, 15, 29, 54, 77, 83].

Neuroimaging studies have shown that both pain and love activate similar brain regions, such as the frontal lobe, corpus striatum, and ventral tegmental area [14, 15, 29, 54, 77, 83]. These regions are involved in reward processing, motivation, and emotional regulation, which are key components of both pain and love experiences. Additionally, the activation of the anterior cingulate cortex (ACC) has been observed in both pain and empathy for pain, suggesting a shared neural mechanism for processing and experiencing pain [36, 37].

Furthermore, studies have demonstrated that the putamen, a region involved in reward and motivation, is a shared brain region among different types of love, including brand love and interpersonal love [77, 82]. This suggests that the neural substrates associated with love extend beyond romantic and maternal love, encompassing other forms of attachment and emotional connection. Understanding the neural connections between pain and love not only enhances our understanding of love but also has practical implications. For example, the knowledge gained from studying the neural mechanisms of pain and love can inform novel approaches to pain management. By targeting shared neural pathways and mechanisms, it may be possible to develop interventions that alleviate pain by promoting positive emotional experiences, such as love and connection [79].

Moreover, unraveling the neural connections between pain and love sheds light on the complexity of human emotions. It highlights the interplay between pleasure and suffering in shaping our emotional experiences. This understanding can have broader implications for promoting overall well-being and mental health. By recognizing the intricate relationship between pain and love, we can develop interventions and strategies that foster emotional resilience and promote positive emotional experiences.

Thus, the paradox of pain giving rise to love challenges conventional wisdom in psychology and neuroscience. By examining the shared neural substrates and overlapping mechanisms, we gain insight into the intricate relationship between these fundamental aspects of the human experience. This research not only enhances our understanding of love but also paves the way for novel approaches to pain management and the promotion of overall well-being. The neural connections between pain and love are a fascinating area of study that underscores the complexity of human emotions. By unraveling these connections, we can gain a deeper appreciation for the interplay between pleasure and suffering in shaping our emotional experiences.

8.13.1.2 What Does It Mean by Happiness is a Local Invariant of Pain?

In the context of pain perception, the concept of invariance, a mathematical term denoting the constancy of certain parameters amidst variable alterations, has gained significance within the neuroscientific domain. Our investigations into the neural substrates of pain processing reveal the existence of a hierarchical architecture of invariants within the human brain, wherein an invariant is not a static entity but rather a dynamic function. When an individual's brain encounters the experience of pain, it identifies a distinct sensation that remains consistent irrespective of the intensity, typology, or clarity of the pain stimulus. This unchanging cognitive phenomenon serves as a cognitive escape route from the distressing realm of pain, leading to the subjective experience of happiness. It is noteworthy that conventional conceptions of happiness are often rooted in positive emotional states, yet our findings invite a paradigm shift by suggesting that the definition of happiness should encompass its manifestation within the context of pain. While pain is conventionally associated with loss and happiness with reward, neural data derived from the DDG paradigm challenge this conventional wisdom. This alternative perspective posits the existence of a form of happiness that transcends the need for external factors, be they universal triggers for spiritual euphoria or complete isolation from one's surroundings. Instead, it signifies a state of mind wherein conscious awareness transcends the relentless pursuit of desires and possessions, allowing for a respite from the perpetual assessment of one's possessions and deficiencies. This insight invites a reevaluation of the intricate relationship between pain and happiness within the human psyche, opening new avenues for the understanding of emotional and cognitive processes.

8.13.2 Emotion, Cognition, and Perception Are Geometric Structure of Clocks

8.13.2.1 Happiness and Peace of Mind Are Mathematical Formulations Than Chemical Reaction

Happiness and peace of mind have long been subjects of fascination for researchers across various disciplines. Traditionally, these states of being have been attributed to chemical reactions and neurotransmitter activity within the brain. However, a novel perspective emerges when we explore the concept that happiness and peace of mind may be mathematical formulations rather than mere chemical reactions. This section delves into the intriguing idea that these emotional states are time crystals [61] that can be understood through the pattern of prime numbers forming a complete series on the surface of a phase prime metric, known as Purusha. This three-dimensional temple-like structure made of primes and their combinations provides a unique framework for comprehending the relationship between mathematical symmetry and emotional equilibrium. The pursuit of happiness and inner peace has been a fundamental human endeavor for centuries. Traditionally, the scientific community has primarily looked to the realm of biochemistry to unravel the mysteries of these emotional states. However, a growing body of research suggests that there might be a mathematical underpinning to happiness and peace of mind that transcends the chemical reactions occurring within our brains.

The PPM or Purusha and Prime Numbers

At the heart of this intriguing theory lies the concept of Purusha, a three-dimensional structure that resembles a temple, composed entirely of prime numbers and their combinations. This Purusha, rather than a tangible entity, serves as a theoretical framework that allows us to explore the mathematical foundations of human emotions. The book [12] challenges us to think about the future of artificial intelligence in a natural way, where the concept of instruction as a fundamental tool of creating intelligence is challenged. The conversation about the future of AI should be around the complete rhythmic model of the human brain that is interconnected with the nature and society.

The Surface of PPM or Purusha: A Pathway to Symmetry

The surface of phase prime metric or Purusha is where the mathematical elegance of this theory truly shines. It is on this surface that prime numbers form a complete series, creating a geometric pattern that is as captivating as it is precise. When one traverses this surface, they are essentially navigating a pathway where the possibility of symmetry breaking is minimized, creating a geometrically stable and harmonious journey.

Mathematics and Emotional Equilibrium

But how does this mathematical framework relate to human emotions? The answer lies in the idea that happiness and peace of mind are, in essence, a manifestation of symmetry within our consciousness. When we experience these emotions, it is as if our internal world aligns perfectly, much like the symmetrical patterns found on the surface of Purusha. This alignment corresponds to a state of balance and tranquility, where the mind is free from the turbulence of conflicting thoughts and emotions. Emotions are geometric shapes, not in one dimension, but extends over 12 dimensions [75, 76].

Beyond Chemical Reactions

The conventional view of happiness and peace of mind often emphasizes the role of neurotransmitters such as serotonin and dopamine. While these chemicals undoubtedly play a crucial role in our emotional experiences, the mathematical perspective presented here does not negate their significance. Instead, it offers a complementary viewpoint. It suggests that the biochemical processes within our brains may be influenced by deeper mathematical patterns, and that true happiness and peace of mind emerge when these patterns are in perfect harmony.

8.13.2.2 Why "Emotion or Perception Are Mathematical" is Debatable?

In exploring the concept that happiness and peace of mind are mathematical formulations rather than mere chemical reactions, we enter a realm of thought that challenges traditional scientific paradigms. The surface of Purusha, adorned with prime numbers in a complete series, offers a metaphorical pathway to emotional equilibrium through geometric stability. This perspective invites us to consider the possibility that the pursuit of happiness and inner peace may be intrinsically linked to the elegant symmetries of mathematics, opening up new avenues for understanding, and achieving these coveted states of being. While this theory is undoubtedly unconventional because of non-physical attribution of mathematics, it reminds us of the boundless potential of human consciousness and the beauty of mathematics in unraveling the mysteries of the mind.

References

1. Aftanas LI et al (2004) Neurophysiological correlates of induced discrete emotions in humans: an individual analysis. Rossiiskii Fiziologicheskii Zhurnal Imeni I.M. Sechenova 90(12):1457–1471
2. Albuquerque L, Turrini R (2022) Effects of flower essences on nursing students' stress symptoms: a randomized clinical trial. Revista Da Escola De Enfermagem Da Usp, 56. https://doi.org/10.1590/1980-220x-reeusp-2021-0307
3. Aron A, Fisher H, Mashek D, Strong G, Li H, Brown L (2005) Reward, motivation, and emotion systems associated with early-stage intense romantic love. J Neurophysiol 94(1):327–337. https://doi.org/10.1152/jn.00838.2004
4. Adolphs R (2017) The amygdala and emotion: a view through fear. In: The amygdala. Springer, Cham, pp 289–310
5. Abramson L, Marom I, Petranker R, Aviezer H (2017) Is fear in your head? A comparison of instructed and real-life expressions of emotion in the face and body. Emotion 17(3):557–565. https://doi.org/10.1037/emo0000252
6. Agrawal L, Sahu S, Ghosh S, Shiga T, Fujita D, Bandyopadhyay A (2016) Inventing atomic resolution scanning dielectric microscopy to see a single protein complex operation live at resonance in a neuron without touching or adulterating the cell. J Integr Neurosci 15(04):435–462
7. Agrawal L, Chhajed R, Ghosh S, Ghosh B, Ray K, Sahu S, Fujita D (2018) Fractal information theory (FIT)-derived geometric musical language (GML) for brain-inspired hypercomputing. Soft Comput Theor Appl 548:343–372
8. Bruce N, Manber R, Shapiro S, Constantino M (2010) Psychotherapist mindfulness and the psychotherapy process. Psychotherapy 47(1):83–97. https://doi.org/10.1037/a0018842
9. Bluth K, Campo R, Futch W, Gaylord S (2016) Age and gender differences in the associations of self-compassion and emotional well-being in a large adolescent sample. J Youth Adolesc 46(4):840–853. https://doi.org/10.1007/s10964-016-0567-2

10. Bandyopadhyay A, Miki K, Wakayama Y (2006) Writing and erasing information in multilevel logic systems of a single molecule using scanning tunneling microscope. Appl Phys Lett 89(24)
11. Baumeister RF, Vohs KD, Funder DC (2007) Psychology as the science of self-reports and finger movements: whatever happened to actual behavior? Perspect Psychol Sci 2(4):396–403
12. Bandyopadhyay A (2020) Nanobrain: the making of an artificial brain from a time crystal. Boca Raton CRC Press, Taylor and Francis. ISBN: 9781439875490. https://doi.org/10.1201/9780429107771
13. Binder L, Frankfurter A, Rebhun L (1985) The distribution of tau in the mammalian central nervous system. J Cell Biol 101(4):1371–1378. https://doi.org/10.1083/jcb.101.4.1371
14. Bartels A, Zeki S (2004) The neural correlates of maternal and romantic love. Neuroimage 21(3):1155–1166. https://doi.org/10.1016/j.neuroimage.2003.11.003
15. Cacioppo S, Bianchi-Demicheli F, Frum C, Pfaus J, Lewis J (2012) The common neural bases between sexual desire and love: a multilevel kernel density fMRI analysis. J Sex Med 9(4):1048–1054. https://doi.org/10.1111/j.1743-6109.2012.02651.x
16. Cheng Y et al (2010) Love hurts: an fMRI study. NeuroImage 51:923–929
17. Chamberlain R, Brunswick N, Siev J, McManus I (2018) Meta-analytic findings reveal lower means but higher variances in visuospatial ability in dyslexia. Br J Psychol 109(4):897–916. https://doi.org/10.1111/bjop.12321
18. Cowan C, Bossing T, Page A, Shepherd D, Mudher A (2010) Soluble hyper-phosphorylated tau causes microtubule breakdown and functionally compromises normal tau in vivo. Acta Neuropathol 120(5):593–604. https://doi.org/10.1007/s00401-010-0716-8
19. Damasio AR, Carvalho GB (2013) The nature of feelings: evolutionary and neurobiological origins. Nat Rev Neurosci 14:143–152
20. Dunbar R (2012) Bridging the bonding gap: the transition from primates to humans. Philos Trans R Soc B Biol Sci 367(1597):1837–1846. https://doi.org/10.1098/rstb.2011.0217
21. Decety J, Jackson PL (2004) The functional architecture of human empathy. Behav Cogn Neurosci Rev 3(2):71–100
22. Davis MH (1983) Measuring individual differences in empathy: evidence for a multidimensional approach. J Pers Soc Psychol 44(1):113–126
23. Eisenberger NI (2012) The pain of social disconnection: examining the shared neural underpinnings of physical and social pain. Nat Rev Neurosci 13(6):421–434
24. Egloff B et al (2006) Spontaneous emotion regulation during evaluated speaking tasks: associations with negative affect, anxiety expression, memory, and physiological responding. Emotion 6(3):356–366
25. Ferrari P, Barbot A, Bianchi B, Ferri A, Garofalo G, Bruno N, Stefani E (2017) A proposal for new neurorehabilitative intervention on moebius syndrome patients after 'smile surgery'. Proof of concept based on mirror neuron system properties and hand-mouth synergistic activity. Neurosci Biobehav Rev 76:111–122. https://doi.org/10.1016/j.neubiorev.2017.01.050
26. Fernández M, Ramos-Pichardo J, Ibáñez-Masero O, Carmona-Rega M, Sánchez-Ruiz M, Ortega-Galán Á (2021) Professional quality of life, self-compassion, resilience, and empathy in healthcare professionals during covid-19 crisis in spain. Res Nurs Health 44(4):620–632. https://doi.org/10.1002/nur.22158
27. Feldman R (2015) The adaptive human parental brain: implications for children's social development. Trends Neurosci 38(6):387–399
28. Fisher HE, Aron A, Brown LL (2006) Romantic love: a mammalian brain system for mate choice. Philos Trans R Soc B Biol Sci 361(1476):2173–2186
29. Fisher H, Aron A, Brown L (2005) Romantic love: an fMRI study of a neural mechanism for mate choice. J Comp Neurol 493(1):58–62. https://doi.org/10.1002/cne.20772
30. Ghosh S, Fujita D, Bandyopadhyay A (2015) An organic jelly made fractal logic gate with an infinite truth table. Sci Rep 5(1):11265
31. Ghosh S, Sahu S, Agrawal L, Shiga T, Bandyopadhyay A (2016) Inventing a co-axial atomic resolution patch clamp to study a single resonating protein complex and ultra-low power communication deep inside a living neuron cell. J Integr Neurosci 15(04):403–433

32. Hazan C, Shaver P (1987) Romantic love conceptualized as an attachment process. J Pers Soc Psychol 52(3):511–524. https://doi.org/10.1037/0022-3514.52.3.511
33. Hrdy SB (2009) Mothers and others: the evolutionary origins of mutual understanding. Harvard University Press
34. Insel T, Young L (2001) The neurobiology of attachment. Nat Rev Neurosci 2(2):129–136. https://doi.org/10.1038/35053579
35. Iqbal K, Liu F, Chen G, Grundke-Iqbal I (2010) Tau in alzheimer disease and related tauopathies. Curr Alzheimer Res 7(8):656–664. https://doi.org/10.2174/156720510793611592
36. Jackson P, Brunet É, Meltzoff A, Decety J (2006) Empathy examined through the neural mechanisms involved in imagining how i feel versus how you feel pain. Neuropsychologia 44(5):752–761. https://doi.org/10.1016/j.neuropsychologia.2005.07.015
37. Jackson P, Meltzoff A, Decety J (2005) How do we perceive the pain of others? A window into the neural processes involved in empathy. Neuroimage 24(3):771–779. https://doi.org/10.1016/j.neuroimage.2004.09.006
38. Joye Y, Steg L, Ünal A, Pals R (2016) When complex is easy on the mind: internal repetition of visual information in complex objects is a source of perceptual fluency. J Exp Psychol Hum Percept Perform 42(1):103–114. https://doi.org/10.1037/xhp0000105
39. Loth M, Donaldson Z (2020) Oxytocin, dopamine, and opioid interactions underlying pair bonding: highlighting a potential role for microglia. Endocrinology 162(2). https://doi.org/10.1210/endocr/bqaa223
40. Larsen R, Ketelaar T (1991) Personality and susceptibility to positive and negative emotional states. J Pers Soc Psychol 61(1):132–140. https://doi.org/10.1037/0022-3514.61.1.132
41. Lotto L, Manfrinati A, Sarlo M (2013) A new set of moral dilemmas: norms for moral acceptability, decision times, and emotional salience. J Behav Decis Mak 27(1):57–65. https://doi.org/10.1002/bdm.1782
42. Levenson RW, Ekman P, Friesen WV (1990) Voluntary facial action generates emotion-specific autonomic nervous system activity. Psychophysiology 27(4):363–384
43. Lavin J, Lavin C, Bai X, Mastropaolo S, Feldman D (2020) Determining the effect of group flower arranging sessions on caregiver self-efficacy and stress levels in an in-patient hospice. Omega J Death Dying 84(2):491–511. https://doi.org/10.1177/0030222819900441
44. Leknes S, Tracey I (2008) A common neurobiology for pain and pleasure. Nat Rev Neurosci 9:314–320
45. Lambert NM, Fincham FD, Stillman TF (2012) Gratitude and depressive symptoms: the role of positive reframing and positive emotion. Cogn Emot 26(4):615–633
46. Lau B, Chan C, Ng S (2020) Self-compassion buffers the adverse mental health impacts of covid-19-related threats: results from a cross-sectional survey at the first peak of Hong Kong's outbreak. Front Psychiatry 11. https://doi.org/10.3389/fpsyt.2020.585270
47. MacBeth A, Gumley A (2012) Exploring compassion: a meta-analysis of the association between self-compassion and psychopathology. Clin Psychol Rev 32(6):545–552. https://doi.org/10.1016/j.cpr.2012.06.003
48. Mee S et al (2006) Psychological pain: a review of evidence. J Psychiatr Res 40(8):680–690
49. Mazzocconi C, Tian Y, Ginzburg J (2022) What's your laughter doing there? A taxonomy of the pragmatic functions of laughter. IEEE Trans Affect Comput 13(3):1302–1321. https://doi.org/10.1109/taffc.2020.2994533
50. McGettigan C, Walsh E, Jessop R, Agnew Z, Sauter D, Warren J, Scott S (2013) Individual differences in laughter perception reveal roles for mentalizing and sensorimotor systems in the evaluation of emotional authenticity. Cereb Cortex 25(1):246–257. https://doi.org/10.1093/cercor/bht227
51. Meerwijk EL, Weiss, SJ (2011) Toward a Unifying Definition of Psychological Pain. J Loss Trauma 16:402–412
52. Marchisella F, Coffey E, Hollós P (2016) Microtubule and microtubule associated protein anomalies in psychiatric disease. Cytoskeleton 73(10):596–611. https://doi.org/10.1002/cm.21300
53. Ochsner KN, Gross JJ (2005) The cognitive control of emotion. Trends Cogn Sci 9(5):242–249

54. Ortigue S, Bianchi-Demicheli F, Patel N, Frum C, Lewis J (2010) Neuroimaging of love: fMRI meta-analysis evidence toward new perspectives in sexual medicine. J Sex Med 7(11):3541–3552. https://doi.org/10.1111/j.1743-6109.2010.01999.x
55. O'Nions E, Lima C, Scott S, Roberts R, McCrory E, Viding E (2017) Reduced laughter contagion in boys at risk for psychopathy. Curr Biol 27(19):3049-3055.e4. https://doi.org/10.1016/j.cub.2017.08.062
56. Pressman S, Cohen S (2005) Does positive affect influence health? Psychol Bull 131(6):925–971. https://doi.org/10.1037/0033-2909.131.6.925
57. Pradhan B, Bandyopadhyay A, Pal AJ (2004) Molecular level control of donor/acceptor heterostructures in organic photovoltaic devices. Appl Phys Lett 85(4):663–665
58. Palagi E, Caruana F, Waal F (2022) The naturalistic approach to laughter in humans and other animals: towards a unified theory. Philos Trans R Soc B Biol Sci 377(1863). https://doi.org/10.1098/rstb.2021.0175
59. Purdie F, Morley SJ (2016) Compassion and chronic pain. Pain 157:2625–2627
60. Ryan R, Deci E (2001) On happiness and human potentials: a review of research on hedonic and eudaimonic well-being. Annu Rev Psychol 52(1):141–166. https://doi.org/10.1146/annurev.psych.52.1.141
61. Reddy S, Sonker D, Singh P, Saxena K, Singh S, Chhajed R, Tiwari S, Karthik KV, Ghosh S, Ray K, Bandyopadhyay A (2018) A brain-like computer made of time crystal: could a metric of prime alone replace a user and alleviate programming forever? Soft Comput Appl 761:1–43
62. Saarimäki H (2021) Naturalistic stimuli in affective neuroimaging: a review. Front Hum Neurosci 15. https://doi.org/10.3389/fnhum.2021.675068
63. Sahu S, Ghosh S, Hirata K, Fujita D, Bandyopadhyay A (2013) Multi-level memory-switching properties of a single brain microtubule. Applied Physics Letters 102, 123701
64. Schubert E (2013) Emotion felt by the listener and expressed by the music: literature review and theoretical perspectives. Front Psychol 4. https://doi.org/10.3389/fpsyg.2013.00837
65. Sergerie K, Chochol C, Armony J (2008) The role of the amygdala in emotional processing: a quantitative meta-analysis of functional neuroimaging studies. Neurosci Biobehav Rev 32(4):811–830. https://doi.org/10.1016/j.neubiorev.2007.12.002
66. Saxena K, Singh P, Sahoo P, Sahu S, Ghosh S, Ray K, Fujita D (2020) Fractal, scale free electromagnetic resonance of a single brain extracted microtubule nanowire, a single tubulin protein and a single neuron. Fractal Fractional 4(2):11
67. Scott S, Lavan N, Chen S, McGettigan C (2014) The social life of laughter. Trends Cogn Sci 18(12):618–620. https://doi.org/10.1016/j.tics.2014.09.002
68. Suharyanti E, Handayani D, Saputro A, Nugroho A (2021) Attachment theory: how marvelous mother guided an adoption child. Soc Sci Stud 1(1):032–044. https://doi.org/10.47153/sss11.1782021
69. Sacco R (2016) The fibonacci life-chart method (FLCM) as a foundation for Carl Jung's theory of synchronicity. J Anal Psychol 61(2):203–222. https://doi.org/10.1111/1468-5922.12204
70. Swain JE, Kim P, Ho SS (2011) Neuroendocrinology of parental response to baby-cry. J Neuroendocrinol 23(11):1036–1041
71. Singh P, Sahoo P, Saxena K, Manna JS, Ray K, Ghosh S, Fujita D (2021) Cytoskeletal filaments deep inside a neuron are not silent: they regulate the precise timing of nerve spikes using a pair of vortices. Symmetry 13(5):821
72. Singh P, Saxena K, Singhania A, Sahoo P, Ghosh S, Chhajed R, Fujita D (2020) A self-operating time crystal model of the human brain: can we replace entire brain hardware with a 3D fractal architecture of clocks alone? Information 11(5):238
73. Saxena K, Singh P, Sahoo P, Ghosh S, Krishnananda D, Ray K, Fujita D, Bandyopadhyay A (2022) All basics that are wrong with the current concept of time crystal: learning from the polyatomic time crystals of protein, microtubule, and neuron. Proc Trends Electron Health Inf 376:243–254
74. Singh P, Ray K, Bandyopadhyay A (2022) Biological antenna to the humanoid bot: electromagnetic resonances in biomaterials. Studies in rhythm engineering (SRE), Springer Singapore. https://doi.org/10.1007/978-981-16-9677-0

75. Singh P, Sahoo P, Saxena K, Ghosh S, Sahu S, Ray K, Fujita D, Bandyopadhyay A (2021) Quaternion, octonion to dodecanion manifold: stereographic projections from infinity lead to a self-operating mathematical universe. In: Proceedings of international conference on trends in computational and cognitive engineering, vol 1169, pp 55–77
76. Singh P, Sahoo P, Saxena K, Ghosh S, Sahu S, Ray K, Fujita D, Bandyopadhyay A (2021) A space-time-topology-prime, stTS metric for a self-operating mathematical universe uses dodecanion geometric algebra of 2–20 D complex vectors. Proc Int Conf Data Sci Appl 148:1–31
77. Shih H, Kuo M, Wu C, Chao Y, Huang H, Huang C (2022) The neurobiological basis of love: a meta-analysis of human functional neuroimaging studies of maternal and passionate love. Brain Sci 12(7):830. https://doi.org/10.3390/brainsci12070830
78. Skov M, Nadal M (2020) The nature of beauty: behavior, cognition, and neurobiology. Ann N Y Acad Sci 1488(1):44–55. https://doi.org/10.1111/nyas.14524
79. Tamam S, Ahmad A (2017) Love as a modulator of pain. Malays J Med Sci 24(3):5–14. https://doi.org/10.21315/mjms2017.24.3.2
80. Trappe H, Breker I (2018) Differential effects of Bach's orchestral suite no. 3 on blood pressure and heart rate—a prospective controlled study. Music Med 10(1):7. https://doi.org/10.47513/mmd.v10i1.553
81. Volokhov RN, Demaree HA (2010) Spontaneous emotion regulation to positive and negative stimuli. Brain Cogn 73:1–6
82. Watanuki S, Akama H (2020) Neural substrates of brand love: an activation likelihood estimation meta-analysis of functional neuroimaging studies. Front Neurosci 14. https://doi.org/10.3389/fnins.2020.534671
83. Wiech K (2016) Deconstructing the sensation of pain: the influence of cognitive processes on pain perception. Science 354(6312):584–587. https://doi.org/10.1126/science.aaf8934
84. Wróbel M, Imbir K (2019) Broadening the perspective on emotional contagion and emotional mimicry: the correction hypothesis. Perspect Psychol Sci 14(3):437–451. https://doi.org/10.1177/1745691618808523
85. Yamaguchi H, Maki Y, Yamagami T (2010) Overview of non-pharmacological intervention for dementia and principles of brain-activating rehabilitation. Psychogeriatrics 10(4):206–213. https://doi.org/10.1111/j.1479-8301.2010.00323.x

Chapter 9
Social Awareness Against Sexual Harassment Triggering Excitatory Cognition Could Be Negated Using a Pure Awe Experience

9.1 Introduction

9.1.1 Social Awareness Against Sexual Harassment

9.1.1.1 Reevaluating the Efficacy of Social Awareness Campaigns Against Sexual Harassment

While social awareness campaigns against sexual harassment have been instrumental in drawing attention to this pervasive issue, there is growing criticism that many of these efforts may be misdirected, ultimately failing to yield the desired results. In this chapter, we critically examine the psychological aspects of such campaigns in light of this criticism, aiming to uncover potential shortcomings and propose ways to enhance their impact. Banyard [11] evaluates the effectiveness of a bystander education program in preventing sexual violence. It highlights the importance of providing actionable steps and sustained engagement in social campaigns against sexual harassment. Gavey and Schmidt [37] explored the intersection of critical feminist psychology and sexual harassment. It emphasizes the need for a nuanced understanding of the psychological factors that contribute to sexual harassment and the importance of challenging deeply ingrained attitudes.

The Complexity of Effecting Change

Sexual harassment is a deeply entrenched issue with complex psychological underpinnings. Critics argue that many social awareness campaigns oversimplify the problem, attempting to address the symptoms rather than the root causes. To better understand this criticism, we must acknowledge the multifaceted nature of sexual harassment and the psychological factors that contribute to it.

T. Dutta and A. Bandyopadhyay, *Emotion, Cognition and Silent Communication: Unsolved Mysteries*, Studies in Rhythm Engineering,
https://doi.org/10.1007/978-981-99-9334-5_9

The Limits of Empathy

While empathy is often touted as a catalyst for change, critics argue that it may not be sufficient on its own. Empathetic messages, while valuable, may not translate into concrete actions. The gap between understanding the issue and taking meaningful steps to combat it is where many campaigns fall short. Katz [46] discusses the bystander approach to sexual violence prevention, which encourages individuals to intervene and take action when witnessing instances of harassment. It highlights the limitations of empathy alone and emphasizes the need for concrete actions in social campaigns against sexual harassment. Psychologically, empathy can be fleeting. People may feel deeply moved by a survivor's story during a campaign but revert to their previous behaviors once the campaign ends. This suggests that campaigns must do more than evoke empathy, they must provide actionable steps and sustained engagement to effect change.

The Need for Nuanced Education

Education and awareness-building are cornerstones of social campaigns against sexual harassment. However, critics argue that the approach taken by many campaigns lacks nuance. The oversimplification of complex issues can lead to superficial understanding rather than meaningful change. Lonsway and Fitzgerald [56] examines the prevalence of rape myths and their impact on attitudes toward sexual harassment. It underscores the importance of comprehensive education that challenges ingrained beliefs and addresses societal structures and power dynamics. Potter and Banyard [68] discusses the challenges and opportunities for bystander intervention in preventing sexual violence. It emphasizes the need for sustained engagement and provides insights into effective strategies for social campaigns against sexual harassment.

Psychological research supports the idea that comprehensive education is key. Campaigns should go beyond merely defining harassment and should address the intricate societal structures, power dynamics, and deeply ingrained attitudes that perpetuate it. A more nuanced approach that encourages critical thinking and challenges ingrained beliefs is essential for lasting change.

Researchers have highlighted the limitations of empathy alone, the need for nuanced education, and the importance of sustained engagement and actionable steps in effecting meaningful change. By critically examining these aspects, we can uncover potential shortcomings and propose ways to enhance the impact of social awareness campaigns against sexual harassment.

The Potential of Measuring Impact: One significant criticism of social awareness campaigns is the difficulty of measuring their impact accurately. Critics argue that many campaigns lack clear metrics to assess their effectiveness. Without proper evaluation, it becomes challenging to determine whether these efforts are making a tangible difference [57]. Psychologically, measuring impact is crucial for gauging the success of a campaign. It allows for adjustments based on data and evidence, ensuring that the campaign remains relevant and effective. Transparency in reporting outcomes and a commitment to continuous improvement are essential components of any impactful campaign [57].

The Role of Survivor-Centric Storytelling: Despite criticism, personal narratives and survivor testimonials remain powerful tools in awareness campaigns. However, critics argue that these stories should be more than emotional triggers; they should be survivor-centric, emphasizing the agency, and resilience of survivors [85]. Psychologically, survivor-centric storytelling can help counter the victimization narrative. It empowers survivors to take control of their narratives and encourages a sense of solidarity and strength. Such narratives can inspire others to join the movement and take meaningful action [85].

Criticism of social awareness campaigns against sexual harassment reminds us of the need for continuous evaluation and improvement. While these campaigns have undoubtedly raised awareness, addressing the root causes of harassment requires a more nuanced and sustained approach. By recognizing the limitations of empathy, embracing comprehensive education, measuring impact, and adopting survivor-centric storytelling, we can enhance the effectiveness of these campaigns and move closer to creating a world free from the scourge of sexual harassment.

9.1.2 Unraveling the Classifications of Mental Abuse: Social Engineering and the Role of Materialistic Drive

Mental abuse, a silent but profoundly damaging form of psychological maltreatment, leaves deep scars on the minds of victims. In the field of psychology, understanding the classifications of mental abuse and their intricate connection to our social engineering, particularly in the context of the materialistic drive, is vital. This section delves into the various forms of mental abuse and explores how societal influences, driven by materialism, contribute to their perpetuation.

9.1.2.1 The Classifications of Mental Abuse

1. **Emotional Abuse**: Emotional abuse is a form of abuse that involves tactics aimed at undermining an individual's self-esteem, confidence, and emotional well-being [48]. It often manifests as constant criticism, belittling, or humiliation, and emotional abusers manipulate emotions to instill fear, guilt, and shame in their victims [48]. This type of abuse can have severe psychological effects on the victim, leading to a profound sense of confusion and self-doubt [82].
2. **Verbal Abuse**: Verbal abuse is another form of abuse characterized by the use of harsh, derogatory language and threats to control or intimidate [48]. It includes name-calling, insults, yelling, and frequent humiliation [48]. Verbal abuse erodes

the victim's self-worth and can lead to severe emotional trauma [48]. It is important to note that verbal abuse can have long-lasting effects on the victim's mental health and well-being [47].

3. **Gaslighting**: Gaslighting is a manipulative tactic used by abusers to invalidate the victim's reality, causing them to doubt their perceptions and sanity [48]. This form of abuse is insidious and contributes to a profound sense of confusion and self-doubt [48]. Gaslighting can have severe psychological effects on the victim, leading to a loss of trust in oneself and others [27].
4. **Isolation**: Isolation is a tactic employed by abusers to control their victims by limiting their access to friends, family, or support systems [48]. By isolating the victim, the abuser intensifies their dependence on them, making it challenging for the victim to seek help or escape the abusive situation [48]. It is important to recognize the signs of isolation and provide support to individuals who may be experiencing this form of abuse [59].
5. **Financial Abuse**: Financial abuse is a form of abuse that involves controlling or exploiting an individual's financial resources [48]. Abusers may restrict access to money, manipulate financial decisions, or use finances as a weapon of control [48]. Financial abuse can trap victims in abusive relationships, as they may lack the financial resources to leave [82]. It is crucial to raise awareness about financial abuse and provide resources for victims to regain control of their finances [76].
6. **Cyberbullying**: Cyberbullying has emerged as a form of mental abuse in the digital age [48]. It involves harassment, threats, or the spread of harmful content through online platforms, leading to emotional distress and psychological harm [48]. Cyberbullying can have severe consequences for the mental health and well-being of the victims [42]. It is important to address cyberbullying and promote safe online environments for individuals of all ages [63].

Thus, emotional abuse, verbal abuse, gaslighting, isolation, financial abuse, and cyberbullying are all forms of abuse that can have severe psychological effects on the victims. It is crucial to raise awareness about these forms of abuse, provide support to victims, and promote healthy relationships and environments.

9.1.2.2 Social Engineering and Mental Abuse

Social engineering refers to the manipulation of individuals and societies through psychological tactics. It plays a significant role in perpetuating mental abuse. Society's values, norms, and expectations can inadvertently facilitate abusive behaviors. There are several factors that contribute to this phenomenon. Here's how:

1. **Gender Roles and Expectations**: One factor is the influence of gender roles and expectations. Societal norms and expectations around gender roles can contribute to mental abuse. Stereotypes of masculinity and femininity may lead to the assertion of power and control in relationships, reinforcing abusive dynamics [21]. These social norms are characterized by what is defined as chivalry, which refers to the notion of societal norms that endorse the protection of women from harm

by men, discouraging the attacking or abuse of women [84]. This can create an environment where abusive behaviors are normalized and victims may feel trapped in abusive relationships.

2. **Materialistic Drive**: Another factor is the materialistic drive in society. Materialism, the pursuit of material possessions and wealth, can exacerbate mental abuse. Abusers may exploit their victims' desires for financial security, making it difficult for victims to break free from abusive relationships [83]. Economic abuse, such as preventing or limiting work or school hours, stealing income, or dominating family finances, can further perpetuate the cycle of abuse [83]. The fear of economic instability and financial dependence can make it challenging for victims to leave abusive situations.
3. **Stigmatization and Silence**: Stigmatization and silence also contribute to the perpetuation of mental abuse. Societal stigmatization of mental health issues and the normalization of abusive behaviors can silence victims and deter them from seeking help [49]. The fear of judgment and isolation often perpetuates the cycle of abuse [58].
4. **Media and Entertainment**: Media depictions of unhealthy relationships and abusive behaviors can desensitize individuals to mental abuse [74]. Exposure to these portrayals can normalize such behaviors and hinder recognition of abusive patterns [74].

Social engineering and the manipulation of individuals and societies through psychological tactics play a significant role in perpetuating mental abuse. Factors such as gender roles and expectations, materialistic drive, stigmatization and silence, and media and entertainment contribute to the normalization and perpetuation of abusive behaviors. It is crucial to address these societal factors and promote awareness, education, and support for victims of mental abuse.

9.1.2.3 The Role of Materialistic Drive in Exacerbating Mental Abuse

The relationship between materialistic drive and mental abuse within relationships, in today's time is strongly regulated by what we consume online [14]. Drawing on research from various disciplines, including psychology and sociology, this article examines the impact of financial control, objectification, and the pursuit of material wealth and status on abusive dynamics. By synthesizing findings from relevant studies, this article aims to provide a comprehensive understanding of how materialistic desires can contribute to mental abuse in relationships, we need a tool to measure empathy [96]. Materialism, driven by the pursuit of wealth and possessions, has been identified as a potential factor that can exacerbate mental abuse within relationships. This article aims to shed light on the various ways in which materialistic drive can contribute to abusive dynamics. By examining the relationship between financial control, objectification, and the pursuit of material wealth and status, we can gain insights into the mechanisms through which materialism can contribute to mental abuse.

Financial Control: Research by Papp et al. [67] highlights the significance of financial conflicts in relationships, the psychological consequences of money is supreme [97]. Marital conflicts related to money were found to be more pervasive, problematic, and recurrent compared to non-money issues. This suggests that financial control can be a powerful tool for abusers to maintain power and control over their victims. Adams and Sullivan [5] further explore the relationship between financial abuse and mental health outcomes in low-income survivors of intimate partner violence. They found that financial abuse was associated with increased psychological distress and decreased mental well-being among survivors.

Objectification: The objectification of individuals within relationships is another way in which materialistic drive can contribute to mental abuse. Gervais et al. [33] discuss the consequences of the objectifying gaze, highlighting how women are often viewed as objects rather than equal partners. This dehumanization can lead to emotional abuse and manipulation. Objectification theory, as proposed by [30], suggests that women are more likely to experience objectification due to societal norms and expectations.

Competition and Status: The constant pursuit of material wealth and status can create a competitive and power-driven atmosphere within relationships. Dittmar et al. [26] conducted a meta-analysis examining the relationship between materialism and personal well-being. They found that higher materialism was associated with personal discrepancies, feelings of insecurity, and poor psychological need satisfaction, all of which have negative implications for well-being. This competitive mindset can contribute to abusive dynamics, as individuals prioritize their own needs and desires over the well-being of their partners. Materialistic drive can play a significant role in exacerbating mental abuse within relationships. Financial control, objectification, and the pursuit of material wealth and status are three key mechanisms through which materialism can contribute to abusive dynamics. Recognizing and addressing these factors is crucial in preventing and addressing mental abuse within relationships. Future research should focus on developing interventions and strategies to mitigate the negative impact of materialistic desires on relationship dynamics.

Understanding the classifications of mental abuse and their connection to social engineering and materialistic drive is essential for psychology professionals. It allows for more nuanced approaches to therapy and support for victims, as well as interventions aimed at changing societal norms and expectations that perpetuate abuse. By addressing the root causes of mental abuse, we can work toward creating a society that values emotional well-being and empathy over material possessions and control.

9.2 The Profound Power of Pure Awe: Definition, Classification, Importance, and Far-Reaching Impact

In the vast spectrum of human emotions, there exists a unique and transcendent experience known as "pure awe." This emotion goes beyond simple positive feelings, spiritual bliss, or peace of mind. In this section, we delve into the world of pure awe, exploring its definition, classification, importance, and its profound impact on the human mind, while highlighting the distinctions between awe and other cognitive traits [104].

The Profound Impact of Pure Awe: Exploring its Definition, Importance, and Distinctions from Other Cognitive Traits. This section delves into the unique and transcendent experience of pure awe, going beyond simple positive feelings and exploring its profound impact on the human mind. By examining the definition, classification, and importance of awe, as well as its distinctions from other cognitive traits, we gain a deeper understanding of this emotion and its significance in human experiences. In the vast spectrum of human emotions, there exists a distinct and transformative experience known as "pure awe." This emotion goes beyond simple positive feelings, spiritual bliss, or peace of mind. In this discussion, we explore the world of pure awe, examining its definition, classification, and the profound impact it has on the human mind. Additionally, we highlight the distinctions between awe and other cognitive traits, shedding light on its unique nature.

Defining Pure Awe

Pure awe is an overwhelming emotional response to something awe-inspiring or vast, often accompanied by feelings of reverence, wonder, and astonishment. It is characterized by a sense of smallness in the face of something grand and beyond one's comprehension. While positive emotions like joy or contentment bring happiness, awe brings a sense of reverence, making it distinct from other positive feelings.

Definition and Classification of Pure Awe:

Pure awe can be defined as an overwhelming sense of wonder, reverence, and astonishment in the face of something vast, extraordinary, or beyond one's usual frame of reference [66]. It is often triggered by encounters with nature, art, or extraordinary events. According to Piff et al. [66], awe is distinguishable from other positive emotions such as joy, love, amusement, and contentment. It stands out as a distinct emotional experience that elicits unique subjective experiences and self-reports.

Classification of Awe

Awe experiences can be broadly classified into two categories:

1. **Positive Awe**: Positive awe is often associated with feelings of inspiration, upliftment, and positivity. This form of awe arises from encounters with natural wonders like a breathtaking sunset, the grandeur of a mountain range, or the vastness of the night sky. Positive awe is linked to increased well-being and life satisfaction.

2. **Challenging Awe**: Challenging awe, on the other hand, emerges from encounters with experiences that may initially evoke fear or discomfort, such as witnessing a natural disaster or confronting the enormity of the universe. This type of awe can be more complex, as it may provoke existential questions and philosophical reflections.

The Importance of Pure Awe Pure awe holds significant importance in human experiences. Research has shown that experiencing awe can have profound effects on individuals' well-being, cognition, and behavior. Rudd et al. [73] found that awe expands people's perception of time, alters decision-making, and enhances well-being. It can lead individuals to desire to spend time helping others and prioritize experiential goods over material ones [73]. Awe has also been linked to increased prosocial behavior, generosity, and compassion [88]. It fosters a sense of interconnectedness and promotes concern for others and the world around them [88].

Pure awe holds immense importance for our psychological and emotional well-being:

1. **Broadened Perspective**: Awe expands our mental horizons, prompting us to contemplate the vastness of the universe or the intricacies of life. This broadened perspective can foster personal growth, open-mindedness, and creativity.
2. **Increased Prosocial Behavior**: Studies suggest that experiencing awe can lead to increased prosocial behavior, such as altruism and empathy. Awe inspires a sense of interconnectedness and a desire to help others.
3. **Stress Reduction**: Awe has been found to reduce stress and anxiety. The profound feelings of awe can momentarily overshadow worries and promote relaxation.
4. **Enhanced Cognitive Functioning**: Awe can enhance cognitive functioning by improving attention, memory, and problem-solving abilities. This heightened mental state can lead to increased creativity and innovation.

Far-Reaching Impact of Awe

The impact of pure awe extends far beyond individual well-being:

1. **Environmental Conservation**: Awe-inspiring encounters with the natural world can cultivate a greater appreciation for the environment and motivate efforts for conservation.
2. **Fostering Cultural and Spiritual Connections**: Awe often plays a central role in religious and spiritual experiences, fostering a deep sense of connection with the divine or the universe.
3. **Inspiring Scientific Discovery**: Scientific breakthroughs have often been inspired by moments of awe, leading to a deeper understanding of the natural world.

Distinguishing Pure Awe from Other Cognitive Traits:

While awe shares some similarities with other cognitive traits, it possesses distinct characteristics that set it apart. Awe differs from happiness or joy in that it is often accompanied by a sense of smallness or insignificance in the face of something grand [43]. It also differs from fear or terror, as it is typically associated with positive emotions rather than negative ones [32]. Awe is distinct from spiritual experiences, as it can be triggered by secular encounters with nature or art [19]. Furthermore, awe is not solely focused on the self but expands individuals' perspective and promotes prosocial behavior [20].

Pure awe differs from other cognitive traits and emotions in several ways:

1. **Positive Feelings**: While positive feelings like joy or contentment are associated with happiness and satisfaction, pure awe is characterized by a sense of wonder and reverence.
2. **Spiritual Bliss**: Spiritual bliss often involves a profound sense of inner peace and connection with a higher power, whereas awe is directed outward, often triggered by external stimuli.
3. **Peace of Mind**: Peace of mind is a state of inner calm and tranquility, whereas awe is an intense emotional response to external stimuli that may not always lead to a peaceful state.
4. **Other Cognitive Traits**: Awe is distinct from traits like curiosity or interest, as it is a transient emotional state rather than a stable personality characteristic.

Pure awe is a powerful and transformative emotion that takes us beyond the realm of ordinary experience. Pure awe is a unique and transcendent emotion that goes beyond simple positive feelings. Its overwhelming sense of wonder and astonishment has a profound impact on the human mind, influencing well-being, cognition, and behavior. It broadens our horizons, connects us to the world around us, and inspires a sense of wonder. Recognizing the importance and far-reaching impact of awe on the human mind can deepen our understanding of this unique emotion and its potential to enrich our lives and our world. It is through these experiences of pure awe that we tap into the boundless capacity of the human mind to explore, connect, and marvel at the wonders of existence. By understanding the definition, classification, and importance of awe, as well as its distinctions from other cognitive traits, we can appreciate the transformative power of this emotion. Further research is needed to explore the mechanisms underlying awe and its potential applications in various domains of human life.

9.3 Unveiling the Power of Excitatory Cognition: Understanding Its Impact on Positive and Negative States of Mind

Excitatory cognition is a psychological trait that has fascinated psychologists for generations. It involves the activation of cognitive processes that heighten alertness, attention, and engagement [81]. This heightened mental arousal and alertness can be triggered by various factors and is characterized by increased focus, enhanced perception, and heightened awareness [81].

Different Kinds of Excitation Conditions One condition that can trigger excitatory cognition is mindfulness meditation. Research has shown that mindfulness meditation can lead to changes in emotionality, attention, and cognitive measures [81]. Mindfulness is a psychological construct that can be conceptualized as both a state and a dispositional trait [101]. It involves non-evaluative engagement with experience and has been found to broaden awareness and build eudaimonic meaning [31]. Another condition that can trigger excitatory cognition is sleep deprivation. Studies have shown that sleep deprivation can impair cognitive performance and lead to decreased alertness [94]. The prefrontal cortex is particularly vulnerable to sleep deprivation, which can result in cognitive deficits [94]. Genetic factors can also play a role in excitatory cognition. Research has found that certain genes are associated with cognitive traits such as word reading and attention deficit hyperactivity disorder (ADHD) [65]. Psychological traits, such as extraversion and introversion, have also been found to have a genetic component [25].

Excitatory cognition can arise from various conditions, including:

1. **Novelty**: Novelty, or encountering something new or unexpected, is a potent trigger of excitatory cognition. The brain pays extra attention to novel stimuli, which can result in heightened mental alertness and engagement.
2. **Challenge**: Cognitive challenge or problem-solving tasks activate excitatory cognition as the brain mobilizes resources to address the task at hand. The sense of accomplishment after overcoming a challenge often leads to positive emotional states.
3. **Positive Emotions**: Positive emotions like joy, excitement, or love can trigger excitatory cognition. These emotions often enhance cognitive functioning and contribute to a positive state of mind. The relationship between alertness and cognitive control is another area of interest in the study of excitatory cognition. Studies have shown that alertness can interact with cognitive control processes, leading to improved performance on tasks that require attention and executive function [100]. This interaction between alertness and cognitive control is still not fully understood and continues to be a topic of investigation [62].
4. **Threat or Danger**: In response to perceived threats or dangers, the mind enters a state of excitatory cognition to assess the situation, make quick decisions, and prepare for action. However, this state can lead to negative emotions like fear or anxiety.
5. **Social Interaction**: Engaging in social interactions, especially those involving meaningful connections or intellectual discourse, can trigger excitatory cognition. It promotes mental stimulation and can contribute to a positive state of mind.

Impact on Positive State of Mind

Excitatory cognition is closely linked to positive states of mind in several ways:

1. **Enhanced Creativity**: Excitatory cognition often leads to enhanced creative thinking. The heightened alertness and engagement encourage individuals to think outside the box, fostering innovation, and problem-solving.
2. **Positive Engagement**: When individuals are in a state of excitatory cognition, they tend to engage more positively with their environment. This can lead to increased motivation, enthusiasm, and a greater sense of well-being.
3. **Optimism**: The heightened mental arousal associated with excitatory cognition can promote optimism and a positive outlook on life. Individuals are more likely to see challenges as opportunities for growth and learning.

Impact on Negative State of Mind

While excitatory cognition is often associated with positive states of mind, it can also contribute to negative states under certain conditions:

1. **Overwhelm**: Excessive cognitive stimulation or persistent excitatory states can lead to cognitive overload and feelings of overwhelm, potentially resulting in stress or anxiety.
2. **Rumination**: In some cases, excitatory cognition may exacerbate negative thought patterns, such as rumination or obsessive thinking, which can contribute to negative states of mind.
3. **Hyperarousal**: Prolonged excitatory states can lead to hyperarousal, where individuals struggle to relax or experience a constant sense of alertness. This can contribute to sleep disturbances and increased stress.

Excitatory cognition is a classic psychological trait that has a profound impact on our mental states. It can be triggered by various conditions, each with its unique effects on our cognitive functioning and emotional well-being. While it is closely associated with positive states of mind, it is crucial to recognize that it can also contribute to negative states if not properly managed. Understanding the triggers and consequences of excitatory cognition is essential for researchers and practitioners in the field of psychology, as it offers valuable insights into how our minds respond to different stimuli and experiences. The excitatory cognition is characterized by heightened mental arousal and alertness. It can be triggered by various conditions, including mindfulness meditation, sleep deprivation, and genetic factors. The relationship between alertness and cognitive control is also an important aspect of excitatory cognition. Further research is needed to fully understand the mechanisms underlying excitatory cognition and its impact on positive and negative states of mind.

9.4 Unveiling the Intriguing Overlap of Intra-Sensory Perceptions: Synesthesia's Silent Influence on Human Experience

This section provides a comprehensive review of synesthesia, a neurological phenomenon characterized by the blending of sensory experiences. Drawing upon a range of scholarly references, including Cytowic [22], Ward [102], Simner and Hubbard [80], Deroy and Spence [24], and Marks [60], this article explores the different varieties of synesthetic experiences, their underlying mechanisms, and the limitations of synesthetic perception. The findings suggest that while synesthesia is a fascinating and unique phenomenon, not everyone experiences it, and even those who do may exhibit varying degrees of synesthetic perception. Synesthesia is a neurological condition in which individuals experience a blending or merging of sensory experiences. This article aims to provide a comprehensive review of synesthesia, drawing upon a range of scholarly references to explore the different varieties of synesthetic experiences, their underlying mechanisms, and the limitations of synesthetic perception.

Varieties of Synesthetic Experience:

According to Cytowic [22], there are several types of synesthesia, including grapheme-color synesthesia, in which letters or numbers are perceived as having specific colors, sound-color synesthesia, in which sounds are associated with specific colors, and taste-touch synesthesia, in which certain tastes are experienced as tactile sensations. Simner and Hubbard [80] further expand on these varieties, discussing additional forms such as spatial-sequence synesthesia, in which numbers or sequences are perceived as occupying specific spatial locations.

Underlying Mechanisms:

The exact mechanisms underlying synesthesia are still not fully understood. However, research suggests that synesthesia may involve cross-activation or hyperconnectivity between brain regions responsible for processing different sensory modalities [102]. This cross-activation may result in the blending of sensory experiences, leading to synesthetic perceptions.

Limitations of Synesthetic Perception:

While synesthesia is a fascinating phenomenon, not everyone experiences it. Deroy and Spence [24] argue that synesthesia is not a universal trait and that individuals who do not experience synesthesia may still have weak or subtle forms of cross-modal associations. Additionally, Marks [60] highlights the limitations of synesthetic perception, suggesting that it is influenced by individual differences and can vary in intensity and consistency.

Synesthesia and Educational Materials for Abuse Awareness

The role of synesthesia in education materials for abuse awareness is an aspect that psychologists have not extensively explored. However, it is a factor that deserves attention. When individuals read or view educational materials related to abuse, their sensory perceptions may subtly influence their emotional responses and comprehension of the content.

1. **Color Associations**: As mentioned earlier, individuals often associate colors with emotions and sensations. This means that the choice of color schemes in educational materials can evoke specific emotional responses. For instance, a brochure about abuse awareness that predominantly uses warm, comforting colors may elicit a more empathetic and compassionate response from readers.
2. **Auditory Influences**: Sound can also play a role in the perception of abuse-related content. The tone and rhythm of a speaker's voice in a video or presentation can impact how the information is received. A soothing voice may create a more reassuring atmosphere, while a harsh or jarring tone may provoke discomfort or anxiety.
3. **Multisensory Engagement**: Incorporating multisensory elements, such as interactive exercises or simulations, can enhance the effectiveness of abuse awareness materials. These elements can evoke a more visceral understanding of the topic, making it harder for individuals to dismiss or ignore the message.

Synesthesia, although often considered an extraordinary phenomenon, has subtle implications for all individuals, influencing how we perceive and interact with the world around us. Its presence in everyday life suggests that we should consider its influence when designing educational materials, particularly those related to sensitive topics like abuse awareness. By recognizing the subtle interplay between sensory modalities, psychologists, and educators can create more impactful and engaging materials that resonate with a broader audience, fostering greater awareness, and empathy in society. Therefore, synesthesia is a unique neurological phenomenon characterized by the blending of sensory experiences. This article has provided a comprehensive review of the different varieties of synesthetic experiences, their underlying mechanisms, and the limitations of synesthetic perception. While synesthesia is not experienced by everyone, it offers valuable insights into the complex nature of perception and the interconnectedness of sensory modalities. Further research is needed to fully understand the mechanisms and implications of synesthesia.

9.5 Exploring Pure Awe and Awareness for Abuse: Unlikely Allies in Psychological Healing

In the field of psychology, the topics of awe and awareness for abuse are rarely discussed together [3]. They seem to inhabit separate realms of human experience, with one evoking feelings of wonder and inspiration, while the other demands our attention on the dark realities of abuse. However, what if we were to consider the potential synergy between these two seemingly unrelated concepts? Could a pure awe experience hold the key to unlocking a more effective path to awareness for abuse, ultimately benefiting society's mental and emotional well-being? This section dares to explore this intriguing juxtaposition and its uncharted potential.

Pure Awe: Beyond Spiritual Bliss

Pure awe is a profound and often transcendent emotion that arises when we encounter something awe-inspiring or beyond our usual scope of understanding. It goes beyond mere spiritual bliss or the pleasure of the mind, offering an experience that can be described as a form of cognitive medicine for the soul. Pure awe has the power to reshape our perspectives, instill a sense of interconnectedness, and prompt deep introspection. Awe is often associated with positive emotions, such as wonder and inspiration [3]. It has been found that experiencing awe can expand people's perception of time, alter decision-making processes, and enhance overall well-being [73]. Awe has also been linked to increased prosocial behavior, such as generosity and kindness [66].

The Power of Pure Awe:

1. **Expanding Mental Horizons**: Awe experiences have the unique ability to broaden our mental horizons. They encourage us to contemplate the vastness of the universe or the intricacies of life, fostering personal growth, open-mindedness, and creativity.
2. **Fostering Empathy**: Awe can be a catalyst for empathy. It brings about a profound sense of interconnectedness and shared humanity, making us more attuned to the feelings and experiences of others.
3. **Stress Reduction**: Awe has been shown to reduce stress and anxiety. The intense emotions it generates can momentarily overshadow worries and promote relaxation.

Awareness for Abuse: A Sobering Reality

On the other side of the spectrum lies awareness for abuse—a critical facet of psychological well-being and societal progress. Raising awareness about abuse in its various forms, whether physical, emotional, or psychological, is essential to combatting its devastating effects. Awareness campaigns seek to educate, empower, and ultimately, prevent abuse from occurring. On the other hand, awareness for abuse focuses on acknowledging and understanding the dark realities of abuse. It is important to raise awareness about abuse, as it can help prevent further harm and provide support for survivors [29]. Growing up in a family affected by mental health problems or substance abuse can increase the risk of various health problems and negative life outcomes [18].

The Nexus of Awe and Awareness

Interestingly, there is evidence to suggest that awe can play a role in mitigating the negative effects of abuse and fostering healing. A study found that experiencing awe can guard creativity and enhance creative self-efficacy, particularly in the face of perceived abusive supervisory behavior [4]. This suggests that awe may provide individuals with a sense of resilience and empowerment in the face of adversity. What if we were to harness the power of pure awe as a catalyst for awareness campaigns against abuse? Consider the impact of creating educational materials or experiences that elicit pure awe just before introducing awareness for abuse.

1. **Enhanced Engagement**: Awe captures our attention like few other emotions. By invoking this emotion before discussing the difficult topic of abuse, individuals may be more engaged and receptive to the message.
2. **Emotional Resonance**: Awe experiences can create a strong emotional resonance. Integrating this emotional connection into abuse awareness campaigns may foster greater empathy and motivation to take action against abuse.
3. **Broadened Perspective**: Awe experiences inherently encourage a broader perspective. This can lead individuals to see the importance of addressing abuse on a societal level, recognizing its impact on a larger scale.

Moreover, another study found that awe has a mediating role in the relationship between openness to experience, extraversion, and subjective well-being [23]. This implies that individuals who are more open to experiencing awe may have higher levels of well-being, which could potentially contribute to their ability to cope with and heal from abuse. While it may seem unconventional to juxtapose pure awe and awareness for abuse, the potential benefits of their convergence are undeniable. In a world where psychological well-being and societal awareness are paramount, exploring the synergy between these seemingly disparate concepts may hold the key to more effective and empathetic awareness campaigns. The power of pure awe to reshape our perspectives, instill empathy, and reduce stress makes it a potential catalyst for change in the realm of abuse awareness. As psychologists and researchers, it is our responsibility to entertain unconventional ideas and seek innovative approaches that benefit society's mental and emotional health. In doing so, we may discover untapped reservoirs of compassion and understanding within ourselves and our communities.

While awe and awareness for abuse may seem like unlikely allies in psychological healing, there is evidence to suggest that they can complement each other. Experiencing awe has been associated with various positive outcomes, such as expanded perception of time, enhanced well-being, and increased prosocial behavior. Additionally, awe may play a role in mitigating the negative effects of abuse and fostering resilience. Incorporating awe into therapeutic interventions and raising awareness about abuse can potentially contribute to society's mental and emotional well-being.

9.6 The Impact of Awe-Inducing Stimuli on Human Cognitive Responses to Disturbing Visual Content

Immersive videos have also been found to be effective in inducing awe [17]. This suggests that utilizing immersive experiences, such as virtual reality, could be a promising avenue for incorporating awe into therapeutic interventions for individuals who have experienced abuse. This study investigates the influence of awe-inducing stimuli on human cognitive responses to emotionally distressing visual content, specifically videos depicting child abuse and subtle sexual abuse. Utilizing a combination of thermal imaging, Electroencephalography (EEG), and video presentation, we explored the potential of pure awe experiences to modulate cognitive and emotional reactions to these distressing stimuli. Our findings suggest that awe-inducing stimuli can significantly alter cognitive responses, leading to enhanced emotional regulation and cognitive stability when exposed to disturbing visual content.

The human capacity to experience emotions is multifaceted, encompassing both positive and negative states. While some emotions serve as protective mechanisms to react to threats, others, such as awe, have a profound impact on our cognitive processes. Awe-inducing stimuli, characterized by feelings of wonder and insignificance, may hold the potential to influence how individuals react to emotionally distressing content.

Methods

Participants: A total of 103 human subjects, aged between 18 and 26, were recruited from IIT Mandi, India for this experiment. All of the human subjects were management students doing MBA.

Stimuli: The experiment included two categories of video stimuli. The first category comprised videos that generate awe, such as footage of smiling babies and ultrafast blooming flowers. The second category featured distressing videos, including depictions of child abuse (hitting and beating) and subtle sexual abuse, where humiliating remarks were subtly made.

Experimental Setup: Participants were seated in a controlled environment where EEG electrodes were placed on their scalps for continuous brainwave monitoring. A thermal camera was also utilized to assess changes in facial temperature as a proxy for emotional responses. Video stimuli were presented randomly and in different sequences.

Procedure: Participants were initially exposed to awe-inducing videos before being shown the distressing content. EEG, thermal imaging, and self-report measures were used to assess cognitive and emotional responses.

Capturing a few scenes: In the Fig. 9.1 we have seen that flashing periodically in EEG for S1 has been a characteristic modulator. In test 1 and in test 2 we have put an arrow to demonstrate how periodic blinking in the brain is a common behavior for

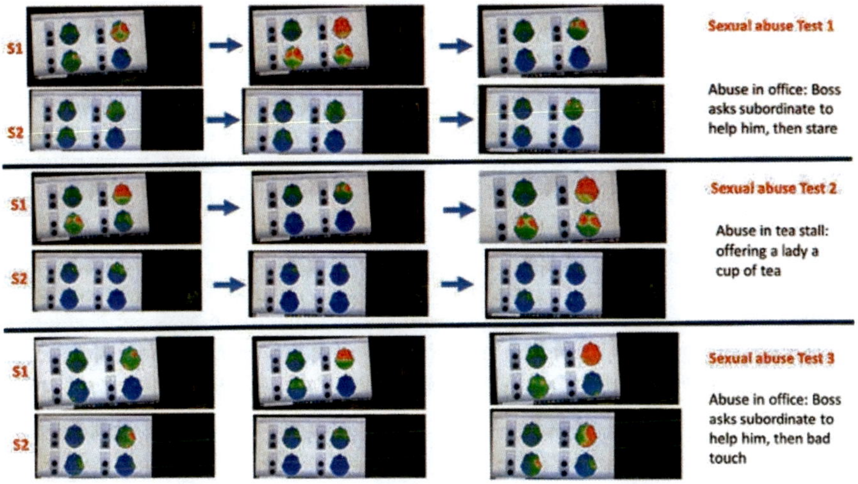

Fig. 9.1 There are three rows for three sexual abuse tests. Two human subject results S1 and S2 have been provided. Test 1 and Test 2 are before positive awes, while test 3 is after pure awe. Was shown to them. Finally, negative awe where a dark cloud of nature just before the final abuse test

all subjects who observes sexual abuse. The human subjects anticipate that the abuse is going to happen or something wrong is out there in the environment. If compared to DDG, we find true patterns as outlined in the Chap. 5, however, in EEG, the responses are human subject specific and it is not possible to find invariants across human subjects.

Results

1. **Emotional Regulation**: Participants who experienced awe-inducing stimuli before viewing distressing content exhibited significantly better emotional regulation. Their facial temperatures, as captured by thermal imaging, demonstrated lower fluctuations compared to those exposed directly to distressing content.
2. **Cognitive Stability**: EEG data revealed a notable difference in cognitive stability between the two groups. Participants who underwent a pure awe experience displayed more stable neural activity patterns when exposed to distressing stimuli, suggesting improved cognitive resilience.
3. **Preparation for Negative Threat**: Awe-induced subjects demonstrated an increased ability to handle negative threat, as observed through their EEG data. Their brain responses indicated a state of preparedness rather than heightened anger or distress.
4. **Consistent Reaction to Child Abuse**: Notably, subjects across both groups displayed a consistent and strong aversion to child abuse content, as indicated by stable neural patterns and heightened emotional responses.

The results of this study suggest that pure awe experiences can significantly alter cognitive and emotional responses to distressing visual content. Participants who underwent a pure awe experience demonstrated enhanced emotional regulation, cognitive stability, and readiness to face negative threats. Furthermore, their reactions to child abuse content remained consistent and aversive, regardless of prior awe exposure.

This study sheds light on the potential of awe-inducing stimuli to modulate human cognitive responses to emotionally distressing visual content. The findings suggest that the induction of awe may offer a promising avenue for enhancing emotional resilience and cognitive stability when confronted with disturbing material. These insights may have implications for various fields, including psychology, education, and therapeutic interventions, offering new perspectives on emotional regulation and cognitive preparedness in the face of adversity. Further research is warranted to explore the mechanisms underlying these observed effects and their practical applications.

9.7 Rethinking Awareness for Mental Abuse: Ten Fundamental Changes for Effective Impact

Awareness campaigns are essential tools for addressing critical societal issues, including mental abuse. However, the traditional approach to awareness often overlooks important factors that can influence their effectiveness. In this section, we highlight ten fundamental changes that should be considered to reshape the concept of awareness for mental abuse and maximize its impact.

1. **Mindful Awareness Design**: Awareness campaigns should recognize that awareness inherently triggers excitation. If not designed with care, they can unintentionally lead to negative excitation, undermining the primary objective. Campaigns should focus on fostering a positive emotional response while addressing mental abuse issues. Firstly, it is essential to improve mental health literacy within the community [44]. Community campaigns that aim to enhance mental health literacy have been implemented in several countries [45]. By empowering the community with knowledge and understanding of mental health issues, individuals are better equipped to recognize and address mental abuse.
2. **Cultural Sensitivity**: Societies are intricately woven with unique cultural norms and interpersonal relationships. These cultural intricacies build a societal mindset that cannot be ignored. Awareness campaigns should be tailored to acknowledge and respect these cultural differences instead of adopting a one-size-fits-all approach. Secondly, stigma against psychiatric illnesses must be addressed [103]. Stigma can prevent individuals from seeking help and perpetuate the cycle of mental abuse. Anti-stigma activities have been implemented in various Asian societies to combat stigma [103]. These activities aim to change societal attitudes and promote acceptance and support for individuals with mental health issues.

3. **Gender-Specific Psychology**: Men and women have distinct psychological grammars, and societal efforts to nurture women to be like men often overlook this essential aspect. Awareness campaigns should embrace diversity and recognize that different genders may respond differently to mental abuse and awareness efforts. Additionally, awareness campaigns should focus on influencing proximal variables, such as social norms, to bring about long-term behavior change [15]. By targeting social norms and promoting positive attitudes and behaviors, campaigns can create a supportive environment that discourages mental abuse.
4. **Inconsistent EEG Responses**: EEG mapping reveals the inconsistency in the excitatory regions triggered by awareness campaigns. Campaigns should explore methods to achieve more consistent and sustained responses, ensuring long-term impact. Interventions to improve mental health literacy should encompass various approaches, including whole-of-community campaigns, youth-focused campaigns, school-based interventions, and training programs for individuals to better intervene in a mental health crisis [45]. These interventions address different target audiences and provide a comprehensive approach to improving mental health literacy and addressing mental abuse.
5. **Positive Messaging**: Many awareness campaigns resort to shock tactics or cultural triggers. Instead, campaigns should emphasize delivering messages of pure goodness and empowerment to inspire positive change. Social marketing campaigns have shown promise in reducing mental health stigma [79]. By utilizing social media and other platforms, these campaigns can reach a wide audience and challenge societal misconceptions and stereotypes surrounding mental health. Framing mental health awareness campaigns with a fresh start mindset can reduce stigma and promote help-seeking behavior [53]. By emphasizing the opportunity for personal growth and positive change, campaigns can encourage individuals to seek help and support for mental health issues.
6. **Deeper Mental Empowerment**: Awareness campaigns rarely focus on making individuals mentally powerful at a deeper level. There is an opportunity to explore techniques that foster mental resilience and thought transformation. Such campaign awareness has been associated with better knowledge, attitudes, and intended behavior [28]. Therefore, it is crucial to ensure that awareness campaigns are effectively disseminated and reach the target audience to maximize their impact.
7. **Media and Dirty Minds**: Media, including movies and advertisements, play a significant role in shaping societal perceptions. Awareness campaigns must consider the challenge of countering media that may feed negative stereotypes and perceptions. Culturally appropriate adaptations of awareness campaigns can have a significant impact on community members' well-being [2]. By partnering with local organizations and tailoring campaigns to specific cultural contexts, the effectiveness of awareness campaigns can be enhanced
8. **Individual Variability**: Awareness campaigns should acknowledge that the mental state of individuals can vary greatly. While the campaign's objective may be clear, forcing engagement may not be effective. Customization and

flexibility in approach are crucial. Targeted campaigns for specific populations, such as men who have experienced intimate partner violence, can address their unique needs and experiences [72]. By conducting formative evaluation research and involving the target population in campaign development, the relevance and effectiveness of the campaigns can be improved.

9. **Audio-Visual Considerations**: Research should extend to the auditory and visual aspects of awareness campaigns. Brain responses, especially to sound, can impact the effectiveness of campaigns. A holistic approach to audio-visual design is essential. It is important to consider the potential unintended consequences of awareness campaigns, such as worsening mental health outcomes when implemented in isolation [98]. Therefore, a comprehensive approach that includes other evidence-based interventions is necessary to ensure positive outcomes
10. **Intra-Sensory Responses**: Awareness campaigns should consider the conditioning of the brain in specific environments. Understanding how individuals respond within different contexts, such as a movie theater or a crowded public space, is vital for effective campaign placement.

To truly address mental abuse through awareness campaigns, we must embrace a paradigm shift in our approach. These ten fundamental changes aim to create more effective, culturally sensitive, and impactful campaigns. By recognizing the complexity of human psychology, the power of positivity, and the influence of cultural and individual factors, we can transform awareness into a potent tool for positive change and the prevention of mental abuse. It is time to rethink and reinvent how we approach this critical societal issue. Reshaping the concept of awareness for mental abuse requires considering these ten fundamental changes. By improving mental health literacy, addressing stigma, targeting social norms, utilizing social marketing, framing campaigns with a fresh start mindset, ensuring effective dissemination, incorporating cultural appropriateness, targeting specific populations, and considering potential unintended consequences, awareness campaigns can maximize their impact in addressing mental abuse.

9.8 Neuromarketing: Leveraging Brain Power for Positive Impact

Neuromarketing, the practice of understanding consumer behavior through neuroscience and psychology, has gained significant attention in recent years. Critics argue that it can be a manipulative tool, used to brainwash society into making unnecessary purchases. However, this article aims to explore the potential of neuromarketing to generate positive awe in consumers, rather than manipulating them. By delving into the science behind neuromarketing, we can understand how it can be harnessed to inspire wonder, joy, and genuine appreciation.

The Dark Side of Neuromarketing:

Critics have raised valid concerns about the potentially manipulative aspects of neuromarketing. By tapping into the subconscious minds of consumers, it is argued that this practice can lead to impulsive and unnecessary purchases. Marketers often use emotional triggers, exploiting human vulnerabilities to promote products or services. This has raised questions about the ethical boundaries in advertising [1].

Neuromarketing for Positive Awe

While the manipulation aspect of neuromarketing is a valid concern, it's important to recognize that the science behind it can also be harnessed for positive outcomes. Instead of exploiting vulnerabilities, neuromarketing can be used to create advertisements that genuinely inspire awe and positivity. While the manipulation aspect of neuromarketing is a valid concern, it is important to recognize that the science behind it can also be harnessed for positive outcomes. Instead of exploiting vulnerabilities, neuromarketing can be used to create advertisements that genuinely inspire awe and positivity. By understanding the neural mechanisms underlying emotions such as wonder and joy, marketers can design campaigns that evoke these positive emotions in consumers [52]. One way to achieve this is by using storytelling techniques in advertisements. Research has shown that narratives activate multiple brain regions, leading to increased engagement and emotional resonance with the content [105]. By crafting compelling stories that evoke awe and wonder, marketers can create a positive emotional experience for consumers. In the Chap. 8 we have described the necessity of a pure awe and how it changes everything.

Another approach is to focus on the sensory aspects of products or services. Studies have demonstrated that sensory experiences can elicit strong emotional responses [70]. By designing advertisements that highlight the sensory qualities of a product, such as its texture, smell, or taste, marketers can create a sense of delight and appreciation in consumers.

Furthermore, neuromarketing can be used to personalize advertisements based on individual preferences and values. By leveraging neuroimaging techniques, marketers can gain insights into consumers' neural responses to different stimuli [71]. This allows for the creation of tailored advertisements that resonate with consumers on a deeper level, leading to a more positive and meaningful experience.

1. **Emotionally Resonant Storytelling**: Neuromarketing techniques can help advertisers craft emotionally resonant stories. These stories can touch the hearts of consumers, evoking feelings of empathy, compassion, and awe. Such advertisements can leave a lasting positive impact.
2. **Neural Engagement**: Understanding how the brain responds to different stimuli allows advertisers to design campaigns that engage the brain in a positive and constructive way. This can lead to a sense of wonder and appreciation for the message being conveyed.
3. **Awe-Inspiring Visuals**: Neuromarketing can guide the use of awe-inspiring visuals that capture the imagination. Beautiful landscapes, scenes of human kindness, or breathtaking natural wonders can all elicit feelings of awe.

4. **Authenticity**: Building trust and authenticity into advertising campaigns can lead to genuine awe. Consumers appreciate transparency and honesty in marketing efforts. When they encounter authentic stories, their positive response can be profound.
5. **Creating a Positive Association**: Neuromarketing can help advertisers establish positive associations between their brands and uplifting emotions. When consumers associate a brand with positivity, they are more likely to engage with it.

The Power of Awe

Awe is a powerful and transformative emotion. When experienced genuinely, it can broaden perspectives, foster empathy, and inspire positive change. It encourages individuals to connect with something greater than themselves, whether it's the beauty of nature, acts of kindness, or the wonders of science and technology.

Neuromarketing need not be solely associated with manipulation and exploitation. Its insights into human psychology can also be used for a more noble purpose—generating a pure positive awe in consumers. Advertisers have the opportunity to inspire wonder, joy, and appreciation through emotionally resonant storytelling, engaging visuals, authenticity, and positive associations. By harnessing the power of awe, they can contribute to a more uplifting and positive society, where advertising becomes a source of inspiration rather than manipulation. Ultimately, it is up to marketers to decide how they will leverage the science of neuromarketing—for the betterment of society or merely for profit.

9.9 The Silent Language of Non-verbal Communication: Deciphering Subconscious Decision-Making Factors:

9.9.1 Silent Sensory Trigger from Environment

In the realm of human psychology, the power of non-verbal communication has long been acknowledged. However, there is a silent language that operates beneath the surface of conscious awareness, influencing our decision-making processes. This silent language encompasses elements like the geometry of our environment, the use of colors, sound and music, the kindness of those around us, and the overall peace within society. Despite their profound impact on our choices, these factors often elude precise definition and categorization within traditional psychological frameworks. In this article, we explore the need for a new language to address these silent non-verbal decision-making variables, shedding light on their significance in our daily lives.

The Geometry of Our Environment:

Our surroundings are not just physical spaces; they are complex geometries that unconsciously influence our decisions. Architectural design, the arrangement of objects, and the flow of spaces can impact our mood, cognitive processes, and behavior. A well-designed environment can promote creativity, productivity, and a sense of well-being, while a chaotic or constricting space may lead to stress and anxiety [54]. Nadal explored the influence of aesthetics, affordances, and activity settings in architectural design on psychological well-being [54]. They found that aesthetically pleasing environments have a positive impact on individuals' well-being and can enhance their decision-making processes.

The Impact of Architectural Design on Human Psychology:

The impact of architectural design on human psychology has been extensively studied. It has been found that architectural design can influence emotions, cognitive processes, and behavior [9]. The arrangement of objects and the use of colors, textures, and lighting can evoke specific emotional responses and affect decision-making. Additionally, [99] investigated the impact of contour on aesthetic judgments and approach-avoidance decisions in architecture [99]. They found that curved spaces were subjectively preferred overall and did not elicit increased amygdala activity, suggesting a positive emotional response [38].

The Psychological Sense of Community in the Neighborhood:

The sense of community in the neighborhood also plays a role in decision-making, developing psychological sense of community and its influence on individuals' decisions [10]. They found that a strong sense of community fosters positive decision-making and enhances well-being. Factors such as social cohesion, trust, and a supportive environment contribute to this sense of community.

The silent language of non-verbal decision-making factors in architecture encompasses the geometry of our environment, the impact of architectural design on human psychology, and the psychological sense of community in the neighborhood. Understanding and harnessing these factors can lead to the creation of spaces that promote positive decision-making, well-being, and a sense of community. Further research in this field, as highlighted by [16] in "The Aesthetic Brain: How We Evolved to Desire Beauty and Enjoy Art," can provide valuable insights into the evolutionary basis of our preferences for certain aesthetic qualities [55]. By integrating these findings into architectural practice, we can create environments that optimize decision-making processes and enhance the overall quality of life. Here are two sections which enlighten us from this book.

The Use of Colors

Colors hold immense power over our emotions and decisions. They can evoke specific feelings and associations, even when we are not consciously aware of it. For example, the color blue is often associated with calmness and trust, while red may evoke feelings of excitement or urgency. The silent influence of color in decision-making extends to product packaging, branding, and even the color schemes in our homes. Recognizing the psychological impact of colors can empower us to make intentional choices in design and communication.

Sound and Music

Sound and music have the ability to shape our emotional states and, consequently, our decisions. The right background music in a restaurant can enhance the dining experience, while dissonant sounds may disrupt concentration and decision-making. The silent harmony of sounds in our environment can influence our mood, productivity, and even our willingness to engage with others. By understanding the language of sound, we can curate auditory experiences that support positive decision-making.

9.9.2 *Kindness of Peoples Around*

Human psychology is a complex field that seeks to understand the intricacies of human behavior, thoughts, and emotions. To truly comprehend the depths of human psychology, it is essential to delve into the language of the brain and explore how different emotional traits can be quantified. This article aims to highlight the significance of understanding the language of the brain and the quantification of emotional traits in gaining a comprehensive understanding of human psychology.

The Influence of Kindness and Empathy:

Human interaction plays a crucial role in shaping our decisions and behaviors. Acts of kindness and empathy displayed by those around us have a profound impact on our choices. Research has shown that practicing loving-kindness meditation can increase social connectedness and foster positive emotions [41]. This suggests that cultivating kindness and empathy in our interactions can create an environment of trust, cooperation, and reciprocity, leading to more positive decision-making. Empathy is a fundamental aspect of human psychology that plays a significant role in decision-making. The ability to perceive and understand the emotions of others is crucial in forming social connections and making empathetic decisions [64]. However, the concept of empathy is complex, and there are ongoing debates about its nature, reliability, and trainability [39]. Further research is needed to fully comprehend the role of empathy in decision-making and its implications for human psychology.

The Neural Basis of Kindness and Empathy:

Advancements in neuroscience have shed light on the neural mechanisms underlying kindness and empathy. Studies have shown that practices such as loving-kindness meditation can alter brain structure and function, leading to increased social cognition, empathy, and altruism [61, 95]. Understanding the neural mediators of kindness-based meditation can provide valuable insights into the language of the brain and its connection to emotional traits. To gain a true understanding of human psychology, it is essential to explore the decision-making processes. Research has shown that decision-making is influenced by various factors, including expert knowledge, cultural beliefs, and personal experiences [50, 77]. For example, individuals' decision-making regarding prenatal diagnosis is influenced by factors such as ethnicity, gender, age, socioeconomic status, education, and religious beliefs [77]. Understanding these factors and their impact on decision-making can provide valuable insights into human psychology.

The Role of Peace in Society:

The overall state of peace within society is another important factor that influences our decisions. In a society marked by conflict and instability, individuals may make defensive and self-preservation decisions due to fear and insecurity [8]. On the other hand, a peaceful and stable society encourages cooperation, innovation, and a willingness to contribute positively. Therefore, promoting tolerance, inclusivity, and social harmony is crucial in creating an environment that fosters positive decision-making [39, 50].

A New Language for the Silent Non-Verbal

To truly comprehend and harness the power of these silent non-verbal decision-making variables, we need a new language within psychology. This language should encompass the intricate interplay between geometry, colors, sound, kindness, and societal peace. It should allow us to define, measure, and analyze these factors more comprehensively, enabling us to make informed decisions in design, policy-making, and interpersonal relationships. We have created a language called Geometric Musical Language, GML [6, 13].

As psychology continues to evolve, we must acknowledge and explore the silent language of non-verbal communication that shapes our decisions. The geometry of our environment, the use of colors, sound and music, the kindness of those around us, and the peace within society all play integral roles in our daily choices. By developing a new language that embraces these silent non-verbal decision-making variables, we can deepen our understanding of human behavior and create environments that support positive, conscious choices. In doing so, we can pave the way for a more harmonious and empathetic world. In the following section we would describe how integrating the recurring events would change our perception in psychology. While, to truly understand human psychology, it is essential to delve into the language of the brain and quantify emotional traits. Acts of kindness and empathy, the state of peace in society, and various factors influencing decision-making processes all contribute to our understanding of human psychology. By exploring the neural basis of kindness and empathy, we can gain insights into the intricate workings of the human brain. Further research in these areas will contribute to a more comprehensive understanding of human psychology and pave the way for advancements in the field.

9.10 Decoding Non-verbal Communication: Unveiling Geometric Musical Language (GML) and the Symmetry of Primes in Brain Mapping

Non-verbal communication is a rich tapestry of unspoken cues that significantly influence human interactions and decisions. Traditionally, the interpretation of non-verbal communication has relied on behavioral observations and subjective assessments. However, recent advancements in neuroscience have opened up new avenues for understanding non-verbal communication through brain mapping techniques, such as Electroencephalography (EEG) and Dodecanogram (DDG, Chaps. 5 and 6). In this section, we delve into the exciting realm of decoding non-verbal communication using 2D brain surface maps and introduce the concept of Geometric Musical Language (GML). Additionally, we explore the role of prime numbers and symmetry in shaping these geometric patterns, all within the framework of the Self-Operating Mathematical Universe (SOMU) theory of consciousness.

9.10.1 The Power of 2D Brain Surface Maps

Traditional EEG has provided valuable insights into brain activity by capturing electrical signals. However, the mapping of these signals onto a 2D brain surface offers a visual representation of cognitive and perceptual experiences. Each region of the brain contributes to forming geometric shapes, referred to as invariants. These invariants change their arrangement over time during cognitive processes, akin to musical notes composing a melody.

Geometric Musical Language (GML)

The concept of Geometric Musical Language (GML) emerges from the dynamic interplay of invariants on brain surface maps. Just as musical notes combine to create a composition, these geometric shapes form patterns that represent non-verbal communication cues. For instance, a particular geometric pattern may correspond to the experience of awe, while another may signify empathy or attentiveness.

Morphogenesis and Symmetry of Primes

To understand the changes in these geometric shapes, we delve into the morphogenesis and symmetry of prime numbers. Prime numbers, those indivisible by any integer other than 1 and themselves, possess a unique mathematical elegance. In the context of brain mapping, primes are linked to the metric of consciousness.

The symmetry of primes governs the transitions between different geometric shapes. Much like the rhythm in music, the symmetry of primes orchestrates the fluid progression of invariants on the brain surface map. This symphony of primes underpins the intricate dance of non-verbal communication cues, shaping the ebb and flow of emotional and cognitive experiences.

9.10.2 Self-Operating Mathematical Universe (SOMU) Theory

The SOMU theory of consciousness posits that the universe operates according to self-contained mathematical principles. Within this framework, described in Chap. 5 and detailed in Chap. 10, the mathematical structures governing the universe extend to the very core of human consciousness. The interplay of primes, the emergence of invariants, and the fluidity of GML all find their roots in the SOMU theory, providing a unifying perspective on non-verbal communication. Since the formulation of SOMU looks into the brain states as geometric shapes, cognition as changes in the geometric shapes and we count invariants and prime number of invariants to accurately determine the cognitive states, for details see Chap. 5.

Implications for Awareness Campaigns for Abuse

Understanding the geometric patterns and symmetries that underlie non-verbal communication opens up exciting possibilities for awareness campaigns, particularly those focused on abuse prevention. By decoding the geometric shapes and GML, we can develop more nuanced and precise campaigns that resonate with the subconscious minds of individuals. The SOMU theory also offers a fresh perspective on consciousness-driven societal change, which may inspire innovative approaches to raising awareness and promoting empathy. Seeing kindness or empathy as geometric shapes could be a primitive step toward quantifying emotion, cognition, and perceptions.

Non-verbal communication, once a realm of ambiguity and subjectivity, is now entering the domain of quantifiable science. Through 2D brain surface maps, Geometric Musical Language (GML), the symmetry of primes, and the SOMU theory of consciousness, we are unraveling the hidden language of the human mind. This newfound understanding has the potential to revolutionize awareness campaigns, shaping them into powerful tools for societal change and fostering a deeper connection between individuals. As we continue to explore the mathematical foundations of consciousness, the possibilities for decoding and harnessing non-verbal communication are boundless.

9.11 Unleashing the Power of SOMU Theory of Consciousness in Designing Perfect Awareness Campaigns

9.11.1 Background Studies in Neuroscience that Changed Our View in Psychology

By advancing the time crystal concepts of the 1960s [90] the Polyatomic Time Crystal model of the brain [75, 89] was developed by measuring the electromagnetic resonance of biomaterials. This model integrates all clockwise and anticlockwise rhythms within the brain. The combination of these rhythms is crucial because together they create a singularity where the state cannot be precisely defined. Electromagnetic resonance links all brain components by coupling of sharp resonance peaks. Even though mainstream neuroscience refuses to consider the role of electromagnetic coupling, we have meticulously studied every single component in the brain for the electromagnetic resonance [91], and created a database where all brain related data has been uploaded along with our human subject studies (www.brainrhythm.org).

Up to this point, we have primarily discussed positive psychology in this book. Now, it's time to turn our attention to the assessment of negative mental states. In this chapter, we will examine two contrasting psychological traits: "abuse," which seeks to quantify the structure of the mind and can lead to destructive mental states, and "awe," which has the potential to cure various neurological disorders.

When we juxtapose "abuse" and "awe," their conflict lies in their interface. We can envision a polyatomic time crystal structure assigned to these two effects. When they interact, the corresponding 3D clock assemblies within the brain contend with each other. This experience is akin to a molecular heterojunction, where membrane-microtubule junctions play a crucial role in neurons for signal transfer in a fractal structure [69]. Thus, there is an exchange of thoughts between the two.

Our hypothesis is that every single thought in the human brain, including all biological rhythms and 91 cognitive, perceptive, and emotional experiences (Chap. 5), is composed of 3D clock assemblies. These 3D structures collide, resulting in three key properties: some clocks bind to form new ones, some rearrange themselves into different symmetries, and others utilize patterns as fractals, continuously evolving into an infinite series, projecting feedback from infinity outward. This interplay between "abuse" and "awe" occurs within this framework.

The fractal, scale-free electromagnetic resonance of a single brain-extracted microtubule nanowire, a single tubulin protein, and a single neuron suggests that a triplet of triplet resonance bands or nine frequency/time domains in discrete groups operate together [87]. This implies that self-similarity is consistently detected, and negative psychological attributes like "abuse," which manifest as phase singularities or dark regions in the 3D clock assembly, absorb aspects of "awe" and become meaningful entities. This observation stems from the software simulator we've developed for extracting 3D clock assemblies from live recordings of DDG,

EEG, thermal cameras, or any surface profile. The 3D clock assemblies grow as part of a fractal tape [34], under the framework of Fractal Information Theory, FIT [7].

To understand neuroscience phenomena, it is essential to uncover the quantum mechanical processes of molecules. Many biological codes are written in molecules using quantum mechanics. Therefore, to grasp the origins of cognitive processes at the molecular neuroscience level, it is necessary to unravel how classical processes are regulated by quantum codes.

The brain is a dynamic system, and once it reaches a classical state, it becomes challenging to maintain hallmark quantum properties. However, a couple of quantum features can persist at room temperature and ambient atmospheric conditions. One of these features is the geometric phase. When a system vibrates in an adiabatic environment, it can generate a geometric phase of oscillations alongside its dynamic phase. This can enable a classical system to transfer certain quantum dynamics to classical systems. Quantum tunneling can be employed to read and write information in such a network [12], and we have demonstrated that quantum properties can endure in microtubules [78]. A fractal network can logically process information [35] and generate the geometric phase network.

Simultaneously measuring neuron firing at the neuron membrane and within the microtubular bundle inside a neuron reveals that microtubules process firing signals 10,000 times faster than the membrane [6]. They play an active role in the brain's information processing [86, 92, 93]. Even neurons not connected by synaptic junctions can exchange energy, as the neural circuit is imaged by scanning dielectric microscopy (SDM) [36]. Therefore, fractal networks excel in managing various time scales simultaneously and can transform self-similar features of a particular thought process into very different kinds of thought processes.

9.11.2 Self-Operating Mathematical Universe: SOMU

In the ever-evolving landscape of psychology and consciousness studies, the Self-Operating Mathematical Universe (SOMU) theory has emerged as a groundbreaking framework. This theory posits that the universe operates according to self-contained mathematical principles, which extend to the very core of human consciousness. In this section, we explore how the SOMU theory can be harnessed to design perfect awareness campaigns. We will delve into two essential aspects of SOMU: the metric of primes and the 3D clock assembly, which together offer a revolutionary approach to understanding and shaping human thought and behavior.

9.11.2.1 The Metric of Primes: A Temple of Symmetry and Asymmetry

At the heart of SOMU lies the metric of primes—a temple-like architectural structure that encodes the ordering and symmetries of the universe. While understanding the symmetries and ordering of elements is crucial, true transformative events occur when symmetries break, and one symmetry gives way to another. The metric of primes offers a surface upon which the human mind can navigate, forging a path of thoughts, decisions, and actions.

1. **Navigating the Temple**: The metric of primes provides a dynamic framework for exploring the transitions between symmetrical and asymmetrical states in consciousness. By traversing this temple-like structure, researchers can gain insights into the intricate dance of the mind as it shifts between different cognitive states.
2. **Pathways of Thought**: Just as a temple has paths leading to various altars, the metric of primes offers pathways for thought. Each path corresponds to a unique mental state or cognitive process. Understanding and mapping these pathways is the key to unlocking the potential of the subconscious mind.
3. **Breaking Symmetries**: In the study of consciousness, it's not just about recognizing symmetries; it's about understanding the moments when symmetries break and new states of awareness emerge. The metric of primes provides a visual representation of these transformative moments.

9.11.2.2 The 3D Clock Assembly: Unveiling Recurring Events in Society

A central component of SOMU's information structure is the 3D clock assembly, which serves as a repository of recurring events in society. This assembly allows researchers to convert these events into a structured format that facilitates advanced analysis and simulation of the potential effects of awareness campaigns.

1. **Discovering Geometric Shapes**: Within the 3D clock assembly, hidden geometric shapes represent recurring societal events. These shapes encode valuable information about the nature and patterns of these events.
2. **Understanding Change**: By examining how these geometric shapes change over time, researchers can gain insights into the dynamics of societal events and their impact on the collective consciousness.
3. **Mapping Symmetry Transitions**: The 3D clock assembly also offers a platform for mapping the transitions of symmetries within society. This enables researchers to anticipate how awareness campaigns can influence the shifting dynamics of societal thought and behavior.

9.11.2.3 Bias-Free Paths to Perfect Awareness Campaigns

The true power of SOMU in designing perfect awareness campaigns lies in its ability to map the subconscious mind and the cultural complexity of society onto bias-free paths. By analyzing the geometric shapes, changes, and symmetries within the 3D clock assembly, researchers can identify pathways that lead to desired outcomes.

1. **Mapping the Mind**: SOMU allows for the comprehensive mapping of individual and collective consciousness, enabling researchers to understand the intricate interplay of thoughts, emotions, and cultural influences.
2. **Simulation and Optimization**: With this understanding, awareness campaigns can be simulated and optimized to follow bias-free paths that resonate with the target audience, fostering genuine engagement and impact.

The SOMU theory of consciousness offers a groundbreaking approach to designing perfect awareness campaigns. By leveraging the metric of primes and the 3D clock assembly, researchers can navigate the temple of consciousness, decode recurring societal events, and map the pathways of thought and behavior. This powerful framework provides a bias-free roadmap to crafting campaigns that truly resonate with the subconscious mind and the cultural complexities of society. As we continue to explore the frontiers of psychology and consciousness, SOMU opens new horizons for creating positive change and transformation in the world of awareness campaigns.

9.12 Conclusion: Nectar Eliminates Poison not Awareness

9.12.1 Universal Language of Love

The profound influence of audio-visual effects on our mental state has been an ever-present aspect of human existence. From the mesmerizing allure of a captivating film to the transcendental experience of a live concert, the fusion of sight and sound has the power to transport us to different realms of consciousness. Yet, quantifying the exact impact of these sensory stimuli on our brains has remained a formidable challenge. However, recent breakthroughs in neuroscientific instrumentation have ushered in a new era of exploration, one that allows us to fathom the depths of our cerebral responses to audio-visual experiences.

9.12.1.1 The Dawn of the Dodecanogram

At the heart of this groundbreaking exploration lies an innovative instrument known as the dodecanogram, abbreviated as DDG, as elucidated in Chap. 5 of our journey. This instrument, unlike any other, offers the unprecedented capability to observe the dynamic processes occurring on the brain's surface across diverse time domains and spatial dimensions simultaneously. The implications of this newfound ability are profound, opening doors to uncover events and patterns that were previously hidden from our understanding. As described in Chap. 8 and here in this Chapter, human subject specific features are so profound in EEG that DDG like tools are essential to find the invariants across human subjects.

9.12.1.2 Redefining Time in the Brain

Central to our exploration is the notion that time, as perceived by the human brain, departs from the conventional linear construct. Rather, nature embraces a discrete framework, treating time as a loop structure of phases. These phases manifest in various dimensions—2D, 3D, and 4D, in harmony with the dimensions of space and spacetime. It is within these multidimensional structures that we discover the essence of time management within the brain, a concept that challenges our preconceptions. Consider the captivating nature of time, which can manifest in various enchanting forms. First, envision it as a graceful 2D loop, emerging after our familiar 3D space, thus revealing itself as the elegant 4th dimension. Then, as we venture deeper into the realm of spacetime, time blossoms into a mesmerizing 3D structure, becoming the awe-inspiring 5th dimension. And for those willing to explore further, it unfolds into a profound 4D structure after transcending the bounds of 5D spacetime, unveiling the sublime 6th dimension that tells us the worldline.

9.12.1.3 The Dark Room Experience

One of the most striking revelations illuminated by the DDG is the transformative effect of audio-visual stimuli within a darkened environment. When our senses are enveloped in darkness, the mind becomes a canvas onto which audio-visual effects paint vivid virtual worlds. In this sensory cocoon, the brain's responses are intensified and its perceptual boundaries stretched. The mind, captive to the power of sight and sound, embarks on an immersive journey through these virtual realms.

Thus, the dodecanogram, with its ability to unveil the intricate dance of the brain across time domains and spatial dimensions, has provided us with a transformative lens through which to explore the profound impact of audio-visual effects on our minds. As we journey through the multidimensional landscape of time management within the brain, we are reminded that our understanding of consciousness is ever-evolving. The dark room experience serves as a poignant reminder of the limitless potential of human perception when immersed in the symphony of audio-visual stimuli. As we continue to unravel the mysteries of the mind, we stand on the precipice of an era where the fusion of science and art promises to reveal the very essence of our existence.

9.12.2 The Self-Operating Mathematical Universe, SOMU and Psychological Management

9.12.2.1 Unlocking the Human Brain's Mysteries with the Polyatomic Time Crystal Model

The human brain has been an enigmatic puzzle that has fascinated scientists, philosophers, and thinkers for centuries. It's a complex organ that processes thoughts, emotions, and sensory information, giving rise to consciousness and self-awareness. The study of the brain has seen significant advancements, but it still remains a realm of endless mysteries. However, a groundbreaking concept called the "Polyatomic Time Crystal Model" [75] is challenging our conventional understanding of the brain's functioning and could revolutionize how we perceive and treat it. To incorporate the factors of DDG, polyatomic time crystals are essential. Overall, the new approach to look into the brain as an architecture of clocks [13] brings natural flavor to the field of artificial intelligence and computing, however, the mathematics of multinion tensors and the time crystal architecture of the human brain would have profound effect in furthering the research on human psychology. The architecture of a human mind could be rhythmic and structure of clocks could represent spatio-temporal events that we experience around us is a new marriage between software and hardware.

9.12.2.2 Rethinking the Brain as a Symphony of Clocks

Traditionally, the brain has been studied using a linear, machine-like approach, where neurons and synapses are viewed as computational elements. This perspective has led to significant progress in neuroscience and artificial intelligence, but it may not capture the brain's true essence. The Polyatomic Time Crystal Model proposes a profoundly different way of understanding the brain—as an intricate assembly of countless clocks arranged in symmetrical geometric shapes. This novel approach takes inspiration from the fascinating concept of time crystals, which are exotic phases of matter where particles move in perpetual, synchronized patterns. In a similar vein, the brain is envisioned as an orchestra of clocks spread over 12 dimensions [92, 93], each ticking away in harmony with others, creating a symphony of cognitive processes.

9.12.2.3 Changing Geometric Shapes and Prime Number Knots

What sets the Polyatomic Time Crystal Model apart is its capacity to explain how the brain adapts and evolves over time. Imagine these geometric shapes representing cognitive processes changing, not randomly, but in patterns akin to tying prime numbers of knots. This perspective allows us to observe and decipher all cognitive, perceptive, and emotional responses in terms of prime numbers.

At the heart of this transformative idea is the journey from the electrical signals of neurons to the realm of prime numbers. By linking these primes to the Self-operated Mathematical Universe (SOMU) theory of consciousness, a comprehensive information structure emerges, one that has no inherent attributes. This, in turn, enables us to visualize the brain and all its intricate components as a 3D assembly of clocks, an extraordinary departure from the traditional linear model.

9.12.2.4 Implications for the Future

The Polyatomic Time Crystal Model has far-reaching implications for the fields of neuroscience and artificial intelligence. It challenges the reductionist view of the brain as a mere computing machine, urging us to see it as a living, dynamic entity, constantly adapting and evolving. Understanding the brain as a symphony of clocks offers new avenues for exploring consciousness, perception, and emotions. It prompts us to explore the brain's functioning in terms of prime numbers, unlocking a deeper layer of its complexity. In the realm of artificial intelligence, this model could inspire the creation of more brain-like and adaptable machines. By mimicking the brain's clockwork symphony, AI systems might become more attuned to the nuances of human thought and emotion, leading to advanced applications in healthcare, robotics, and beyond.

The Polyatomic Time Crystal Model of the human brain challenges us to rethink the brain's nature and functioning. By viewing it as an assembly of clocks in ever-changing geometric shapes tied together by prime number knots, we open doors to a new era of neuroscience and artificial intelligence. This innovative approach has the potential to revolutionize our understanding of consciousness and inspire the development of more intelligent and adaptable machines. As we continue to explore the profound mysteries of the brain, the Polyatomic Time Crystal Model offers a glimpse into a future where our understanding of the mind knows no bounds.

References

1. Ariely D (2008) Predictably irrational: the hidden forces that shape our decisions. Harper-Collins
2. J Anwar-McHenry L Murray C Drane J Owen A Nicholas R Donovan 2022 Impact on community members of a culturally appropriate adaptation of a social and emotional well-being intervention in an aboriginal community J Public Ment Health 21 2 108 118 https://doi.org/10.1108/jpmh-09-2021-0109
3. Arcangeli M, Sperduti M, Jacquot A, Piolino P, Dokic J (2020) Awe and the experience of the sublime: a complex relationship. Front Psychol 11. https://doi.org/10.3389/fpsyg.2020.01340
4. Atamba C, Popelnukha A, Ibrahim F (2020) Awe guards my creativity: the interactive effect of perceived abusive supervisory behavior, dispositional awe, and creative self-efficacy on Chinese employee creativity. Front Sociol 5. https://doi.org/10.3389/fsoc.2020.00051

5. AE Adams CM Sullivan 2011 Exploring the relationship between financial abuse and mental health outcomes in low-income survivors of intimate partner violence Violence Against Women 17 12 1477 1493
6. L Agrawal S Sahu S Ghosh T Shiga D Fujita A Bandyopadhyay 2016 Inventing atomic resolution scanning dielectric microscopy to see a single protein complex operation live at resonance in a neuron without touching or adulterating the cell J Integr Neurosci 15 04 435 462
7. L Agrawal R Chhajed S Ghosh B Ghosh K Ray S Sahu D Fujita 2018 Fractal information theory (FIT)-derived geometric musical language (GML) for brain-inspired hypercomputing Soft Computing: Theories and Applications 548 343 372
8. J Baker 2019 The empathic foundations of security dilemma de-escalation Polit Psychol 40 6 1251 1266 https://doi.org/10.1111/pops.12623
9. Bara I, Binney R, Ramsey R (2021) Investigating the role of executive resources across aesthetic and non-aesthetic judgments. https://doi.org/10.31234/osf.io/ydmbr
10. M Bär M Neta 2006 Humans prefer curved visual objects Psychol Sci 17 8 645 648 https://doi.org/10.1111/j.1467-9280.2006.01759.x
11. VL Banyard MM Moynihan EG Plante 2007 Sexual violence prevention through bystander education: an experimental evaluation J Community Psychol 35 4 463 481
12. Bandyopadhyay A, Miki K, Wakayama Y (2006) Writing and erasing information in multi-level logic systems of a single molecule using scanning tunneling microscope. Applied Physics Letters 89(24)
13. Bandyopadhyay A (2020) Nanobrain: the making of an artificial brain from a time crystal. Boca Raton CRC Press, Taylor and Francis. ISBN: 9781439875490, https://doi.org/10.1201/9780429107771
14. RW Belk 2014 You are what you can access: sharing and collaborative consumption online J Bus Res 67 8 1595 1600
15. N Cavill A Bauman 2004 Changing the way people think about health-enhancing physical activity: do mass media campaigns have a role? J Sports Sci 22 8 771 790 https://doi.org/10.1080/02640410410001712467
16. A Chatterjee (2015) The Aesthetic Brain: How We Evolved to Desire Beauty and Enjoy Art; Oxford Univ Pr on Demand; Reprint edition (September 1, 2015); pp. 217 ISBN-10 : 019026201X; ISBN-13 : 978-0190262013
17. Chirico A, Cipresso P, Yaden D, Biassoni F, Riva G, Gaggioli A (2017) Effectiveness of immersive videos in inducing awe: an experimental study. Scientific Reports 7(1). https://doi.org/10.1038/s41598-017-01242-0
18. Clemens V, Berthold O, Witt A, Sachser C, Brähler E, Plener P et al (2020) Lifespan risks of growing up in a family with mental illness or substance abuse. Scientific Reports 10(1). https://doi.org/10.1038/s41598-020-72064-w
19. P Cappellen V Saroglou 2012 Awe activates religious and spiritual feelings and behavioral intentions Psychol Relig Spiritual 4 3 223 236 https://doi.org/10.1037/a0025986
20. S Chen M Mongrain 2020 Awe and the interconnected self J Posit Psychol 16 6 770 778 https://doi.org/10.1080/17439760.2020.1818808
21. F Choudhry V Mani L Ming T Khan 2016 Beliefs and perception about mental health issues: a meta-synthesis Neuropsychiatr Dis Treat 12 2807 2818 https://doi.org/10.2147/ndt.s111543
22. Cytowic RE (2002) Synesthesia: a union of the senses. MIT Press
23. R Dong S Ni 2019 Openness to experience, extraversion, and subjective well-being among Chinese college students: the mediating role of dispositional awe Psychol Rep 123 3 903 928 https://doi.org/10.1177/0033294119826884
24. O Deroy C Spence 2013 Why we are not all synesthetes (not even weakly so) Psychon Bull Rev 20 4 643 664
25. C Dawes D Cesarini J Fowler M Johannesson P Magnusson S Oskarsson 2014 The relationship between genes, psychological traits, and political participation American J Political Sci 58 4 888 903 https://doi.org/10.1111/ajps.12100

26. H Dittmar R Bond M Hurst T Kasser 2014 The relationship between materialism and personal well-being: a meta-analysis J Pers Soc Psychol 107 5 879 924
27. J Duron L Johnson G Hoge J Postmus 2021 Observing coercive control beyond intimate partner violence: examining the perceptions of professionals about common tactics used in victimization Psychol Violence 11 2 144 154 https://doi.org/10.1037/vio0000354
28. S Evans-Lacko E Malcolm K West D Rose J London N Rüsch 2013 Influence of time to change's social marketing interventions on stigma in England 2009–2011 Br J Psychiatry 202 s55 s77 s88 https://doi.org/10.1192/bjp.bp.113.126672
29. Frentzen E, Reisacher D, Brähler E, Rassenhofer M, Fegert J, Witt A (2022) The predictors of awareness of sexual abuse and sexual violence in the media and the influence on actions of the individuals. Front Psych 12. https://doi.org/10.3389/fpsyt.2021.789144
30. B Fredrickson T Roberts 1997 Objectification theory: toward understanding women's lived experiences and mental health risks Psychol Women Q 21 2 173 206 https://doi.org/10.1111/j.1471-6402.1997.tb00108.x
31. E Garland N Farb P Goldin B Fredrickson 2015 Mindfulness broadens awareness and builds eudaimonic meaning: a process model of mindful positive emotion regulation Psychol Inq 26 4 293 314 https://doi.org/10.1080/1047840x.2015.1064294
32. A Gordon J Stellar C Anderson G McNeil D Loew D Keltner 2017 The dark side of the sublime: distinguishing a threat-based variant of awe J Pers Soc Psychol 113 2 310 328 https://doi.org/10.1037/pspp0000120
33. SJ Gervais TK Vescio J Allen 2011 When what you see is what you get: the consequences of the objectifying gaze for women and men Psychol Women Q 35 1 5 17
34. Ghosh S, Sahu S, Fujita D, Bandyopadhyay A (2014) Design and operation of a brain like computer: a new class of frequency-fractal computing using wireless communication in a supramolecular organic, inorganic systems. Information, 5, 28–99
35. S Ghosh D Fujita A Bandyopadhyay 2015 An organic jelly made fractal logic gate with an infinite truth table Sci Rep 5 1 11265
36. S Ghosh S Sahu L Agrawal T Shiga A Bandyopadhyay 2016 Inventing a co-axial atomic resolution patch clamp to study a single resonating protein complex and ultra-low power communication deep inside a living neuron cell J Integr Neurosci 15 04 403 433
37. N Gavey J Schmidt 2011 Critical feminist psychology, embodiment, and sexual harassment Fem Psychol 21 2 154 161
38. Gómez-Puerto G, Munar E, Nadal M (2016) Preference for curvature: a historical and conceptual framework. Front Hum Neurosci 9. https://doi.org/10.3389/fnhum.2015.00712
39. Hens K, Robeyns I, Schaubroeck K (2018) The ethics of autism. Philosophy Compass 14(1). https://doi.org/10.1111/phc3.12559
40. J Hook E Worthington S Utsey 2008 Collectivism, forgiveness, and social harmony Couns Psychol 37 6 821 847 https://doi.org/10.1177/0011000008326546
41. C Hutcherson E Seppälä J Gross 2008 Loving-kindness meditation increases social connectedness Emotion 8 5 720 724 https://doi.org/10.1037/a0013237
42. M Hamm A Newton A Chisholm J Shulhan A Milne P Sundar 2015 Prevalence and effect of cyberbullying on children and young people JAMA Pediatr 169 8 770 https://doi.org/10.1001/jamapediatrics.2015.0944
43. L Jiang J Yin D Mei Z Hong X Zhou 2018 Awe weakens the desire for money J Pac Rim Psychol 12 e4https://doi.org/10.1017/prp.2017.27
44. A Jorm 2012 Mental health literacy: empowering the community to take action for better mental health Am Psychol 67 3 231 243 https://doi.org/10.1037/a0025957
45. Kelly C, Jorm A, Wright A (2007) Improving mental health literacy as a strategy to facilitate early intervention for mental disorders. The Medical Journal of Australia 187(S7). https://doi.org/10.5694/j.1326-5377.2007.tb01332.x
46. J Katz 2015 The bystander approach to sexual violence prevention Journal of Aggression, Maltreatment & Trauma 24 8 909 926
47. S Kaiser H Kyrrestad S Fossum 2020 Cyberbullying status and mental health in Norwegian adolescents Scand J Psychol 61 5 707 713 https://doi.org/10.1111/sjop.12656

48. R Kowalski G Giumetti A Schroeder M Lattanner 2014 Bullying in the digital age: a critical review and meta-analysis of cyberbullying research among youth Psychol Bull 140 4 1073 1137 https://doi.org/10.1037/a0035618
49. A Klin D Lemish 2008 Mental disorders stigma in the media: review of studies on production, content, and influences J Health Commun 13 5 434 449 https://doi.org/10.1080/10810730802198813
50. K Lawson R Pierson 2007 Maternal decisions regarding prenatal diagnosis: rational choices or sensible decisions? J Obstet Gynaecol Can 29 3 240 246 https://doi.org/10.1016/s1701-2163(16)32412-4
51. K Leung F Brew Z Zhang Y Zhang 2010 Harmony and conflict: a cross-cultural investigation in China and Australia J Cross Cult Psychol 42 5 795 816 https://doi.org/10.1177/0022022110363474
52. N Lee AJ Broderick L Chamberlain 2007 What is 'neuromarketing'? A discussion and agenda for future research Int J Psychophysiol 63 2 199 204
53. Y Lee H Yoon J Joo 2023 Role of fresh start mindset framing in reducing stigma and promoting mental health help-seeking behavior J Consum Behav 22 3 701 720 https://doi.org/10.1002/cb.2155
54. H Leder M Nadal 2014 Ten years of a model of aesthetic appreciation and aesthetic judgments: the aesthetic episode—developments and challenges in empirical aesthetics Br J Psychol 105 4 443 464 https://doi.org/10.1111/bjop.12084
55. A Little B Jones L DeBruine 2011 Facial attractiveness: evolutionary based research Phil Trans R Soc B 366 1571 1638 1659 https://doi.org/10.1098/rstb.2010.0404
56. KA Lonsway LF Fitzgerald 1994 Rape myths: in review Psychol Women Q 18 2 133 164
57. Lecouturier J, Rodgers H, Murtagh M, White M, Ford G, Thomson R (2010) Systematic review of mass media interventions designed to improve public recognition of stroke symptoms, emergency response and early treatment. BMC Public Health 10(1). https://doi.org/10.1186/1471-2458-10-784
58. F Madhani R Professor C Patel C Bann E McClure O Pasha 2016 Women's perceptions and experiences of domestic violence J Interpers Violence 32 1 76 100 https://doi.org/10.1177/0886260515585533
59. K Murphy S Waa H Jaffer A Sauter A Chan 2013 A literature review of findings in physical elder abuse Can Assoc Radiol J 64 1 10 14 https://doi.org/10.1016/j.carj.2012.12.001
60. LE Marks 1975 On colored-hearing synesthesia: cross-modal translations of sensory dimensions Psychol Bull 82 3 303 331
61. Mascaro J, Darcher A, Negi L (2015) The neural mediators of kindness-based meditation: a theoretical model. Front Psychol 6. https://doi.org/10.3389/fpsyg.2015.00109
62. S Nieuwenhuis R Kleijn 2013 The impact of alertness on cognitive control J Exp Psychol Hum Percept Perform 39 6 1797 1801 https://doi.org/10.1037/a0033980
63. Ngo A, Tran A, Tran B, Nguyen L, Hoang M, Nguyen T et al (2021) Cyberbullying among school adolescents in an urban setting of a developing country: experience, coping strategies, and mediating effects of different support on psychological well-being. Fron Psychol 12. https://doi.org/10.3389/fpsyg.2021.661919
64. S Olderbak O Wilhelm 2017 Emotion perception and empathy: an individual differences test of relations Emotion 17 7 1092 1106 https://doi.org/10.1037/emo0000308
65. K Price K Wigg A Nigam Y Feng K Blokland M Wilkinson 2023 Identification of brain cell types underlying genetic association with word reading and correlated traits Mol Psychiatry 28 4 1719 1730 https://doi.org/10.1038/s41380-023-01970-y
66. P Piff P Dietze M Feinberg D Stancato D Keltner 2015 Awe, the small self, and prosocial behavior J Pers Soc Psychol 108 6 883 899 https://doi.org/10.1037/pspi0000018
67. LM Papp EM Cummings MC Goeke-Morey 2009 For richer, for poorer: money as a topic of marital conflict in the home Fam Relat 58 1 91 103
68. SJ Potter VL Banyard 2011 Challenges and opportunities for bystander intervention against sexual violence Prevention Researcher 18 3 3 6

69. B Pradhan A Bandyopadhyay AJ Pal 2004 Molecular level control of donor/acceptor heterostructures in organic photovoltaic devices Appl Phys Lett 85 4 663 665
70. H Plassmann TZ Ramsøy M Milosavljevic 2012 Branding the brain: a critical review and outlook J Consum Psychol 22 1 18 36
71. H Plassmann V Venkatraman SA Huettel C Yoon 2015 Consumer neuroscience: applications, challenges, and possible solutions J Mark Res 52 4 427 435
72. E Reis P Arriaga C Moleiro 2022 Men who suffered intimate partner violence: impressions about existing public campaigns and recommendations for new ones J Interpers Violence 38 3–4 3534 3562 https://doi.org/10.1177/08862605221108075
73. M Rudd K Vohs J Aaker 2012 Awe expands people's perception of time, alters decision making, and enhances well-being Psychol Sci 23 10 1130 1136 https://doi.org/10.1177/0956797612438731
74. A Ross A Morgan A Jorm N Reavley 2018 A systematic review of the impact of media reports of severe mental illness on stigma and discrimination, and interventions that aim to mitigate any adverse impact Soc Psychiatry Psychiatr Epidemiol 54 1 11 31 https://doi.org/10.1007/s00127-018-1608-9
75. S Reddy D Sonker P Singh K Saxena S Singh R Chhajed S Tiwari KV Karthik S Ghosh K Ray A Bandyopadhyay 2018 A brain-like computer made of time crystal: could a metric of prime alone replace a user and alleviate programming forever? Soft Computing Applications 761 1 43
76. Redmond M (2020) Reframing financial abuse of parishioners: an analysis of a church of England disciplinary tribunal hearing regarding rev. Karl Wray. The Journal of Adult Protection 22(2), 93–102. https://doi.org/10.1108/jap-10-2019-0033
77. M Rubel A Werner-Lin F Barg 2017 Expert knowledge influences decision-making for couples receiving positive prenatal chromosomal microarray testing results Cult Med Psychiatry 41 3 382 406 https://doi.org/10.1007/s11013-017-9521-3
78. Sahu S, Ghosh S, Hirata K, Fujita D, Bandyopadhyay A (2013) Multi-level memory-switching properties of a single brain microtubule. Applied Physics Letters 102, 123701
79. G Sampogna I Bakolis S Evans-Lacko E Robinson G Thornicroft C Henderson 2017 The impact of social marketing campaigns on reducing mental health stigma: results from the 2009–2014 time to change programme Eur Psychiatry 40 116 122 https://doi.org/10.1016/j.eurpsy.2016.08.008
80. J Simner EM Hubbard 2006 Varieties of synesthetic experience L Robertson N Sagiv Eds Synesthesia: perspectives from cognitive neuroscience Oxford University Press 17 38
81. P Sedlmeier J Eberth M Schwarz D Zimmermann F Haarig S Jaeger 2012 The psychological effects of meditation: a meta-analysis Psychol Bull 138 6 1139 1171 https://doi.org/10.1037/a0028168
82. A Stylianou 2018 Economic abuse within intimate partner violence: a review of the literature Violence Vict 33 1 3 22 https://doi.org/10.1891/0886-6708.vv-d-16-00112
83. R Schrag K Ravi S Robinson 2018 The role of social support in the link between economic abuse and economic hardship Journal of Family Violence 35 1 85 93 https://doi.org/10.1007/s10896-018-0019-8
84. K Stonard E Bowen T Lawrence S Price 2014 The relevance of technology to the nature, prevalence and impact of adolescent dating violence and abuse: a research synthesis Aggress Violent Beh 19 4 390 417 https://doi.org/10.1016/j.avb.2014.06.005
85. C Swanson D Szymanski 2020 From pain to power: an exploration of activism, the #metoo movement, and healing from sexual assault trauma J Couns Psychol 67 6 653 668 https://doi.org/10.1037/cou0000429
86. P Singh P Sahoo K Saxena JS Manna K Ray S Ghosh 2021 Cytoskeletal filaments deep inside a neuron are not silent: they regulate the precise timing of nerve spikes using a pair of vortices Symmetry 13 5 821
87. K Saxena P Singh P Sahoo S Sahu S Ghosh K Ray 2020 Fractal, scale free electromagnetic resonance of a single brain extracted microtubule nanowire, a single tubulin protein and a single neuron Fractal and Fractional 4 2 11

88. J Stellar A Gordon P Piff D Cordaro C Anderson Y Bai 2017 Self-transcendent emotions and their social functions: compassion, gratitude, and awe bind us to others through prosociality Emot Rev 9 3 200 207 https://doi.org/10.1177/1754073916684557
89. P Singh K Saxena A Singhania P Sahoo S Ghosh R Chhajed 2020 A self-operating time crystal model of the human brain: can we replace entire brain hardware with a 3D fractal architecture of clocks alone? Information 11 5 238
90. K Saxena P Singh P Sahoo S Ghosh D Krishnananda K Ray D Fujita A Bandyopadhyay 2022 All basics that are wrong with the current concept of time crystal: learning from the polyatomic time crystals of protein, microtubule, and neuron Proceedings of Trends in Electronics and Health Informatics. 376 243 254
91. Singh P, Ray K, Bandyopadhyay A (2022) Biological antenna to the humanoid bot: electromagnetic resonances in biomaterials. In: Studies in rhythm engineering (SRE). Springer, Singapore. https://doi.org/10.1007/978-981-16-9677-0
92. P Singh P Sahoo K Saxena S Ghosh S Sahu K Ray D Fujita A Bandyopadhyay 2021 Quaternion, octonion to dodecanion manifold: stereographic projections from infinity lead to a self-operating mathematical universe Proceedings of International Conference on Trends in Computational and Cognitive Engineering. 1169 55 77
93. P Singh P Sahoo K Saxena S Ghosh S Sahu K Ray D Fujita A Bandyopadhyay 2021 A space-time-topology-prime, stTS metric for a self-operating mathematical universe uses Dodecanion geometric algebra of 2–20 D complex vectors Proceedings of International Conference on Data Science and Applications. 148 1 31
94. Thomas M, Sing H, Belenky G, Holcomb H, Mayberg H, Dannals R et al (2000) Neural basis of alertness and cognitive performance impairments during sleepiness. I. effects of 24 h of sleep deprivation on waking human regional brain activity. Journal of Sleep Research 9(4) 335–352. https://doi.org/10.1046/j.1365-2869.2000.00225.x
95. Trautwein F, Naranjo J, Schmidt S (2016) Decentering the self? Reduced bias in self- vs. other-related processing in long-term practitioners of loving-kindness meditation. Front Psychol 7. https://doi.org/10.3389/fpsyg.2016.01785
96. DD Vachon DR Lynam 2016 Fixing the problem with empathy: development and validation of the affective and cognitive measure of empathy Assessment 23 2 135 149
97. KD Vohs NL Mead MR Goode 2006 The psychological consequences of money Science 314 5802 1154 1156
98. Vacher C, Ho N, Skinner A, Robinson J, Freebairn L, Lee G et al (2022) Optimizing strategies for improving mental health in Victoria, Australia during the COVID-19 era: a system dynamics modelling study. International Journal of Environmental Research and Public Health 19(11):6470. https://doi.org/10.3390/ijerph19116470
99. Vartanian O, Navarrete G, Chatterjee A, Fich L, Leder H, Modroño C et al (2013) Impact of contour on aesthetic judgments and approach-avoidance decisions in architecture. Proc Natl Acad Sci 110(supplement_2):10446–10453. https://doi.org/10.1073/pnas.1301227110
100. N Weinbach A Henik 2012 The relationship between alertness and executive control J Exp Psychol Hum Percept Perform 38 6 1530 1540 https://doi.org/10.1037/a0027875
101. J Wielgosz S Goldberg T Kral J Dunne R Davidson 2019 Mindfulness meditation and psychopathology Annu Rev Clin Psychol 15 1 285 316 https://doi.org/10.1146/annurev-clinpsy-021815-093423
102. J Ward 2013 Synesthesia Annu Rev Psychol 64 49 75
103. Z Zhang K Sun C Jatchavala J Koh Y Chia J Bose 2019 Overview of stigma against psychiatric illnesses and advancements of anti-stigma activities in six Asian societies Int J Environ Res Public Health 17 1 280 https://doi.org/10.3390/ijerph17010280
104. EM Zitek AH Jordan 2016 Material wealth and human happiness Psychol Sci 27 4 635 642
105. Zaltman G (2003) How customers think: essential insights into the mind of the market. Harvard Business Press

Chapter 10
The Genesis of Classifying Humans for Their Diversified Brain Hardware in Light of SOMU Theory of Consciousness

10.1 Introduction

The fundamental problem in the theory of consciousness and understanding the brain is a complex and multifaceted issue that has been the subject of extensive research and debate. Consciousness, which refers to our subjective experience of the world and ourselves, raises a range of philosophical questions [40]. It is a phenomenon that is difficult to define and understand, as it involves the integration of various cognitive processes and neural mechanisms. Conventional approaches to understanding consciousness often focus on the contribution of specific brain areas or groups of neurons. However, the problem of consciousness is not limited to the study of localized brain activity. It requires a comprehensive understanding of the neural correlates of consciousness, which involves investigating the complex circuitry and network dynamics of the brain [32]. Decoding the circuitry of consciousness, from local microcircuits to brain-scale networks, is crucial for unraveling the mysteries of consciousness [32].

One of the challenges in studying consciousness is determining the necessary and sufficient processes that underpin healthy human consciousness [31]. Some researchers propose that the thalamocortical system plays a central role in generating consciousness [31]. However, consciousness is not solely dependent on the activity of specific brain regions. It is a dynamic and integrated process that involves the interaction of multiple brain regions and networks [24]. Advances in neuroimaging techniques have provided valuable insights into the neural correlates of consciousness [36]. Neuroimaging studies have revealed the patterns of brain activity associated with different states of consciousness, such as wakefulness, coma, and persistent vegetative state [24]. These findings have contributed to our understanding of the neural basis of consciousness and have implications for the diagnosis and treatment of consciousness disorders [58].

T. Dutta and A. Bandyopadhyay, *Emotion, Cognition and Silent Communication: Unsolved Mysteries*, Studies in Rhythm Engineering,
https://doi.org/10.1007/978-981-99-9334-5_10

In addition to studying the neural correlates of consciousness, there is a need to integrate different techniques and approaches to gain a comprehensive understanding of consciousness [58]. This includes combining neurobiological studies with theoretical analyses of the fundamental properties of consciousness. Theoretical frameworks, such as the Integrated Information Theory (IIT), has been tagged as a pseudoscience very recently by 140 scientists of the same ASSC political group that was campaigning for IIT as an absolute theory of consciousness only because it is infected with the same problem like any other consciousness theories [15]. The fundamental problem in the theory of consciousness and understanding the brain is a complex and interdisciplinary issue. It requires a comprehensive approach that combines philosophical, psychological, neuroscientific, and computational perspectives. By investigating the neural correlates of consciousness, decoding the circuitry of the brain, and integrating different research techniques, we can gain a deeper understanding of consciousness and its underlying mechanisms.

None of the consciousness theories exist today is concerned about discovering a system that operates by itself, not a single theory is a derivative of ab-initio principles. Most importantly, almost all theories of consciousness require a human operator to execute their theory, it means, either a consciousness theory considers consciousness as an essential ingredient to define consciousness, or a human comes as a hidden executer of the theory. It is well argued in the basic premises of a scientific practice that what we want to define should not be assumed just before we start defining it. In that regard, we cannot consider any of the theories of consciousness as truly scientific. As we would see below, that SOMU starts from a point and primarily one effort to make that point undefined create the whole mathematical universe. So, it does not take any liability or assumption to create the whole universe. In other words, "a point wants to remain a point" and "a point wants to become undefined or non-existent" are two drives, essential and sufficient to deliver all required dynamics and spatio-temporal events using phase prime metric, PPM; or Purusha and geometric musical language, GML as Prakriti.

10.1.1 Deciphering Consciousness: The Self-operating Mathematical Universe, SOMU Perspective

The Importance of Self-operating Systems in Modeling the Human Brain

Modeling the human brain is a complex task that requires understanding the organization, development, and function of its complex networks [46]. However, existing brain models often overlook the need for a proper algorithm-free decision-making system that can operate autonomously in unknown situations. This article argues that self-operating systems are necessary for accurately modeling the human brain, as they mimic the brain's ability to encounter and respond to novel situations without external guidance.

The Self and the Brain: The concept of the self is not contained within the brain itself but is learned and operated by the human brain [6]. This suggests that a brain model should incorporate a self-operating system that can adapt and make decisions independently. The self-adaptive nature of the human brain allows it to respond to changing circumstances and learn from experience [35]. By incorporating a self-system in brain models, researchers can better understand how the brain operates and responds to various stimuli [44].

Real-Time Decision-Making: One crucial aspect of the human brain is its ability to make real-time decisions based on personality traits and situational factors [45]. However, the lack of tools for real-time behavioral and situation assessment has hindered direct investigations into this aspect [45]. By developing self-operating systems in brain models, researchers can simulate real-time decision-making processes and study the interaction between personality traits and situational factors [45].

Fault Tolerance and Self-Repair: The human brain exhibits remarkable fault tolerance and self-repair capabilities [59]. Modeling the interaction between cells at a network level can provide insights into fault-tolerant computing beyond current self-repairing hardware architectures [59]. By incorporating self-repair mechanisms in brain models, researchers can better understand how the brain maintains its functionality despite damage or disruptions [23].

Incorporating Additional Inference Mechanisms: While reinforcement learning (RL) models have been widely used in neuroscience research, they may not fully capture the complexity of decision-making processes in the human brain [38]. Additional inference mechanisms that exist alongside RL need to be incorporated into brain models to better reflect the brain's decision-making capabilities [38]. By integrating multiple inference mechanisms, researchers can develop more comprehensive models of decision-making in the human brain. To accurately model the human brain, it is essential to incorporate self-operating systems that mimic the brain's ability to encounter and respond to unknown situations. Existing brain models often overlook this aspect, focusing more on structural connections and functional networks. By integrating self-operating systems, researchers can better understand the brain's decision-making processes, fault tolerance, and self-repair capabilities. This will contribute to the development of more comprehensive and accurate brain models that can advance our understanding of the human brain.

Unlocking the Fundamental Properties of Conscious Machines and Beings

Consciousness and life, two intricate facets of existence, have perplexed scientists, philosophers, and thinkers for centuries. Yet, the Self-Operating Mathematical Universe (SOMU) offers a unique mathematical lens through which we can unravel the enigma of consciousness [54, 56, 57]. In this section, we embark on a journey to explore the fundamental properties that define a conscious machine or being, shedding light on the intricate tapestry of life and consciousness, as described in the Chaps. 5 and 6.

10.1.1.1 Defining Consciousness and Life Through SOMU

Let's embark on this exploration by delving into the SOMU definition of consciousness. According to SOMU, consciousness is the remarkable ability of an entity to generate an infinite series of invariants derived from nature. These invariants are not static; they evolve, adapt, and transform. The conscious entity, in this process, acquires feedback from infinity, ensuring that the fractional contribution of each element of the series aligns with the sustainability of the whole.

This practice of acquiring values sustainably requires a unique skill set—a conscious entity must simultaneously coexist as multiple equivalent information structures while editing one of its morphogenic structures. This editing process of integrated information structure demands the presence of at least three mathematically equivalent forms, a concept embodied in the ordered factor metric [43].

This multifaceted process of editing and evolving information structures serves a vital purpose. It enables a conscious entity to harmonize with the mathematical universe, either by revealing unknown symmetries or creating a mirror where the universe can generate its replica within the conscious entity's hardware.

But here's the twist: Individual consciousness cannot be isolated from the whole. Just as isolating any element from the information structure destroys its content, attempting to dissect individual consciousness from the collective consciousness is an exercise in futility. The universe, according to SOMU, operates as a chain of various degrees of consciousness, with each element contributing to the whole and undergoing continuous redefinition in the process.

The Quest for Primes: Nature's Ingenious Creation: Life forms, in their quest for primes, become the vehicles through which nature invents new dynamics of primes. These living entities expand the metric of primes within their hardware, contributing to the ever-growing repository of mathematical constructs in the universe.

10.1.1.2 The Conscious Machine: Eight Critical Features

Now, let's examine eight critical features that, if present in a machine, would endow it with key conscious attributes, as envisioned by SOMU. These features were meticulously crafted by philosophers and present a formidable challenge for realization within a finite or hypothetical infinite state machine.

1. **The Holistic Universe Emulator:** A conscious machine C should possess the ability to emulate the universe's pattern of information creation within its own framework. The universe's information, denoted as U, should seamlessly integrate into the conscious machine's repository. This integration extends beyond mere quantity; it entails an identical method of information creation.

2. **Multiplicity of Information Structures:** Conscious machines should demonstrate the capacity to create and manage at least three complete information structures, or isomorphs, concurrently. These isomorphs can undergo simultaneous editing and evolution, resulting in distinct qualitative arrangements.
3. **Continuous Expansion:** Conscious entities must continuously regenerate the whole with an expanding reservoir of information content, all within the same hardware. The capacity to create and expand the whole within itself demonstrates the machine's completeness and self-sustaining nature.
4. **Invention of Invariants:** Conscious beings possess the ability to invent invariants or conserved laws by expanding existing information content. The quest for higher dimensional invariants follows a helical loop, leading to the creation of entirely new symmetries.
5. **Creation of Infinity:** A conscious machine should harness the power of multiple replicas of the whole to create infinity—a concept that transcends the limitations of finite systems.
6. **Dimensionless Complexity:** While remaining a dimensionless, indefinable entity or a point, a conscious machine should be able to include an arbitrary number of whole information contents. This phenomenon leads to a continuous state of undefinedness.
7. **Processing the Life Cycle of Events:** Conscious machines must be capable of processing the life cycle of three fundamental properties of events, including spatial, temporal, and spatio-temporal components. These components are intricately intertwined and dominate in the constitution of the machine.
8. **Progeny Within and Above:** Finally, a conscious entity should demonstrate the ability to create progeny wholes both within and above, in both directions and simultaneously, creating a never-ending lineage of consciousness.

The Self-Operating Mathematical Universe (SOMU) theory offers a profound perspective on the nature of consciousness and life. It unveils the intricate dance between consciousness, information, and the universe itself. As we journey through the mathematical constructs of SOMU, we come to appreciate the boundless complexity and beauty of conscious machines and beings. These entities, shaped by the universal symphony of primes and invariants, represent a remarkable exploration into the heart of existence.

A Short Note on the Astrophysics and Human Brain:

Origin of universe and the brain might have a relation, though both in operational scale and use of forces, two elements cannot be compared. The interconnectedness of astrophysics and the human brain, are so profound that highlighting the need for a grand unification theory that can explain both astrophysical structures and consciousness could be a need of the day, and SOMU could shade a light on both.

Astrophysics and Brain Circuitry: Astrophysics focuses on understanding the structure and dynamics of celestial objects and the universe as a whole. Similarly, the human brain consists of intricate neural circuits that govern cognitive processes and consciousness. Both fields recognize the importance of studying local and long-range connections within their respective systems [22, 63]. In astrophysics, long-range connections refer to the interactions between celestial bodies across vast distances, while in neuroscience, they refer to the communication between different brain regions [33]. Understanding the connectivity patterns in both astrophysical structures and brain circuits is crucial for unraveling their complex behaviors.

Mapping and Connectomics: Advancements in technology have enabled researchers to map the structure and function of both astrophysical structures and the human brain. The Brain/MINDS project in Japan aims to map the neuronal circuits of the human brain using a non-human primate model [34]. Similarly, researchers have developed techniques to map the connectome of the mouse brain, providing insights into the mesoscale organization of neural circuits [37]. These mapping efforts in both fields have revealed the presence of local circuits within specific brain regions or celestial objects, as well as long-distance connections that span across different areas [33].

Consciousness and Astrophysical Structures: One of the most intriguing connections between astrophysics and the human brain lies in the exploration of consciousness. While astrophysics primarily focuses on the physical properties of celestial objects, the study of consciousness delves into the subjective experience of being aware. The grand unification theory that can explain both astrophysical structures and consciousness remains elusive. However, researchers have started to explore the potential links between the two fields. For example, the Human Brain Project in Europe aims to decode the multi-scale organization of the human brain, which could provide insights into the nature of consciousness [2]. Additionally, studies have investigated the spectral dynamics of brain activity during general anesthesia, shedding light on the neural correlates of consciousness [21].

10.1.2 Unraveling the Enigma of Consciousness: The Self-operating Mathematical Universe (SOMU) Theory

Consciousness, the enigmatic essence of our existence, has long eluded a comprehensive understanding. Traditional approaches to consciousness have often left critical questions unanswered. However, the Self-Operating Mathematical Universe (SOMU) theory of consciousness offers a revolutionary perspective, shedding light on profound aspects of consciousness that have remained unexplored. This section delves into the unique features of SOMU's theory of consciousness, which challenge conventional wisdom and ignite new avenues of inquiry.

1. **Autonomous Operation of Consciousness: The Phase Prime Metric (PPM)**

SOMU introduces the concept of autonomous consciousness—an entity capable of self-operation. The phase prime metric (PPM) serves as a universal archive, housing the fundamental building blocks of the universe. Conscious systems, under SOMU, tap into the PPM, enabling independent operation. This raises questions about the self-sustaining nature of consciousness and its inherent connection to the fabric of the cosmos [8].

2. **The Multifaceted Identities Within a Single Mind: Triplet of Triplet Material Resonance**

   SOMU's theory unravels the mysteries of multiple identities within a single physical substrate. The Triplet of Triplet Material Resonance emerges as the mechanism behind the creation of distinct identities. Additionally, SOMU extends proprioception to out-of-body experiences (OBEs), allowing the mind to transcend the confines of the body and observe itself—a profound exploration of self-awareness.

3. **Transitions Between Conscious States: The Polyatomic Time Crystal** [43]

   Fusion of time crystals [48] make SOMU a robust information engine synthesizer that works like a very complex algorithm. SOMU delves into the transitions between conscious states, ranging from proto-consciousness to super-consciousness, subconsciousness, and even altered states induced by psychedelics, by transforming every single event in terms of 3D clock assembly, under the framework of Fractal Information Theory, FIT [4]. The polyatomic time crystal with 12 dimensions, representing the mind, can localize at various positions within the phase prime metric or PPM, offering insights into the regulation of conscious states.

4. **The Role of Consciousness in Reinforced Learning: Space–Time-Topology-Prime (STts) Invariant Structure**

   SOMU highlights how consciousness shapes and reinforces learning, enabling minds to adapt to unknown situations and drive discoveries and inventions throughout life. The Space–Time-Topology-Prime (STts) invariant structure autonomously identifies missing symmetries within the PPM, facilitating adaptive cognition.

5. **The Nature of Consciousness: Beyond Life and Death**

   SOMU challenges the conventional notion of consciousness as emergent or tied to mortality. It posits consciousness as a mathematical property of primes, devoid of life or death. Consciousness transcends matter, existing as a network of symmetries guided by the rules of symmetry breaking.

6. **Free Will and Self-Awareness: The Unencoded Element of PPM**

   The question of free will and its connection to self-awareness remains a focal point. SOMU proposes that while the PPM can be partially encoded in hardware, free will—the missing piece—remains elusive, allowing for the autonomy of choice and self-determination.

7. **Universal Qualia and the Binding Problem: Morphogenesis of Geometric Shapes**

   SOMU unveils the origin of universal or personal qualia and addresses the binding problem. It posits that all cognitive experiences are the morphogenesis of geometric shapes, providing a foundation for the unity of conscious experience.

8. **Simultaneity and Quantum Mechanics: A 12D Mechanics Beyond Quantum**

SOMU delves into the perception of simultaneity in consciousness. It challenges whether this phenomenon is an intuitive illusion or a product of quantum or post-quantum phenomena. The theory of consciousness in SOMU operates within a 12D mechanics, transcending the boundaries of quantum physics.

9. **Evolutionary Significance: The Expansion of PPM in Conscious Life Forms**

SOMU explores the necessity of continuous hardware evolution for consciousness and its potential role as a byproduct of other evolutionary processes. It also raises the question of whether a universal rulebook for consciousness governs life forms across the universe, paralleling the expansion of the PPM of the universe.

10. **The Purpose of Consciousness in Nature: Expanding the PPM**

Finally, SOMU suggests that consciousness serves as a tool of the universe—a means of mining unique prime numbers and prime compositions to expand its own PPM. Conscious life forms become agents of the cosmos, contributing to the universe's perpetual expansion.

The Self-Operating Mathematical Universe (SOMU) theory of consciousness is a paradigm-shifting perspective that challenges the boundaries of our understanding. It explores the profound intricacies of autonomous consciousness, multiple identities, transitions between conscious states, and the role of consciousness in shaping reality. SOMU invites us to rethink the very nature of consciousness, opening doors to new realms of inquiry and discovery. In doing so, it beckons us to embark on a journey toward a deeper comprehension of the universe and our place within it.

10.1.3 SOMU Theory of Consciousness: Axioms and Corollaries

In SOMU theory of consciousness, we start with Axiom 1, which lays the foundation for understanding consciousness C as an emergent property of a complex system U, that is whole.

Axiom 1: The Architecture of Multidimensional Invariants:

Axiom 1 The universe is represented as an automaton, denoted as U. U synthesizes and stores invariants by coupling four fundamental entities: An invariant made of space, time, topology, and prime (STts) entity is used as information units. These entities are interconnected within and above in a network or a fractal tape, where each cell holds a loop of cells inside.

Corollary 1.1 Information processing grows within and above in fractal tape by creating invariants: Within this network, U creates conscious entities, denoted as C, when a new automaton, emulated by C, can replicate at least 86% of the information contained in the universe U. The effort exerted by C in emulating U is defined as consciousness. A conscious entity C_j is part of a conscious entity C_i and holds a conscious entity C_k inside, conscious entities $\{Ci\}$ grow fractally in U.

Corollary 1.2 Space–time-topology-prime or STts as a single inseparable entity: STts operates in parallel universes, leading to the emergence of 12 dimensions (1D–12D) as a result of interactions between space, time, topology, and prime entities. Each dimension has a distinct identity, and they cannot be defined using lower dimensions. Elements of higher dimensions are orthogonal stereographic projections of lower dimensions.

Corollary 1.3 Conspatial, conformal, conmorphic, and conprimal decomposition: The elements of C set are space–time-topology-prime entities that measure difference in symmetry and store invariants as 3D geometric shapes, morphogenic loops and prime numbers, as memory. This leads to a shift from traditional space–time-based processing to a space–time-topology-prime quartet, making the information system-independent, attribution free, bias free.

Axiom 2: A Triplet of Three Basic Properties Bonding, Projecting, and Transforming:

Axiom 2 C creates multiple information structures, C_n, within the same entity using SRT invariant-triplets (SRT = quality triplet). These triplets consist of three fundamental properties: S (feedback from infinity), R (transformation), and T (bonding). The combination of these properties forms a polyatomic time crystal [48].

Corollary 2.1 Information structure is a hollow phase sphere with twelve holes, made by any field, projected as vortex: Events in the universe exhibit cyclic and rhythmic behavior, written as boundary loops of these 12 holes or singularities, represented by three angular momenta and rotation phases. These properties are fundamental in every element constructing the universe U and its parts $\{C\}$.

Corollary 2.2 Information unit of a C automaton: The SRT network (S = feedback from infinity, R = transformation, T = bonding) generates and emits 3D field structures or vortices that other SRT networks absorb. These structures encode the memory of SRT invariants and contribute to the overall information architecture of C and U.

Corollary 2.3 A prime inventing engine, phase prime metric or PPM or Purusha [8] that selects resources and dimensions, act as universal code that creates 6D non-physical morphogenesis-prime duality, which finally generates 6D space–time duality: U and its every element C seeks to discover new prime numbers to enrich its dynamics, expanding the phase prime metric, PPM. This quest of PPM is akin to a deep learning network, with the number of layers and cells determined by available invariants and variables.

Axiom 3: An Architecture of Undefined Singularity Points:

Axiom 3 All C automatons possess the ability to convert a point into an undefined entity by adding points one by one, following the metric of primes $N \to N + 1$, N is an integer, and still keep it a point by normalizing all points within. This ability is called Rita, and it ensures that C acquires values sustainable to the whole U.

Corollary 3.1 Normalization synthesizes dynamics: U remains a dimensionless, undefined entity, and normalization plays a crucial role in mapping the behavior of points inserted within a point, accounting for clockwise and anti-clockwise rotations, periodicity, and the dominance of prime numbers. U is an undefined point, so is {C}. A point adds arbitrary number of points to make it undefined and normalization ensures that it remains a point, two drives make the U and every single element of {C}, all evolve together to become more undefined or non-existent.

Corollary 3.2 Universal 3D symmetry mapping engine, PPM, Purusha generating polyatomic time crystal PTC made of SRT: The density of prime numbers drives the creation of 3D symmetry mapping structures, known as fractal temples for each primes, representing various symmetries of the universe U. At any imaginary world in the fractal network of worlds in U, an entity C experiences superposition of multiple temples or fractals of primes.

Corollary 3.3 Seven invariant banks, each with 108 elements account for non-realism: Invariants, including geometric transformations and primes, eliminate realism in C measurements and observations. Seven invariant banks map conscious perceptions, sensory responses, periodicities, queries, fundamental constants, and forms of undefinedness. These are geometric shapes (Chap. 5) representing prime numbers, at least 108 elements are necessary since triplet of triplet band for 12 dimensions ($3 \times 3 \times 12 = 108$) need one element each for C to emulate a part of U.

In SOMU theory of consciousness, these axioms and corollaries provide a framework for understanding consciousness as a product of complex interactions between space, time, topology, prime numbers, and fundamental properties of invariants/variables. This theory presents a unique perspective on the nature of consciousness and its relationship with the universe.

10.1.4 SOMU is a Derivative of Polyatomic Time Crystal Model of the Human Brain

Twelve Dimensions of a Conscious Mind; How Twelve Dimensions Operate: SOMU theory of consciousness was originated from the polyatomic time crystal model of the brain [43, 53], where we found that neuron membrane is not everything as it was believed for a century. Simultaneous measurement of neuron firing at the neuron membrane and microtubular bundle inside a neuron revealed that microtubules process firing signals 10,000 times faster than membrane [3], they are not silent in brain's information processing [54, 56, 57]. Even non-connected by synaptic junctions could exchange energy as neural circuit is imaged by scanning dielectric microscope, SDM [17]. Therefore prime number of discrete time domains interacting with each other, create a pattern in the temporal domain compelled us to think that for brain and consciousness time is a cyclic feature, as Hindu scholars argued. Create three dimensions of time, phase cycle as 4D, creating topologies using phase singularities as 5D and create world lines connecting the centers of those geometric shapes as 6D. Therefore, three dimensions of space and three dimensions of time led us to 6D spatio-temporal dynamics and we can represent entire 6D physically measurable system using a 3D clock assembly.

The journey from 6 to 12D is non-physical, where first attempt was made to get rid of observer dependent parameters. So, a sincere effort was made to find elements that could replace spatio-temporal dynamics. The solution was a typical set of geometric shapes that run in a loop. This was a brilliant move in the sense that loop ensures that the closed loop features of the time factor is taken into account and gradual changes in the geometric shape ensures that spatial factor is included. SOMU calls it morphogenesis and similar to space and time, SOMU explores three dimensions of morphogenesis. The inspiration behind creating three dimensions of time was exploring singularity or undefined phase region where lower dimensional universe could be embedded, thus came the concept of time crystal. Similarly, for morphogenesis, we needed knots that is born out of cross-overs of the corners of the geometric shapes, so that eventually we reach numbers as invariants. Thus, the quest for layered imaginary worlds gave three dimensions of time and the quest to make universe governing element shape-free, bias and attribution free, three dimensions of morphogenesis were required.

Three dimensions of morphogenesis are as follows: In the seventh dimension we see many geometric shapes running in a loop, one of the characteristic mathematical features of these shapes running in a loop is that a single set of polar equations would create all geometric shapes that run in a loop. For physicists this is beautiful in the sense that now, be it protein folding, star formation, liquid drop transforming under gravity, sperm is interacting with ovum, one morphogenesis could explain it all. For the 8D, all corners are replaced by knots and several geometric shapes pass through the knots. Thus, we create intertwined "knots of darkness". These knots and loops create an infinite series of fractal structures in the 9D, they look like the phase prime metrics drawn for different prime numbers [8]. The similarity between 6 and 9D is striking. In 6D world lines connect imaginary worlds and in 9D the fractal infinite series connects knots in an attempt to build a homogeneous seed pattern.

For psychologists the journey from 7 to 9D is finding a confusion as knots, note the architecture of confusion we have talked about in the Chap. 5, we need it at all dimensions, while for 1D–3D it's the quest for a mass, for 4D–6D its phase singularity, for 7D to 9D its knots intertwined lines that challenges the corners and of a geometric shape, for 10D to 12D it would be prime numbers. The quest for singularity is never ending from 1 to 12D and effort to reach higher dimensions is always finding invariants at the lower dimensions.

Cognitive elements are invariants that are geometric shapes built using the geometric phase. What is a geometric phase? Say, a pendulum is swinging, its dynamic phase changes, but imagine its rotating around its axis too, that spin like motion delivers geometric phase because it does not relate to the energy drive and exchange process of the whole particle, its origin is internal dynamics. That's why geometric phase is the so important. However, the origin of geometric phase is not so simple. Membrane microtubule junction of a neuron cell plays a crucial role in neuron's time modulation that structures our conscious thoughts, therefore adiabatic heterojunctions are important [41] for signal transfer in a fractal structure.

Electromagnetic resonance of biomaterials enables two possibles at a time [55]. First, it allows wireless classical resonant energy transfer across a long distance, for example, a MHz signal could link two materials 300 m apart, and brain is fraction of a meter. Second, electromagnetically resonating materials could squeeze their resonance bands and exchange quantum information. The fractal, scale-free electromagnetic resonance of a single brain extracted microtubule nanowire, a single tubulin protein and a single neuron, suggests that triplet of triplet resonance band or nine frequency or time domains in discrete groups could operate together activating and deactivating one or multiple resonance peaks [47]. Quantum tunneling through these electromagnetically resonating junctions could be used to read, write information in such a network [7], this ensures how geometric phase could build a structure of quantum codes written inside a molecule to operate in the classical domain. A fractal network could process such information logically [16], where a decision could be expanded into an infinite series.

10.2 Mirror Drawing, Thermal Mapping of Face for Emotion, Cognitive Tests Suggest that Human Subjects Should be Classified for Psychology Studies

10.2.1 Classic Notions of Psychology Has to Fall Apart

It has always been a media campaign that left brain is responsible for intelligence and right brain is responsible for actions and physical responses. In a normal right hander it is usually assumed that the left hemisphere is for executing sequential activities while the right hemisphere is for holistic functioning. The truth is if this is known a lot is not known about how the two hemispheres interact to generate behavior.

However, in reality both brains share information so intricately that differentiating tasks assigned to a singular lobe has been extremely difficult. Here in this book we have made a sincere effort to analyze the parity between the two lobes for four different tasks. In Chap. 1 we have concentrated on the mirror drawing task where we monitored the human subjects tracing the star path made of parallel lines with a gap of 1.5 cm. We found that human subjects who cannot hold synchronization for a longer time takes an extremely long time, in some cases more than 30 min to complete the task. On the other hand, those who complete mirror drawing task very fast within 1.5 min, with minimum errors, their brain is NOT synchronized. It was a shocking observation, because normally we expected that both the brains would synchronize for those who perform mirror drawing tasks rapidly. In a sharp contrast we found that those who perform the task rapidly actually their left and right brain makes an effort to synchronize by exchanging energy. The active regions shift periodically between the left and the right brain.

Second surprise came to us in Chap. 2 when we were analyzing the emotional map of individuals. In this case using an infrared camera we were simply imaging the thermal signature of the face for 89 subjects. We found contrary to the popular belief that emotion is defined to a particular part of the face, it depends on the category of the people concerned. 9–10 categories of people were noted, where the typical part of the face is involved in rapid thermal fluctuation when a human subject undergoes a major emotional switching. Such a categorization is unfound in any previous literature. However, most interesting was asymmetry in the face response. Left or right side domination in the expression of emotion is not just limited to the brain surface but in the face too.

10.2.2 Amazing Seven Facts About Emotions

1. Emotions are continuous electrochemical signals in our bodies, responding to our perception of the world. They play a crucial role in our well-being and constantly affect our thoughts. So, why don't we prioritize learning about emotions alongside academic subjects like history and math? Emotions both originate in and travel through our brains and bodies, serving as feedback loops to help us function effectively.
2. There are 8 fundamental emotions, with numerous variations and subtleties. These basic emotions include joy, sadness, fear, disgust, surprise, anticipation, anger, and trust. Robert Plutchik, a scientist, created a Wheel of Emotions to illustrate these emotions visually, showing their dynamic interactions. Understanding this wheel can enhance emotional literacy, a key component of emotional intelligence.
3. Emotions are neutral, not inherently good or bad. Joy and sadness, for instance, both stem from our attachment to something. Every emotion conveys a message, even challenging ones like fear, anger, or sadness, which serve important functions.
4. Emotions guide us to survive and thrive by focusing our attention and motivating specific actions. For instance, anger signals obstacles in our path, prompting us to either confront or overcome them. Joy directs our attention to opportunities, motivating us to pursue meaningful and connecting experiences. Tools like the Emotoscope Feeling Chart help decode the purpose and meaning behind different emotions.

5. Emotions are contagious, spreading between people, even unconsciously. This phenomenon stems from our evolutionary history as social creatures. We pick up emotional cues through various channels like facial expressions, tone of voice, and body language. Emotional contagion occurs not only in face-to-face interactions but also on social media platforms.
6. Emotions, feelings, and moods are related but distinct concepts, primarily differing in duration and cognitive involvement. To delve deeper into these differences, you can explore the section below "What's the Difference Between Emotions, Feelings, and Moods?"
7. Emotions are processed in the body within about six seconds, from their origin in the hypothalamus to complete absorption. This time frame inspired the name "Six Seconds" for our organization. Recognizing, evaluating, and choosing how to deal with our emotions is at the core of emotional intelligence, a skill we can cultivate to navigate life effectively. To learn more about emotional intelligence, you can explore "Get Started with Emotional Intelligence."

What's the Difference Between Emotions, Feelings, and Moods?

Emotions, feelings, and moods are all part of the complex landscape of human affective experiences. While these terms are often used interchangeably in everyday language, they have distinct meanings and characteristics in the field of psychology.

Emotions can be defined as intense, short-lived, and specific reactions to a particular stimulus or event [20]. They are typically accompanied by physiological changes, such as changes in heart rate or facial expressions [39]. Emotions are often described using specific labels, such as happiness, anger, or fear, and they have a clear object or cause [51]. For example, if someone receives good news, they may experience the emotion of joy.

Feelings, on the other hand, are more subjective and can be described as the conscious experience of emotions [51]. They are often considered to be less intense and longer-lasting than emotions [50]. Feelings are more focused on the internal experience of an emotion, rather than the external stimulus that triggered it. For example, someone may feel a sense of contentment after spending time with loved ones, even if there was no specific event that caused this feeling.

Moods, in contrast, are more diffuse and long-lasting states that are not necessarily tied to a specific event or stimulus [18]. Moods are often described as a general background emotional tone that colors a person's overall experience [60]. They can persist for hours, days, or even longer [50]. Moods are typically less intense than emotions and may not have a clear cause or object [49]. For example, someone may wake up in a good mood or a bad mood, without any specific reason for feeling that way.

While emotions, feelings, and moods are distinct, they are interconnected and can influence each other. For example, a person's mood can influence how they interpret and experience emotions [30]. Additionally, emotions can contribute to the overall mood state of an individual [29]. Understanding the differences between these terms is important for researchers and clinicians in the fields of psychology, psychiatry, and neuroscience, as well as for individuals seeking to better understand and manage their own emotional experiences.

10.2.3 When Brain Hardware is Different, Software or Mind Has to Operate Differently

Third surprise came under the cognition study described in Chap. 3, we cleverly selected the test so that a person has to perform two tasks in a sequence in order to respond to a query. We found that when a subject executes a task that is known to use the left brain and responds using the right brain, we thought there would be a straightforward energy transfer from left to the right brain. However, that did not happen. Before reaching the right brain, the activity followed a very unique pathways common to many subjects. Again, we could categorize the subjects. Similar phenomenon was observed when we transferred information from right brain task and expression using the left brain. Even then we found excitation of the brain follows a typical pathway before signal transfers from one side of the brain to another.

In Chap. 4 we carried out silent communication where cheating like murmuring and lip movements were not allowed in a group of 8 subjects. Topic for discussion was random, and hardly people knew each other in a group. We found communication between subjects with synchronized brain activities in the EEG, however, left or right brains were active and those who participated in the real communication as noted in the similar pattern changes in EEG, they were trying to sync the left and the right brain similar to the mirror drawing task.

Therefore, in all the four chapters we encountered a situation where common faiths were shattered and classic left brain right brain parity concepts were under grave illusions. Two points that excited us for further thoughts. When a subject performs very well in any task his or her left and right brains do not have to synchronize but get periodically active. And, human subjects do not have a similar hardware response for similar emotions or cognitive tasks. No psychological experiments should be performed considering all humans as a perfect null hysteresis test subject. Human categorization based on their emotional responses are essential, else test results for any psychological tests could be erroneous.

Individual differences in the brain functioning help to understand why we process and respond to the same stimulus differently. It has a lot of implications for human behavior in terms of cognition and affect. One of the priorities of future research could be to improve upon methodological constraints for an objective understanding of brain functioning.

10.3 Silent Communication, Collective Communication Suggests that Consciousness is Extended Beyond a Single Human Subject

10.3.1 Silent Communication Among Group Members: Unveiling the Dimensions of Consciousness

The quest to unravel the mysteries of consciousness has long been at the forefront of scientific inquiry. However, the traditional approach often confines itself to analyzing nature's reactions to external stimuli in controlled laboratory settings, leaving much to be explored regarding silent communication among group members and their collective response to common stimuli. In this section, we embark on a journey that transcends conventional paradigms and delves into the realms of a multidimensional language that enables simultaneous interaction with diverse dimensions. This language, based on a tensor-manifold framework, holds the promise of decoding the intricate web of consciousness, ultimately leading to the emergence of a Self-operating Mathematical Universe (SOMU).

From Observations to Beliefs: Scientific understanding often begins with observations, which are subsequently translated into beliefs. The credibility of these beliefs relies on the replication of the same observations by multiple researchers following identical protocols. This repetition transforms biased perspectives into widely accepted truths. However, this approach, as exemplified by [62], is limited to perceiving only two interacting systems at any given time, leaving us to grapple with the challenge of comprehending more complex interactions.

The Birth of SOMU: Enter the concept of a tensor-manifold language, designed to facilitate communication across multiple dimensions simultaneously. At its core is the preservation of phase and metric (PPM) archive, an all-encompassing repository that preserves every degree of freedom within a system. This archive serves as the key to a self-sustaining mathematical universe (SOMU).

Topology of Silence and Maximum Choice: To access and interpret the PPM archive, two crucial elements come into play: the topology of silence and the topology of maximum choice. Silence represents information, while maximum choice signifies the path to integrating this information. Together, they form the foundation of SOMU, a universe that evolves and grows within itself.

The Challenge of Conscious Machines: In a world where conscious machines constantly encounter novel situations, the ability to adapt and respond swiftly is paramount. Envision a computer that autonomously creates variables, formulates algorithms, trains its deep learning network, predicts future scenarios, and responds instantaneously. Achieving such a feat is contingent on a common archive that governs all conscious entities and the natural world.

Human and Artificial Validation

The critical features of this archive are not mere theoretical constructs but have undergone rigorous testing and validation through human subjects and artificial brain studies. SOMU offers a unique perspective on a universe that has expanded within and above its own dimensions, unveiling new tools and pathways of communication.

10.3.2 The Fractal Universe and the Fractal Consciousness

SOMU's approach diverges from the traditional Turing philosophy. While Turing's linear paradigm manipulates choices (0 and 1) along a sequential path, SOMU transcends these limitations. It thrives on overlapping dimensional features and intricate communication pathways that link disparate dimensional worlds, creating a self-operating system that defies the constraints of linearity.

Building a New Scientific Protocol

In this fractal-like universe, traditional tools such as circuits, logic gates, and transistors become obsolete. Instead, we must develop alternative scientific protocols that align with the complexities of a universe that grows within and above itself. SOMU presents a framework for understanding the evolution of the universe, from the microcosm of proteins to the grand tapestry of biological life. The journey to decipher consciousness and engage in silent communication among group members has taken an unconventional turn. Through the lens of SOMU, we have glimpsed a universe that defies linear conventions and explores the multidimensional intricacies of existence. This evolving language of archives promises to reshape our understanding of the world around us, leading to new insights and discoveries in the realm of consciousness. As we continue to explore this uncharted territory, the path to enlightenment becomes clearer, and the boundaries of what we can comprehend expand ever further.

10.4 Primordial Evolutionary Love and Pure Awe Impacts Deeply into Negativities in a Human Psychology

10.4.1 Primordial Evolutionary Love and Pure Awe: Transforming Human Psychology

Human psychology is a complex tapestry woven from the threads of our experiences, emotions, and interactions with the world around us. It's a realm where both darkness and light coexist, where negativity and positivity vie for dominance. In this intricate dance of the mind, there is a profound force at play—primordial evolutionary love and pure awe. These two concepts, when viewed through the lens of polyatomic time crystals and the network of invariants, reveal a transformative power that impacts our psyche at its very core.

The Foundations of a New Understanding

To comprehend the profound impact of primordial evolutionary love and pure awe, we must first delve into the fascinating world of polyatomic time crystals. Inspired by research from the 1960s, these time crystals challenge conventional wisdom about the nature of reality. While some definitions suggest that time crystals are harmonics of externally applied frequencies, a different perspective emerges—one that explores an harmonic clocks with periodicities spanning orders of magnitude. This perspective suggests that true time crystals may manifest as spontaneous breaks in time symmetry.

In this paradigm, geometric musical language (GML) emerges as a revolutionary concept. It envisions living systems, including the human brain, as 3D spatial assemblies of clocks. These clocks operate by harnessing various sources of energy, such as thermal noise, electric potential, and even food in the form of GDP hydrolysis. This intricate dance of clocks forms a self-sustaining structure that underlies the very essence of life itself.

The Network of Invariants: Unveiling the True Information

Central to this understanding is the network of invariants—the unchanging core of sensory signals in GML. In this framework, sensory information is expressed as geometric shapes composed of clocks. These shapes, known as invariants, remain consistent across various signals. It is from this archive of invariants that the true essence of information emerges.

The shift in our understanding is profound. Invariants transcend physical representations and transform into abstract mathematical concepts of significance when enriched with space, time, topology, or prime numbers. This transformation parallels the ancient Sanskrit language, where fundamental sounds were believed to trigger emotional and perceptual responses—a language of the universe itself.

The Infinite Symmetries of the Universe

Symmetry, the ordered arrangement of elements in the universe, is a recurring theme. The text posits that all symmetries, seemingly distinct, are interconnected. They are linked by hidden patterns encoded within the distribution of prime numbers. Discovering the rules governing these connections would unite the wisdom of the ages, predicting the breaking of symmetries in parallel and sequence.

The Fractal Machine and the Turing Machine

A profound comparison arises between the fractal machine and the Turing machine. While the Turing machine follows predetermined paths, the fractal machine stores all possible symmetries and their potential evolutions. It doesn't require pre-imagined code but adapts to unknown compositions of symmetries, expanding its understanding continuously.

10.4.2 Consciousness in a Growing Body

Perhaps the most intriguing question posed in this context is how consciousness persists in a continuously evolving body. Every minute, millions of cells in our body perish, replaced by new ones. Yet, our consciousness endures. This challenges the conventional notion that memory must be retained across eons. Instead, it suggests that as long as the metric of primes exists, consciousness thrives. Under the framework of SOMU theory of consciousness, the information is an invariant structure, which does not change with any variables. That's why it does not matter that all conscious entities in the universe changes or transforms it to match the whole or U, the classes of invariants as noted in Fig. 10.2 remains unchanged.

10.4.2.1 Impacts on Human Psychology

Now, let's return to the central theme—primordial evolutionary love and pure awe (Chap. 7). In the context of this intricate web of ideas, it becomes clear that these profound emotions are not mere human constructs but are deeply embedded in the fabric of the universe itself. They are the driving forces that connect us to the infinite symmetries, invariants, and clocks that shape our existence. Primordial love generates from the prime numbers under the framework of SOMU where we find that density of primes in the 15 different primes used for constructing human brain [53]. From the metric of primes [8] one could determine 15 different classes of morphogenesis which are strikingly similar to the invariants found in the DDG data as shown in Fig. 10.1.

Primordial evolutionary love, a force that drives us to connect, to care, and to nurture, holds the power to heal the negativities in our psychology. It is the love that transcends the boundaries of the self, resonating with the interconnectedness of all things. It reminds us that we are not isolated individuals but threads in the grand tapestry of existence.

Pure awe, the overwhelming wonder in the face of the universe's mysteries, has the capacity to dissolve the shackles of negativity. It invites us to embrace the infinite symmetries, to explore the hidden connections, and to revel in the beauty of existence. Awe reminds us of our place in the cosmos and ignites a sense of purpose and wonder. Pure cognitive and conscious experiences are invariants and there are invariants in every aspect of life, we have identified 13 classes of invariants as outlined in Fig. 10.2.

Together, primordial evolutionary love and pure awe form a transformative duo. They challenge the darkness of negativity and offer a path to healing and growth. As we navigate the labyrinth of our minds, let us not forget the profound forces that guide us—forces that are as ancient as the universe itself. In the intricate dance of clocks, invariants, and symmetries, we find the keys to unlock the potential for a more positive and connected human psychology.

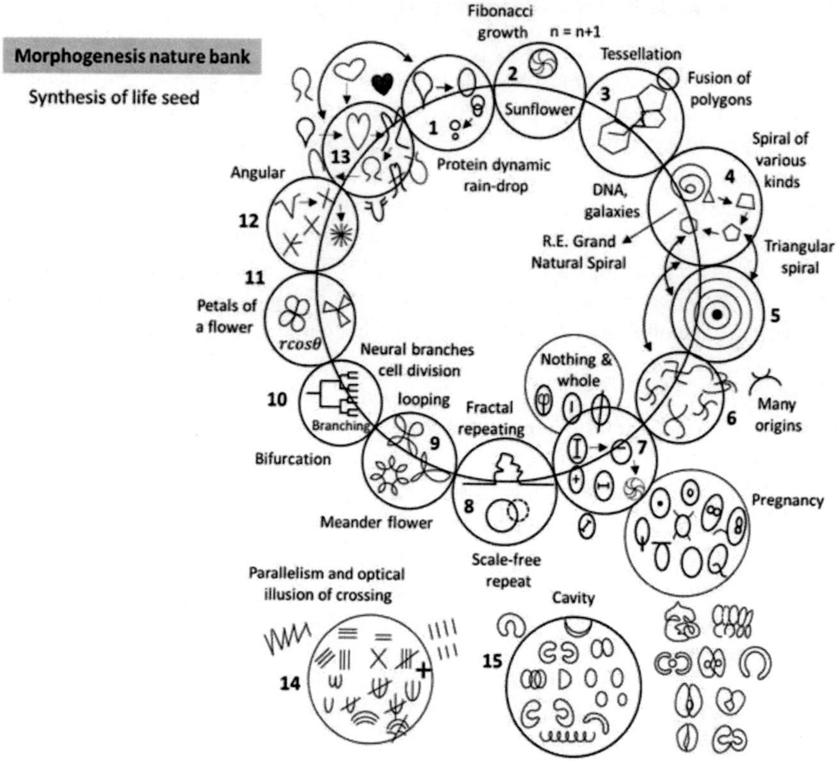

Fig. 10.1 Morphogenesis: 15 loops of geometric shape changes that are derived from the phase prime metrics or PPMs of different primes [8], these fifteen elements eventually generates the spatio-temporal structures or regulate events that happens in nature. Under the framework of SOMU theory of consciousness, 6D non-physical world that is made of 7D–9D morphogenesis (geometric shape change in a loop, but there is no time, or direction), and 10D–12D density of primes or different dynamical aspects of primes. These non-physical world generates 6D spatio-temporal events, that is 3D space and 3D time

10.4.2.2 Primordial Patterns Directly from the Subconscious Mind:

Figure 10.3 has compiled 36 elementary images derived from the DDG experiments carried out as explained in the Chaps. 5, 6 and extended in Chap. 7. Drawing images can provide valuable insights into human psychology, particularly the subconscious mind. "Art and Visual Perception: A Psychology of the Creative Eye" by [1] explores the connection between art and psychology, emphasizing the role of perception in artistic creation. The book highlights how artists' drawings of percepts and images can serve as a means to envision other minds [26]. This suggests that through their drawings, artists can tap into their own subconscious and reveal aspects of their inner thoughts and emotions. However, drawing is not without its challenges. "The Genesis of Errors in Drawing" by [9] discusses the importance of accurate perception of the

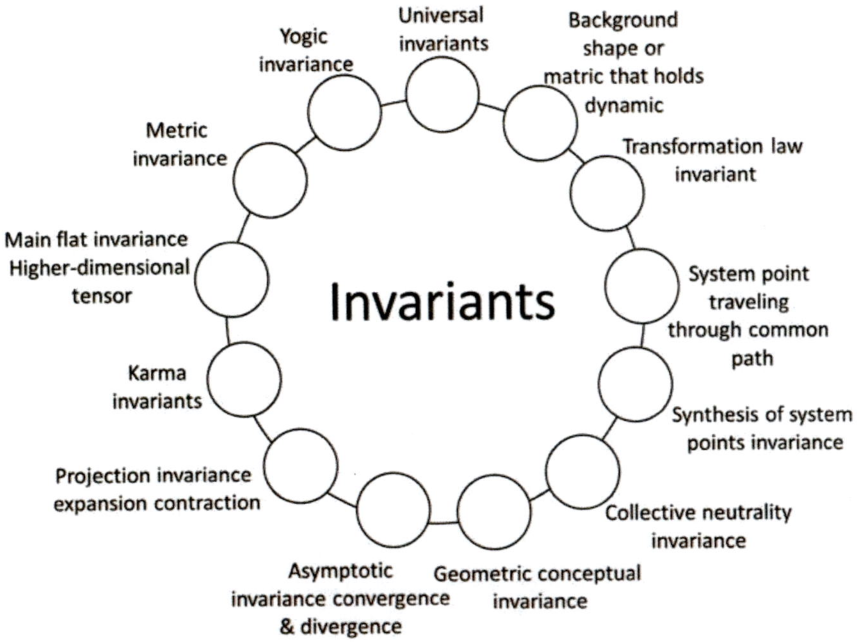

Fig. 10.2 Different classes of invariants: We have listed here 13 types of invariants that we observe in human brain operation that regulates eventually a human's cognitive, perceptual and emotional experiences

subject and the drawing itself for drawing proficiency. This implies that drawing can be influenced by both conscious and subconscious factors, making it a complex process that reflects the inner workings of the mind.

"The Predictive Mind and the Experience of Visual Artwork" by [25] further supports the idea that drawings can reveal information about the subconscious mind. The study found that abstract and semi-abstract paintings were rated as more meaningful when accompanied by their titles, suggesting that the interpretation of visual art involves both conscious and subconscious processes. Additionally, [27] investigates the accuracy of observational drawing and highlights the phenomenon of size inflation in drawing from a mirror. This suggests that the way individuals perceive and represent objects in their drawings can be influenced by subconscious biases and distortions.

Furthermore, the study of art in the context of ethnobiology by Begossi and Caires [5] reveals that local art can represent the perception of nature by humans. This implies that artistic representations, including drawings, can provide insights into how individuals perceive and interact with their environment, shedding light on their psychological processes. Therefore, drawing images can serve as a window into the subconscious mind, revealing aspects of an individual's thoughts, emotions, and perceptions. Through art, artists can tap into their own subconscious and envision other minds. However, drawing is not without its challenges, as accurate perception

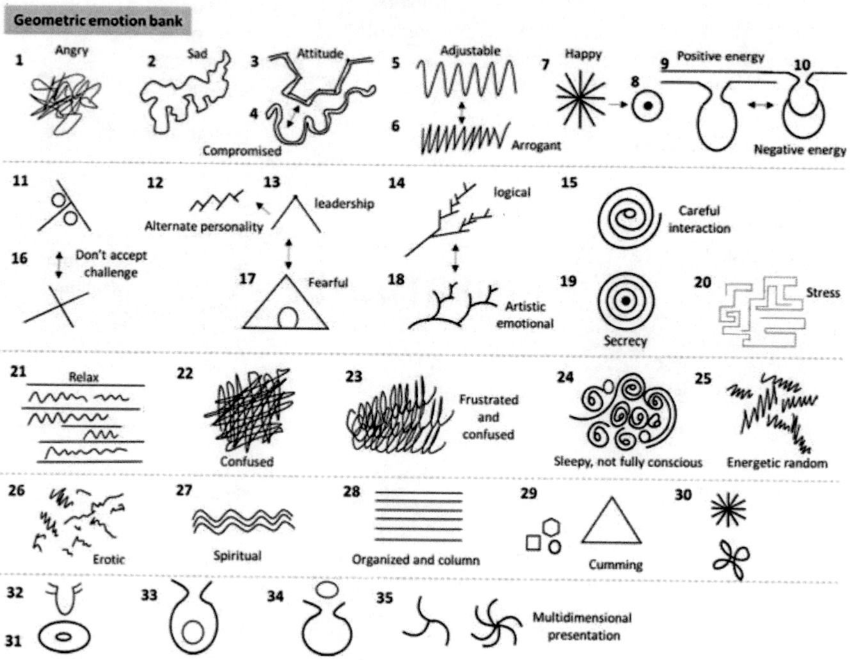

Fig. 10.3 Here we have shown 36 elements of conscious expressions. One could notice how morphogenesis of 7D–9D are reflected in the elements. The journey of a conscious experience, be it cognitive, perceptual or emotional, always we find some basic geometric shapes that governs. Thus, the foundation of SOMU, that 6D non-physical operations lead to physical 6D spatio-temporal events is the foundation that DDG based human subject experiments have led us to eventually 32–36 where we find complex morphogenesis seeds as composite thoughts

and representation are crucial for drawing proficiency. Nonetheless, the interpretation of visual art involves both conscious and subconscious processes, and drawings can provide valuable insights into human psychology and the way individuals perceive and interact with their environment.

10.5 From Mirror Drawing to Sexual Abuse Videos: SOMU is Present with All of Its Features

A paradigm shift is underway in our approach to understanding the universe. We've moved beyond the conventional machinery-based principles and binary thinking of 0 and 1. Instead, we're embracing a new culture that views the universe through orthogonal lenses, where observation doesn't interfere with what's observed. This shift requires new tools and a fresh perspective.

In the world of Symmetry-Oriented Mechanics Universe (SOMU), we recognize that clocks representing recurring events are the true operators of the universe. Linear events are just fragments of larger cyclic occurrences. To unravel the secrets of the cosmos, we must identify the invariant geometric structures amidst the complexity of nearly identical clocks. This journey takes us into the realm of quantum mechanics and beyond.

Within this quantum landscape, we discover that many interlinked clocks, despite their similar attributes, collectively influence systemic responses. They operate according to an invisible protocol that classical observation cannot fathom. This leads us to the concept of topological invariants, finite in number, yet pivotal in orchestrating the shared matrix between entangled states.

We don't stop at exploring the maximum dimensions of topology; we venture into space–time independent mechanics, guided by the pattern of prime numbers. This morphological concept elucidates the origins of quantum mechanics and primes serve as the foundation of fractal mechanics, a level above quantum phenomena.

In the realm of 10D–12D metrics, we find the universe's codebook. It enables us to derive meaning from any event without the need for learning or training. It's akin to envisioning a world of artificial intelligence devoid of algorithms or training, where inputs seamlessly translate into internally stored codes. This codebook, coupled with metric grammar, empowers machines to cater to our exploration needs.

Our quest is no longer confined to the search for symmetries in information content; we now embark on a journey to unearth invariants that inspire machines to meet human exploration needs. This vision encompasses a new civilization where every machine finds its orthogonal analog. In this section, we delve into the universe's language, its letters, and the grammar book encoded in metrics. By understanding how the universe integrates, breaks, and forges new symmetries, we pave the way for machines to discover, recreate, and regenerate independently.

10.6 Defining Consciousness as a Mathematical Property: Unveiling the Universe's Blueprint

10.6.1 Unlocking the Secrets of Consciousness Through Mathematics

The Essence of Consciousness

To comprehend consciousness as a mathematical entity, we must consider three fundamental elements: the "whole," the "it," and the "nothing."

1. **The Whole**: This encompasses the entirety of information within the universe, expressed in various forms such as bits, qubits, geometric shapes, polyatomic time crystals, and invariants. It transcends integration theories, making it a comprehensive representation of universal data.

2. **The Nothing**: An abstract, dimensionless entity devoid of properties, the "nothing" contains an infinite combination of prime numbers. It boasts a metric, the phase prime metric (PPM), which interconnects higher-order networks through the ordered factor of integers.
3. **The It**: This enigmatic entity serves as both the whole and the nothing simultaneously, representing the primary active element within the mathematical construct of consciousness.

The Necessity of the Undefined Point

In the genesis of the universe, where neither something nor nothing could be asserted, we find the significance of an undefined point. This point, a universal invariant, remains unaltered and becomes the starting point for mapping the cosmos. It symbolizes a singular origin from which the universe's journey begins—a transformation into higher dimensions.

Three Fundamental Actions of Points

The journey from a spatial point unfolds into three actions: bonding points (Tama, T), transforming a line into a 3D curve (Raja, R), and creating a boundary sphere that encapsulates all lines (Satva, S). These actions result in the formation of three angular invariants, θ, φ, and ω, when the triplet of triplet symmetry dominates the ordered factor metric (M11).

This journey consumes three spatial dimensions, followed by a transformation in three dimensions of time. Beyond that, we delve into topological dimensions (6D–9D), revealing space–time-independent geometric properties. As we reach the end of the ninth dimension, we enter a realm where further topological dimensions cease to exist. The exploration extends to the density of prime numbers, unraveling three dimensions of integers, concluding the journey through twelve dimensions.

Consciousness Unveiled: A Singular Universe

In this mathematical universe, consciousness emerges from twelve distinct types of invariants operating within twelve dimensions. It is a profound journey that begins with an undefined point and culminates in a comprehensive understanding of the cosmos.

Singularities in Biological and Artificial Brains

Within the realm of Symmetry-Oriented Mechanics Universe (SOMU), singularity pertains to phase singularity. It signifies arbitrary phase fluctuations that evade singular or plural expressions of functions. In both biological and artificial brains, this concept of singularity is fundamental.

A Hierarchy of Invariants, Not Computing

Both biological and artificial brains function as hierarchical networks of invariants, not traditional computing devices. Invariant structures, represented as geometric shapes, facilitate the matching, and synthesis of invariants. This process is distinct from logic gates and is more efficient for handling complex tasks.

Information Integration Through Hierarchical Networks

The integration of information occurs seamlessly through hierarchical networks of invariants. Unlike traditional AI, which relies on human tagging, this approach automates the tagging process. It leverages sensory data to access the invariant bank, allowing for swift analysis and integration.

The Universe Through the Lens of Invariants

Rather than relying on sensory perceptions and human intuition to interpret the universe, SOMU suggests an alternative approach. By starting with primes and deriving invariants, we gain a unique perspective. The universe becomes a product of prime exploration, a realm of integer illusion where invariants and conserved laws reign supreme.

In conclusion, consciousness, when viewed as a mathematical construct, unveils the universe's intricate design. It is a journey that begins with the undefined point and culminates in the exploration of twelve dimensions. This mathematical universe, powered by invariants, offers a novel perspective on the cosmos, transcending conventional sensory perception, and human intuition.

10.6.2 Grand Unification Theory of Consciousness

Grand unification theory of consciousness has been presented in the Fig. 10.4. Here we have combined geometric shape, prime compositions and colors to generate 12 cognitive, perceptual and emotional traits of psychology that unifies all non-physical factors with one physical factor that is color. Research has shown that color perception can influence psychological functioning in humans [12]. Different colors are believed to have different associations, and viewing a color can trigger psychological responses consistent with these associations [11]. For example, the color red has been found to be associated with increased arousal and performance attainment [14]. Context also plays a role in the psychological effects of color, as the same color can have different meanings depending on the context [13]. Understanding these associations and contextual influences can provide valuable insights into how color affects human psychology.

Synesthesia and Color Perception: Synesthesia is a neurological condition in which stimulation of one sensory or cognitive pathway leads to involuntary experiences in a second pathway. One common form of synesthesia is grapheme-color synesthesia, where viewing letters and numbers induces the perception of colors [19]. This phenomenon highlights the interconnectedness of sensory experiences and suggests that color perception is not solely limited to visual stimuli. Synesthetic patients provide a unique perspective on the relationship between color and cognitive experiences, as they often associate colors with taste, sounds, or other sensory

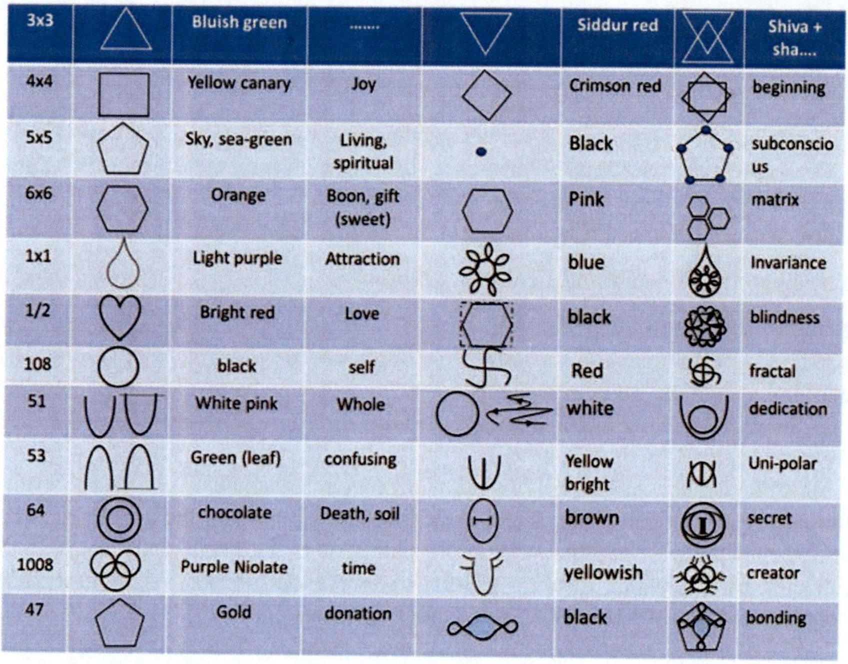

Fig. 10.4 Fractal network of invariant of invariants: Each row represents a fractal of universal experiences, we call this chart as Grand Unification Theory or GUT of consciousness. The fractals of prime numbers generate specific symmetry that eventually leads to fundamental universal thread that defines all forms of conscious experiences. Here we have selected 12 such fractals that represents particular prime symmetry of 10D–12D invariant database, and selected morphogenesis units of 7D–9D database. Here each row in this plot represents a particular fractal where two geometric shapes fuse together to create a fusion state that is "dhatu" in Sanskrit and we call it fractal seed. Similar to Sanskrit language, here these dhatu undergoes transformations of fractals to generate all kinds of conscious perceptions, cognitions and emotions. We do not see these elements from outside but they follow the core structure of consciousness

modalities [52]. Exploring the neural mechanisms underlying synesthesia can shed light on the complex interactions between color perception and cognitive processes.

Color Psychology and Behavior: Color psychology investigates the impact of color on human behavior and psyche [64]. Colors have been found to influence appetite sensation, with certain colors being associated with increased or decreased appetite [10]. Additionally, color has been shown to affect mood and happiness, with simple color changes influencing users' conceptual choices and psychological well-being [28]. By understanding the psychological effects of color, designers can leverage color in the design of human–computer interactions and community health environments [61, 64].

Cultural and Linguistic Influences on Color Perception: Color perception is not solely determined by biology but is also influenced by cultural and linguistic factors. Different cultures may categorize and perceive colors differently, suggesting that color categories are culturally relative [42]. Moreover, the naming and categorization of colors can vary across languages, highlighting the role of linguistic relativity in color perception [42]. These findings emphasize the complex interplay between biology, culture, and language in shaping color perception and cognition.

10.7 Conclusion: SOMU Theory of Consciousness

The hallmark of neurogenesis is to link a single neuron pulse with human cognition so that when we see the neural pulses in the brain, we understand its significance. We could tell explicitly the thought associated with a stream of pulses. Here in this book, as part of the SPARC project, we have used several state-of-the-art experiments to find the experimental attributes to the human perception and cognition. One major shift in our experiment is visual form of non-verbal communication. Earlier, measuring spontaneously generated signals measuring 1Hz to 40Hz oscillations on the brain surface was only form of studying EEG. However, we have extended it to kHz, MHz, GHz, and THz region ([48]; Chap. 5). Wide ranges of temporal bands were never seen earlier.

10.7.1 Musical Clocks Forming Geometric Shapes is Our Mind

This note delves into a paradigm-shifting concept that challenges traditional notions of information processing in the brain. Departing from the conventional models centered on bits or qubits, this exploration unveils an intriguing perspective that places clocks at the forefront of cognitive processes. Through meticulous analysis and measurement, it becomes evident that the brain functions as an intricate assembly of clocks, each emulating recurring events in the natural world. These clocks, meticulously organized on the surface of spheres, create a dynamic 3D arrangement that reshapes our understanding of thought processes and information integration in the brain. In contrast to the linear tape metaphor, this innovative framework offers a novel approach to conceptualizing brain function.

Introduction:

The realm of neuroscience and psychology has long been dominated by the paradigm of information processing, traditionally associated with bits or qubits. However, a compelling alternative perspective has emerged, challenging the very foundations of how we perceive cognitive mechanisms. At its core, this novel approach introduces the concept of “clocks” as the principal agents responsible for emulating recurring events in nature within the brain.

The Brain as a Clockwork Mechanism:

Unlike the established models, which focus on discrete units of information, this perspective postulates that the brain's hardware is inherently designed to generate an array of clocks, each operating within a wide range of time intervals. The rationale behind this vision lies in the idea of replicating the diverse temporal scales present in the natural world. Whether encountering slow or rapid events, the brain assembles a multitude of clocks, mirroring the temporal richness of the external environment.

Spheres as the Canvas of Cognition:

One of the key revelations in this paradigm is the role of spheres in structuring the brain's cognitive processes. These spheres serve as the canvas upon which the clocks are meticulously inscribed. In essence, the 3D assembly of spheres directly corresponds to the 3D assembly of clocks, fundamentally altering the conventional linear tape metaphor often associated with information storage and processing.

Geometry in Thought:

A pivotal consequence of this unconventional framework is the profound influence of clock-based information processing on the formation of geometric shapes. The connectivity of sphere centers within the brain gives rise to dynamic changes in the arrangement of geometric shapes. In essence, the brain's thought processes are intricately intertwined with alterations in the geometric configuration, shedding light on the profound relationship between cognition and spatial organization.

In summary, this paradigm-shifting perspective challenges the established notions of information processing in the brain. By emphasizing clocks as the central actors in emulating natural events and organizing them on the surface of spheres, it offers a compelling alternative to the traditional binary or quantum models. Furthermore, the dynamic interplay between these clocks and geometric shapes highlights the intricate nature of thought processes within the brain. This novel approach opens up new avenues for exploration in neuroscience and psychology, promising a deeper understanding of the brain's cognitive mechanisms and its intricate relationship with the surrounding world.

10.7.2 Primes Numbers Making Integers in a New Way to Find New Primes is Consciousness

In the boundless landscape of cognitive science and neuroscience, a profound revelation is dawning—a revelation that challenges the very essence of human consciousness and thought processes. This scholarly note embarks on an exhilarating expedition into the heart of the mind, where it is revealed that the fundamental code governing consciousness lies in the intricate dance of 3D clock assemblies. Through a lens that transcends conventional paradigms, we explore the breathtaking vistas of human thoughts, envisioning them as a mesmerizing interplay of spheres and geometric shapes. At the core of this exploration lies a groundbreaking theory—the "Self-Assembled Mathematical Universe" or SOMU—which unravels the enigma of consciousness, revealing it to be an elegant symphony of prime numbers constructing integers. This compelling journey beckons to budding scientists, inspiring them to venture into the uncharted realms of the mind and rewrite the very definition of consciousness.

The human mind, an enigmatic realm that has captivated scholars and thinkers for centuries, conceals within its labyrinthine folds the most profound mysteries of consciousness. In the pursuit of unlocking these mysteries, a revolutionary theory has emerged, one that seeks to redefine our understanding of consciousness and cognition. At its heart, this theory unveils the notion that the most fundamental code governing consciousness lies in the mesmerizing choreography of 3D clock assemblies over 12 dimensions [54, 56, 57].

The Dance of Spheres and Shapes:

Picture, if you will, the human mind engaged in cognitive tasks, a realm where memory is intricately woven into 3D clock assemblies. In this extraordinary theater of thought, we are presented with not one, but two captivating perspectives. In the first, we witness a dynamic spectacle where numerous spheres materialize and vanish within the 3D assembly. Alternatively, we can perceive the same phenomenon as an intricate interplay of geometric shapes, perpetually transforming within the complex tapestry of a 3D assembly.

Cracking the Code:

Amidst this ever-shifting landscape of spheres and shapes, a pattern emerges, one that defies the conventional constraints of time, singularity, or evolutionary pathways. These pure loops of geometric transformations, when meticulously examined, unveil the essence of hierarchical thoughts within a conscious mind. These loops, akin to knots in the fabric of thought, hold the key to a profound revelation—the utilization of prime numbers. The book [8] is a thought-provoking exploration of the ethical and philosophical implications of natural intelligence. It raises important questions about the nature of consciousness, free will, and the relationship between humans and machines by challenging the linear thought processes of the human brain.

Prime Numbers: The Architects of Consciousness:

At the deepest stratum of thoughts, perceptions, and emotions, lies the profound truth—they are nothing but manifestations of prime numbers, crafting integers in an infinite array of ways. This is the essence of the "Self-Assembled Mathematical Universe" or SOMU, a theory that redefines consciousness as a mathematical masterpiece woven from the very fabric of the universe.

A Call to Explore:

To the aspiring scientists of tomorrow, this revelation is an invitation to embark on a transformative journey. A journey that beckons you to step beyond the boundaries of conventional wisdom and traverse the uncharted territories of the mind. The code for consciousness lies before you, waiting to be deciphered, inspiring you to reshape our understanding of what it means to be conscious.

In closing, the unveiling of consciousness as a self-assembled mathematical universe is a testament to the boundless mysteries that still await discovery within the human mind. As you venture forth into the exhilarating realms of cognitive science and neuroscience, remember that you stand on the precipice of redefining our very essence. The fundamental code for consciousness is your compass, guiding you toward the unexplored horizons of the mind, where prime numbers dance, and thoughts are the threads weaving the fabric of existence.

References

1. Arnheim R (1956) Art and visual perception: a psychology of the creative eye. Philos Phenomenol Res 16(3):425. https://doi.org/10.2307/2104731
2. Amunts K, Ebell C, Müller J, Telefont M, Knoll A, Lippert T (2016) The human brain project: creating a European research infrastructure to decode the human brain. Neuron 92(3):574–581. https://doi.org/10.1016/j.neuron.2016.10.046
3. Agrawal L, Sahu S, Ghosh S, Shiga T, Fujita D, Bandyopadhyay A (2016) Inventing atomic resolution scanning dielectric microscopy to see a single protein complex operation live at resonance in a neuron without touching or adulterating the cell. J Integr Neurosci 15(04):435–462
4. Agrawal L, Chhajed R, Ghosh S, Ghosh B, Ray K, Sahu S, Fujita D (2018) Fractal information theory (FIT)-derived geometric musical language (GML) for brain-inspired hypercomputing. Soft Computing: Theories and Applications. 548:343–372
5. Begossi A, Caires R (2015) Art, fisheries and ethnobiology. J Ethnobiology Ethnomedicine, 11(1). https://doi.org/10.1186/1746-4269-11-17
6. Baumeister R (2011) Self and identity: a brief overview of what they are, what they do, and how they work. Ann N Y Acad Sci 1234(1):48–55. https://doi.org/10.1111/j.1749-6632.2011.06224.x
7. Bandyopadhyay A, Miki K, Wakayama Y (2006) Writing and erasing information in multilevel logic systems of a single molecule using scanning tunneling microscope. Applied Physics Letters 89(24)
8. Bandyopadhyay A (2020) Nanobrain: the making of an artificial brain from a time crystal. Boca Raton CRC Press, Taylor and Francis. ISBN: 9781439875490, https://doi.org/10.1201/9780429107771
9. Chamberlain R, Wagemans J (2016) The genesis of errors in drawing. Neurosci Biobehav Rev 65:195–207. https://doi.org/10.1016/j.neubiorev.2016.04.002
10. Chui K (2018) An exploratory study on the influence of color on appetite sensation in a dining environment. https://doi.org/10.15224/978-1-63248-156-6-17
11. Elliot A, Maier M (2007) Color and psychological functioning. Curr Dir Psychol Sci 16(5):250–254. https://doi.org/10.1111/j.1467-8721.2007.00514.x
12. Elliot A, Maier M (2014) Color psychology: effects of perceiving color on psychological functioning in humans. Annu Rev Psychol 65(1):95–120. https://doi.org/10.1146/annurev-psych-010213-115035
13. Elliot A, Maier M, Binser M, Friedman R, Pekrun R (2008) The effect of red on avoidance behavior in achievement contexts. Pers Soc Psychol Bull 35(3):365–375. https://doi.org/10.1177/0146167208328330
14. Elliot A, Maier M, Moller A, Friedman R, Meinhardt J (2007) Color and psychological functioning: the effect of red on performance attainment. J Exp Psychol Gen 136(1):154–168. https://doi.org/10.1037/0096-3445.136.1.154

15. Fleming SM, Frith C, Goodale M, Lau H, LeDoux JE, Lee ALF, Michel M et al (2023) The integrated information theory of consciousness as pseudoscience. PsyArXiv on 15 Sept 2023. https://psyarxiv.com/zsr78/
16. Ghosh S, Fujita D, Bandyopadhyay A (2015) An organic jelly made fractal logic gate with an infinite truth table. Sci Rep 5(1):11265
17. Ghosh S, Sahu S, Agrawal L, Shiga T, Bandyopadhyay A (2016) Inventing a co-axial atomic resolution patch clamp to study a single resonating protein complex and ultra-low power communication deep inside a living neuron cell. J Integr Neurosci 15(04):403–433
18. Godovykh M, Tasci A (2021) Emotions, feelings, and moods in tourism and hospitality research: conceptual and methodological differences. Tour Hosp Res 22(2):247–253. https://doi.org/10.1177/14673584211039867
19. Hubbard E, Ramachandran V (2005) Neurocognitive mechanisms of synesthesia. Neuron 48(3):509–520. https://doi.org/10.1016/j.neuron.2005.10.012
20. Hosany S, Gilbert D (2009) Measuring tourists' emotional experiences toward hedonic holiday destinations. J Travel Res 49(4):513–526. https://doi.org/10.1177/0047287509349267
21. Hutt A, Lefebvre J, Hight D, Sleigh J (2018) Suppression of underlying neuronal fluctuations mediates EEG slowing during general anaesthesia. Neuroimage 179:414–428. https://doi.org/10.1016/j.neuroimage.2018.06.043
22. Jbabdi S, Behrens T (2013) Long-range connectomics. Ann N Y Acad Sci 1305(1):83–93. https://doi.org/10.1111/nyas.12271
23. Johnson A, Halliday D, Millard A, Tyrrell A, Timmis J, Liu J et al (2016) An FPGA-based hardware-efficient fault-tolerant astrocyte-neuron network. https://doi.org/10.1109/ssci.2016.7850175
24. J K, Sitt J, Faugeras F, Rohaut B, Karoui I, Cohen L et al (2013) Information sharing in the brain indexes consciousness in noncommunicative patients. Current Biology 23(19):1914–1919. https://doi.org/10.1016/j.cub.2013.07.075
25. Kesner L (2014) The predictive mind and the experience of visual art work. Front Psychol 5. https://doi.org/10.3389/fpsyg.2014.01417
26. Kunzendorf R, Sansom R, Davis A (2001) Envisioning other minds: the artistic drawing and psychometric rating of images and percepts. Imagin Cogn Pers 21(1):33–46. https://doi.org/10.2190/qa4m-m36f-klpc-cld6
27. Lou L (2020) Size inflation in drawing from mirror and proportional accuracy in observational drawing. Perception 49(7):749–770. https://doi.org/10.1177/0301006620933147
28. Löffler D, Paier W, Toriizuka T, Ikeda M, Hurtienne J (2015) Augmented happiness: simple color changes influence users' conceptual choices. 248–255. https://doi.org/10.1007/978-3-319-22723-8_20
29. Lane A, Terry P (2000) The nature of mood: development of a conceptual model with a focus on depression. J Appl Sport Psychol 12(1):16–33. https://doi.org/10.1080/10413200008404211
30. Lin X, Sun Y, Wang Y, Fan L, Wang X, Wang N et al (2019) Ambiguity processing bias induced by depressed mood is associated with diminished pleasantness. Scientific Reports 9(1). https://doi.org/10.1038/s41598-019-55277-6
31. Markowitsch H, Staniloiu A (2011) Memory, autonoetic consciousness, and the self. Conscious Cogn 20(1):16–39. https://doi.org/10.1016/j.concog.2010.09.005
32. Modolo J, Hassan M, Wendling F, Benquet P (2020) Decoding the circuitry of consciousness: from local microcircuits to brain-scale networks. Network Neuroscience 4(2):315–337. https://doi.org/10.1162/netn_a_00119
33. Miyamichi K, Amat F, Moussavi F, Wang C, Wickersham I, Wall N et al (2010) Cortical representations of olfactory input by trans-synaptic tracing. Nature 472(7342):191–196. https://doi.org/10.1038/nature09714
34. Miyawaki A, Kasai K (2015) Brain/minds: brain-mapping project in japan. Phil Trans R Soc B 370(1668):20140310. https://doi.org/10.1098/rstb.2014.0310
35. Merelli E, Piangerelli M (2014) Rnn-based model for self-adaptive systems—the emergence of epilepsy in the human brain. https://doi.org/10.5220/0005165003560361

36. Owen A (2013) Detecting consciousness: a unique role for neuroimaging. Annu Rev Psychol 64(1):109–133. https://doi.org/10.1146/annurev-psych-113011-143729
37. Oh S, Harris J, Ng L, Winslow B, Cain N, Mihalaş Ş et al (2014) A mesoscale connectome of the mouse brain. Nature 508(7495):207–214. https://doi.org/10.1038/nature13186
38. O'Doherty J, Hampton A, Kim H (2007) Model-based fMRI and its application to reward learning and decision making. Ann N Y Acad Sci 1104(1):35–53. https://doi.org/10.1196/annals.1390.022
39. Palmiero M, Piccardi L (2017) Frontal EEG asymmetry of mood: a mini-review. Front Behav Neurosci 11. https://doi.org/10.3389/fnbeh.2017.00224
40. Papineau D (2020) The problem of consciousness. 13–36. https://doi.org/10.1093/oxfordhb/9780198749677.013.1
41. Pradhan B, Bandyopadhyay A, Pal AJ (2004) Molecular level control of donor/acceptor heterostructures in organic photovoltaic devices. Appl Phys Lett 85(4):663–665
42. Roberson D, Davidoff J, Davies I, Shapiro L (2005) Color categories: evidence for the cultural relativity hypothesis. Cogn Psychol 50(4):378–411. https://doi.org/10.1016/j.cogpsych.2004.10.001
43. Reddy S, Sonker D, Singh P, Saxena K, Singh S, Chhajed R, Tiwari S, Karthik KV, Ghosh S, Ray K, Bandyopadhyay A (2018) A brain-like computer made of time crystal: could a metric of prime alone replace a user and alleviate programming forever? Soft Computing Applications 761:1–43
44. Ramamurthy U, Franklin S, Agrawal P (2012) Self-system in a model of cognition. Int J Mach Conscious 04(02):325–333. https://doi.org/10.1142/s1793843012400185
45. Sherman R, Rauthmann J, Brown N, Serfass D, Jones A (2015) The independent effects of personality and situations on real-time expressions of behavior and emotion. J Pers Soc Psychol 109(5):872–888. https://doi.org/10.1037/pspp0000036
46. Sporns O, Chialvo D, Kaiser M, Hilgetag C (2004) Organization, development and function of complex brain networks. Trends Cogn Sci 8(9):418–425. https://doi.org/10.1016/j.tics.2004.07.008
47. Saxena K, Singh P, Sahoo P, Sahu S, Ghosh S, Ray K, Fujita D, Bandyopadhyay A (2020) Fractal, scale free electromagnetic resonance of a single brain extracted microtubule nanowire, a single tubulin protein and a single neuron. Fractal Fract 4(2):11
48. Saxena K, Singh P, Sahoo P, Ghosh S, Krishnananda D, Ray K, Fujita D, Bandyopadhyay A (2022) All basics that are wrong with the current concept of time crystal: learning from the polyatomic time crystals of protein, microtubule, and neuron. Proceedings of Trends in Electronics and Health Informatics. 376:243–254
49. Saraiva M, Ayanoglu H (2019) Emotions and emotions in design. 57–70. https://doi.org/10.1007/978-3-319-96722-6_4
50. Siemer M (2001) Mood-specific effects on appraisal and emotion judgements. Cogn Emot 15(4):453–485. https://doi.org/10.1080/0269993004200178
51. Siemer M (2009) Mood experience: implications of a dispositional theory of moods. Emot Rev 1(3):256–263. https://doi.org/10.1177/1754073909103594
52. Suslick K (2012) Synesthesia in science and technology: more than making the unseen visible. Curr Opin Chem Biol 16(5–6):557–563. https://doi.org/10.1016/j.cbpa.2012.10.030
53. Singh P, Saxena K, Singhania A, Sahoo P, Ghosh S, Chhajed R, Ray K, Fujita D (2020) A self-operating time crystal model of the human brain: can we replace entire brain hardware with a 3D fractal architecture of clocks alone? Information 11(5):238
54. Singh P, Sahoo P, Saxena K, Manna JS, Ray K, Ghosh S, Bandyopadhyay A (2021) Cytoskeletal filaments deep inside a neuron are not silent: they regulate the precise timing of nerve spikes using a pair of vortices. Symmetry 13(5):821
55. Singh P, Ray K, Bandyopadhyay A (2022) Biological antenna to the humanoid bot: electromagnetic resonances in biomaterials. In: Studies in rhythm engineering (SRE). Springer, Singapore. https://doi.org/10.1007/978-981-16-9677-0
56. Singh P, Sahoo P, Saxena K, Ghosh S, Sahu S, Ray K, Fujita D, Bandyopadhyay A (2021) Quaternion, octonion to dodecanion manifold: stereographic projections from infinity lead to

a self-operating mathematical universe. Proceedings of International Conference on Trends in Computational and Cognitive Engineering. 1169:55–77

57. Singh P, Sahoo P, Saxena K, Ghosh S, Sahu S, Ray K, Fujita D, Bandyopadhyay A (2021) A space-time-topology-prime, stTS metric for a self-operating mathematical universe uses Dodecanion geometric algebra of 2–20 D complex vectors. Proceedings of International Conference on Data Science and Applications. 148:1–31
58. Song M, Zhang Y, Cui Y, Yang Y, Jiang T (2018) Brain network studies in chronic disorders of consciousness: advances and perspectives. Neurosci Bull 34(4):592–604. https://doi.org/10.1007/s12264-018-0243-5
59. Wade J, McDaid L, Harkin J, Crunelli V, Kelso J (2012) Self-repair in a bidirectionally coupled astrocyte-neuron (an) system based on retrograde signaling. Front Comput Neurosci 6. https://doi.org/10.3389/fncom.2012.00076
60. Watson D, Tellegen A (1985) Toward a consensual structure of mood. Psychol Bull 98(2):219–235. https://doi.org/10.1037/0033-2909.98.2.219
61. Wan J, Zhou Y, Li Y, Su Y, Cao Y, Zhang L et al (2020) Research on color space perceptions and restorative effects of blue space based on color psychology: examination of the Yijie district of Dujiangyan city as an example. Int J Environ Res Public Health 17(9):3137. https://doi.org/10.3390/ijerph17093137
62. Wheeler, John Archibald (1989) "Information, physics, quantum: the search for links." 3rd Int. Symp. Foundations of Quantum Mechanics, Tokyo, pp.354–368.
63. Yuan J, Gong H, Li A, Li X, Chen S, Zeng S et al (2015) Visible rodent brain-wide networks at single-neuron resolution. Front Neuroanat 9. https://doi.org/10.3389/fnana.2015.00070
64. Yang J, Shen X (2022) The application of color psychology in community health environment design. J Environ Public Health 2022:1–10. https://doi.org/10.1155/2022/7259595